D1300654

Invitation to Complex Analysis

The Random House/Birkhäuser Mathematics Series:

INVITATION TO COMPLEX ANALYSIS
by R. P. Boas

BRIDGE TO ABSTRACT MATHEMATICS: MATHEMATICAL PROOF AND STRUCTURES
by Ronald P. Morash

ELEMENTARY NUMBER THEORY
by Charles Vanden Eynden

INTRODUCTION TO ABSTRACT ALGEBRA
by Elbert A. Walker

Invitation to Complex Analysis

RALPH PHILIP BOAS

Professor of Mathematics, Emeritus
Northwestern University

The Random House/Birkhäuser Mathematics Series

Random House
New York

INVITATION TO COMPLEX ANALYSIS *Ralph Philip Boas*

First Edition

98765432

All inquiries should be addressed to Random House, Inc., 201 East 50th Street, New York, N.Y. 10022. Published in the United States by Random House, Inc., and simultaneously in Canada by Random House of Canada Limited, Toronto.

Library of Congress Cataloging-in-Publication Data
Boas, Ralph Philip.
Invitation to complex analysis.

(The Random House/Birkhäuser mathematics series)
Bibliography: p.
Includes index.
1. Functions of complex variables. I. Title.
II. Series.
QA331.B644 1987 515.9 86-31369
ISBN 0-394-35076-6

Manufactured in the United States of America by Halliday Lithograph, Hanover, MA.

to Mary

Preface

This book is intended as a textbook for a first serious (senior or graduate level) course on complex analysis; but it is also intended to be useful for independent study. It assumes that its readers have had a course in advanced calculus or introductory real analysis (the material in the first part of my *A Primer of Real Functions* is more than enough background).

I have adopted a rather informal style, in the hope that students can actually read the text; in particular, there are no numbered equations. I hope that you will not mistake informality, and avoidance of special symbols and abbreviations, for lack of rigor. I have deliberately not tried to present every result in the greatest possible generality. I have also taken a rather cavalier attitude toward topological properties that I do not know how to prove without a large amount of apparatus (the Jordan curve theorem, for example). I have not tried to arrange the material in strict logical order; and I have strayed beyond the confines of my field whenever I thought that a digression would be interesting. Several topics are unusual in a book at this level, but they are interesting and no more difficult than the standard material. Sections 15C, 17D–F, 18B–C, 20D–E, 21, and all of Chapter 5 are supplementary to the main line of the book and can be omitted from a short course. On the other hand, the topics in Chapter 5 can, if desired, be taken up earlier: Section 30 after Section 25; Sections 31 and 32 any time after Section 10; Section 33 after Section 16; and Section 34 after Section 14.

Rather few illustrative examples are worked out in the text. Instead, I usually present them as exercises, with solutions at the back of the book. By encouraging students to work through the exercises themselves, instead of passively reading solved problems, I hope to get the students actively involved in the subject. For teachers who want to assign exercises that do not have solutions in the book, there are supplementary problems at the ends of most sections; an answer pamphlet for these exercises is available.

I believe that, for beginners, pure theory is sterile unless it can be seen to be used for something. Technique *is* important. If you are going to use complex analysis, for example to evaluate integrals or describe fluid flows, you need to be able to do these things correctly. An experienced scientist will not be seriously led astray through carelessness, because the real world is rather unforgiving. I have, however, seen students badly confused by trying to apply half-understood mathematics. I have confined my examples of applications to a few that are simple enough to be understood easily. Complex analysis has many applications—that is one reason for its popularity and importance—but many applications are either very technical or demand so much background that they do not fit into a short book. I have discussed the application of conformal mapping to potential problems in some detail, but only after developing enough background so that they can be understood.

Since I think of Cauchy's theorem as central, I get to it as early as possible, pausing only long enough to introduce the principal elementary functions in order to have something to integrate. The properties of functions as mappings are postponed so as not to introduce too many different ideas at once.

I have not tried to provide historical comments; the books by Burckel and Remmert provide better historical material than I could possibly collect myself.

Since my involvement with the subject now goes back more than half a century, I find it impossible to give credit where it is due for much of my material, almost none of which is original.

Superscript numbers in the text refer to the notes at the ends of main sections; books in the list of references are cited by authors' names.

Many people have contributed helpful comments. I am especially indebted to H. P. Boas, M. L. Boas, R. B. Burckel, S. M. Cargal, K. F. Clancey, G. B. Folland, S. Hellerstein, S. G. Krantz, A. M. Trembinska, and the students in my class in complex analysis in 1983–1984.

R. P. Boas
Evanston, Illinois
November 1986

To the student

You may think of this book as a guided tour through parts of a picturesque country, where you will see some of the famous scenery, as well as a few less notable but still attractive sights. You will not climb any mountains or visit large industrial plants, but you will be invited to get out of the bus from time to time and hike around the countryside on your own. You will also acquire facility with the native language.

To drop the metaphor, if you are reading the book by yourself you can omit starred sections without losing continuity. If you find a section becoming too difficult, you can usually skip to the next section without missing very much. I do assume that you are familiar with calculus in two dimensions; and also with basic facts about sets, limits, and convergence. If you have missed some of this material, you may have to take some theorems on faith, but you should still be able to appreciate what they say.

Some of the exercises are significant theorems that have simple proofs, some are auxiliary results, and some are merely for practicing techniques. Starred exercises are either peripheral or are expected to be difficult. You will get much more out of the book if you try to do the exercises for yourself. Solutions are given at the back of the book (except for the Supplementary Exercises), but you may be able to find better solutions for yourself.

I have tried to make the book easy to read. There are few special symbols and abbreviations, since I think that most people read words more easily than symbolism. The mathematical difficulty increases somewhat toward the end of the book.

Complex analysis was originally developed for the sake of its applications; I have tried to show what some of the applications are. However, the subject now has an independent and active life of its own, with many elegant and even surprising results. I hope that you will come to appreciate its intrinsic fascination as well as its power.

R. P. Boas

Contents

Invitation to Complex Analysis

CHAPTER 1

From Complex Numbers to Cauchy's Theorem

1 *Complex Numbers*

1A. Why complex numbers?

The main emphasis of this book is on complex analysis—the use of complex numbers and complex-valued functions—as a subject that is interesting in itself and also has practical applications. The applications could be treated without actually using complex numbers, but complex analysis certainly makes them seem simpler and more comprehensible, as I hope you will see.

1B. Definitions

In elementary algebra, complex numbers appear as expressions $a + bi$, where a and b are ordinary real numbers and $i^2 = -1$. One often writes $i = \sqrt{-1}$, without worrying very much about which of the (presumably) two square roots is intended. Complex numbers are manipulated by the usual rules of algebra, with the convention that i^2 is to be replaced by -1 wherever it occurs.

If you do enough algebra with complex numbers, you will sooner or later want to write expressions whose meanings are not obvious. For instance, since

$$\left(\frac{1+i}{\sqrt{2}}\right)^2 = \left(\frac{-1-i}{\sqrt{2}}\right)^2 = i,$$

is $\sqrt{i}$ to be $(1+i)/\sqrt{2}$ or $-(1+i)/\sqrt{2}$? Questions like this troubled the mathematicians of three centuries ago, but today we realize that symbols are meaningless until we define them. We begin by defining complex numbers themselves. (There are alternative definitions, which you might find more instructive: See Exercises 1.24 and 1.26.)

We define a **complex number** to be an **ordered pair** of real numbers. "Ordered" means that the pair consisting of first x and then y is different from the pair consisting of these numbers in the opposite order (unless $x = y$); the first pair is denoted by (x, y). Instead of (x, y) we also write $x + iy$, and we often denote the pair (x, y) by z or some other single letter. If we are dealing with two complex numbers, we are likely to call them $z_1 = (x_1, y_1)$ and $z_2 = (x_2, y_2)$, or $z = (x, y)$ and $w = (u, v)$, or $z = (x, y)$ and $\zeta = (\xi, \eta)$. Two ordered pairs (x, y) and (u, v) are equal if and only if $x = u$ and $y = v$.

Addition of ordered pairs is defined as component-wise addition, $(x, y) + (u, v) = (x + u, y + v)$. **Multiplication** of two complex numbers is defined by $(x, y) \cdot (u, v) = (xu - yv, xv + yu)$. This multiplication is (of course) suggested by the formal computation $(x + iy)(u + iv) = xu + iyu + ixv + i^2yv = xu - yv + i(xv + yu)$. Multiplication of a complex number (x, y) by a real number c is defined by $c \cdot (x, y) = (cx, cy)$ (see Exercise 1.2).

It is easy to prove that addition and multiplication of complex numbers have the expected properties of being commutative, associative, and so on. You may want to verify a few of these properties for yourself.

We shall *identify* the complex number $(x, 0)$ with the real number x; that is, when we write x alone, it stands for $(x, 0)$.

EXERCISE 1.1 Show that the complex number $i = (0, 1)$ has the property $i^2 = (-1, 0) = -1$.

If x and y are real numbers, Exercise 1.1 shows that $(x, 0) + (0, 1)(y, 0) = (x, y)$, which says that $(x, y) = x + iy$, as we expected.

EXERCISE 1.2 Show that $c(x, y) = (c, 0)(x, y)$, so that multiplication by a real number is consistent with complex multiplication.

A number iy with y real and not zero is called a **pure imaginary**. The only satisfactory general term for $x + iy$, $y \neq 0$, is "not real"; don't say "complex" for this, because the real numbers are also complex numbers.

If we look at the complex numbers merely as ordered pairs, we observe that the ordered pairs (x, y) also form the ordinary Euclidean plane, or more precisely the Cartesian product of two real lines. Hence it is permissible to think of the complex numbers as points in a plane; indeed, it is often very helpful to use this interpretation. If we do this, we call the plane the **complex plane**.[1]* It becomes the Euclidean plane after we define the distance between (x, y) and (u, v) to be $\{(x - u)^2 + (y - v)^2\}^{1/2}$. With this distance the complex plane becomes a metric space. The only axiom for a metric space that requires

* Superscript numbers refer to the notes at the end of the section.

any effort to verify is the triangle inequality, which we shall check a little later (Exercise 1.10).

1C. The point at infinity

It is often useful to think of the complex plane as containing, besides the points (x, y) that we already have, a "point at infinity." To see how to provide such a point, consider a sphere, of diameter 1, tangent to the plane at the origin, then project the sphere onto the plane from the north pole N by drawing a line from N to each point of the plane, and make that point correspond to the point where the line intersects the sphere a second time (see Fig. 1.1). This is a **stereographic projection**; it projects meridian semicircles on the sphere onto rays from the origin of the plane.

It should be clear that every point on the sphere except N corresponds to just one point of the plane. The correspondence between the sphere and the plane will be one-to-one if we adjoin to the plane a single "ideal" point ∞ ("infinity") that is to correspond to N. We think of ∞ as being outside every disk in the plane. Notice that there is just *one* point at infinity, and all straight lines in the plane go through it; parallel lines in the plane correspond to circles on the sphere that are tangent to each other at N. The sphere is the Riemann sphere, and the complex plane with ∞ adjoined is the **extended plane**. When we need to distinguish the plane without ∞ from the extended plane, we call

FIGURE 1.1

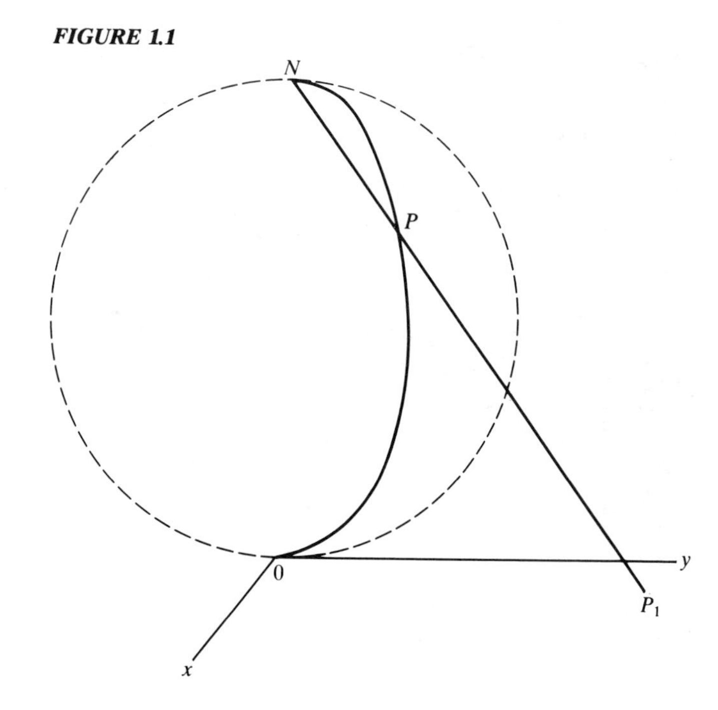

the former the **finite plane**. The sphere, being a subset of Euclidean three-space, is a metric space, the distance being the length of the chord connecting the points. If we take this length as a new distance between the images of the same points in the plane, we obtain a metric for the extended plane.

The notation $z \to \infty$ means the same thing as $x^2 + y^2 \to \infty$. We can calculate with ∞ by using the conventions $a \cdot \infty = \infty$ if $a \neq 0$; $a/0 = \infty$ if $a \neq 0$ or ∞; $a/\infty = 0$ if $a \neq \infty$; $a \pm \infty = \infty$ if $a \neq \infty$. Expressions like $\infty + \infty$, $\infty - \infty$, and $0 \cdot \infty$ should be avoided; they can occur only as abbreviations for limiting processes.

1D. Terminology of the complex plane

A certain amount of terminology is essential. The **conjugate** of $z = x + iy$ is $\bar{z} = x - iy$. (The notation z^* is used instead of $\bar{z}$ by people who want to use the bar to indicate an average.) The **real part** of z is x and is denoted by $\operatorname{Re} z$ [or $\operatorname{Re}(z)$ if z is a complicated expression]. This seems reasonable enough, but the **imaginary part** of z is y, NOT iy (which may seem unreasonable, but is the convention). The notation is $y = \operatorname{Im} z$. Thus we have $z + \bar{z} = 2 \operatorname{Re} z$, and similar formulas. Geometrically, z and $\bar{z}$ are reflections of each other with respect to the real axis; z and $-z$ are reflections of each other with respect to the origin (see Figure 1.2).

Division of complex numbers If $c + di \neq 0$, we define $(a + bi)/(c + di)$ to be the complex number $p + qi$ such that $(p + qi)(c + di) = a + bi$. In practice, we find quotients by writing

$$\frac{a + bi}{c + di} = \frac{a + bi}{c + di}\,\frac{c - di}{c - di} = \frac{ac + bd + i(cb - ad)}{c^2 + d^2}.$$

That is, we multiply the numerator and denominator of the fraction by the conjugate of the denominator. For a formal proof, we observe that the

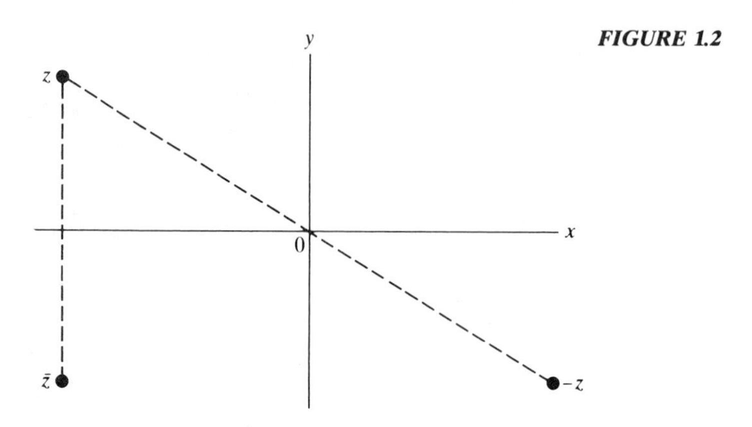

FIGURE 1.2

equation $(p+qi)(c+di)=a+bi$ is equivalent to

$$(pc-qd)+i(pd+qc)=a+bi,$$

that is, to the pair of equations

$$pc-qd=a,$$
$$pd+qc=b,$$

which can be solved for p and q if (and only if) $c^2+d^2 \neq 0$.

EXERCISE 1.3 Verify that $z+\bar{z}=2\operatorname{Re} z$ and $z-\bar{z}=2i\operatorname{Im} z$; also that the conjugates of sums, products, and quotients are the sums, products, and quotients of the conjugates of the numbers involved.

By using Exercise 1.3, we can calculate the real or imaginary part of a complex number without having to find the other part. For example,

$$\operatorname{Re}\left(\frac{2+3i}{3-4i}\right)=\frac{1}{2}\left(\frac{2+3i}{3-4i}+\frac{2-3i}{3+4i}\right)=\frac{(2+3i)(3+4i)+(2-3i)(3-4i)}{2(3+4i)(3-4i)}$$
$$=\frac{-6}{25}.$$

EXERCISE 1.4 Find the real and imaginary parts of $(i+2)/(i-2)$; $(1+i\sqrt{3})^2$; $1/(1-i)$; $(2-3i)/(3i+2)$.

The product $z\bar{z}=x^2+y^2$ can be interpreted geometrically as the square of the distance from the point z to 0. We denote it by $|z|^2$. Here $|z|$ is the **absolute value** of z. It is often also called the **modulus** of z, especially if you are speaking and want to save breath by abbreviating it to "mod."

EXERCISE 1.5 Show that $|zw|=|z|\,|w|$, that $|\bar{z}|=|z|$, that $|z/w|=|z|/|w|$; and that $|\operatorname{Re} z|\le|z|$, with equality only if z is real.

EXERCISE 1.6 Find the absolute values of the numbers in Exercise 1.4. [It is usually easier to find $|z|$ as $(z\bar{z})^{1/2}$ than as $\{(\operatorname{Re} z)^2+(\operatorname{Im} z)^2\}^{1/2}$.]

A word of caution: The absolute value of a real number is sometimes defined as the number "without its sign." This will not do for *complex* numbers: For example, the absolute value of $-2i$ is 2, not $2i$.

We can interpret $|z-w|$ as the Euclidean distance between the points z and w. Consequently, geometric figures that are defined in terms of distances can be described by equations or inequalities between absolute values. For example, the equation $|z|=2$ says that the distance from the point z to the point 0 is always equal to 2; in other words, $|z|=2$ is the equation of a circle with center at 0 and radius 2. Similarly, the inequality $|z|<2$ describes an open disk with center at 0 and radius 2. (See Fig. 1.3, p. 6.)

We could get the same conclusion by writing the first equation as $x^2+y^2=4$ and recognizing this as the equation of a circle; but it is better to think directly

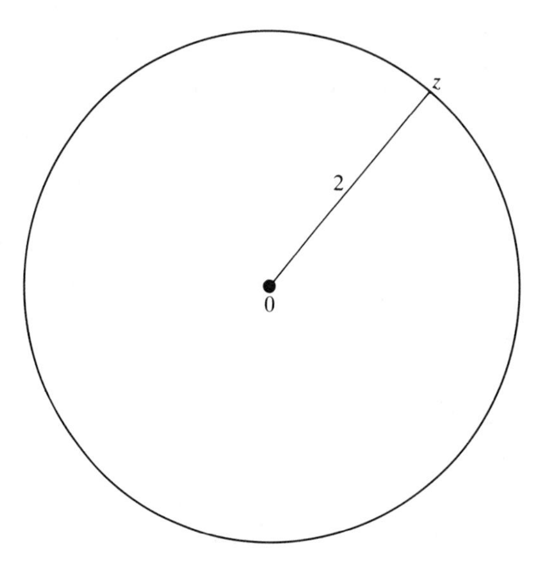

FIGURE 1.3

about the geometric meaning of equations and inequalities whenever possible.

The equation $|z| = 1$ describes the **unit circle**; the **unit disk** is either $|z| < 1$ (open), or $|z| \leq 1$ (closed). The inequality $\operatorname{Im} z > 0$ defines the **upper half plane**; $\operatorname{Re} z > 0$ defines the **right-hand half plane**.

EXERCISE 1.7 Describe geometrically the sets of points determined by the relations

(*a*) $\lvert z\rvert < 2$	(*b*) $\lvert z\rvert \geq 3$	(*c*) $\lvert z - 2 + 3i\rvert < 5$
(*d*) $\operatorname{Re} z > 1$	(*e*) $\operatorname{Re}[(2+3i)z] > 0$	(*f*) $\operatorname{Im} z \geq \operatorname{Re} z$
(*g*) $\lvert z - i\rvert + \lvert z - 1\rvert = 2$	(*h*) $\lvert z - i\rvert = \lvert z + 1\rvert$	(*i*) $\operatorname{Re} z = \lvert z - 2\rvert$

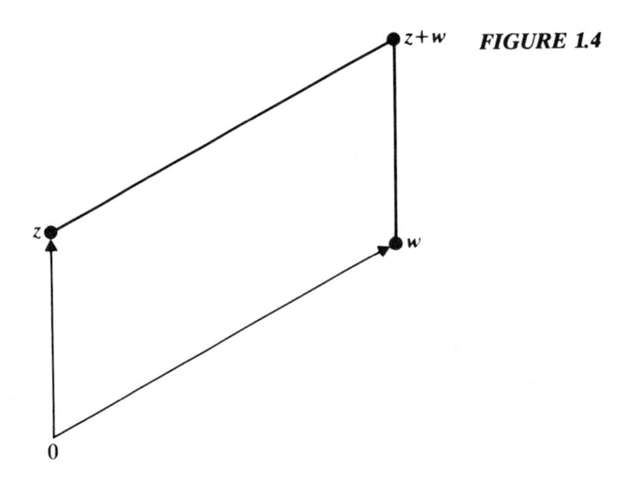

FIGURE 1.4

EXERCISE 1.8 Show that if we identify the complex number z with the vector from 0 to z, then addition of complex numbers corresponds to addition of vectors by the parallelogram law (see Fig. 1.4).

A **neighborhood** of a point usually means an open disk centered at that point; more loosely, it can mean any open set that contains the point. (Of course, such a set does contain an open disk centered at the point.) A **neighborhood of ∞** is the exterior of a closed disk (centered anywhere).

Supplementary exercises

1. Find the real parts, imaginary parts, and absolute values of

(*a*) $\dfrac{i-7}{6+2i}$ (*b*) $\dfrac{1}{(3i+2)(2i+3)}$ (*c*) $(2+i\sqrt{5})^2$

(*d*) $(2i-3)^3$ (*e*) $\dfrac{i-1}{i+1}$

2. Describe geometrically the sets of points determined by the relations

(*a*) $|z+i|<5$ (*b*) $|2i-z|=|z+1+3i|$

(*c*) $2>\operatorname{Re} z>-3$ (*d*) $\operatorname{Re}[(3-4i)z]>0$

(*e*) $(\operatorname{Im} z)^2 \le \operatorname{Re} z$ (*f*) $2\operatorname{Re} z<|z|^2$

1E. Polar coordinate representation of complex numbers

Complex numbers can be located by using the polar coordinates of the corresponding points in the plane. If $x=r\cos\theta$ and $y=r\sin\theta$, then[2] $z=r(\cos\theta+i\sin\theta)$ and $r=|z|$. There are, of course, many angles θ for the same point z; any correct one is an **argument**[3] of z, and is denoted by $\arg z$. We do not define $\arg 0$. If we need to, we can restrict the values of θ so that $\arg z$ becomes single-valued, that is, a function; usually, but not always, we take $\arg z$ to have its **principal value**, $-\pi<\arg z\le\pi$. It is useful to think of the principal value of $\arg z$ as the angle above the "horizon" $0x$ at which z is seen by an observer at 0 (see Fig. 1.5, p. 8), and so $\arg(z-a)$ (principal value) is the angle from the ray in the positive real direction at a to the ray $0z$.

EXERCISE 1.9 Find the absolute values and principal angles of the following complex numbers, and plot them in the complex plane: $3-4i$; $-4-11i$; $2.236+3.142i$; $(1+i)/(1-i)$; $(2+i)/(3-i)$; $-1+i\sqrt{3}$; $\cos(3\pi/2)+i\sin(3\pi/2)$; $2[\cos(7\pi/6)+i\sin(7\pi/6)]$.

EXERCISE 1.10 (**Triangle inequality**) Show that $|z+w|\le|z|+|w|$, with equality if and only if z and w are on the same ray through the origin. Draw a figure to see why this is called the triangle inequality. Deduce the following variant of the triangle inequality: $|z-w|\ge\big||z|-|w|\big|$.

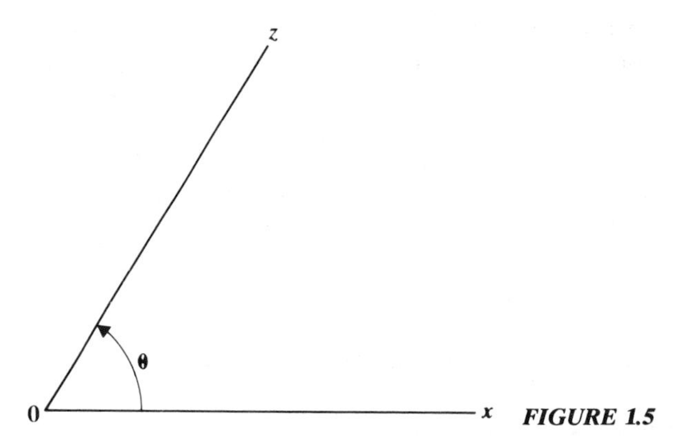

FIGURE 1.5

It follows by induction from Exercise 1.10 that $|z_1 + z_2 + \cdots + z_n| \le |z_1| + |z_2| + \cdots + |z_n|$. However, the following proof establishes this for all n at once, and determines the condition for equality.

Write $\sum z_k$ in polar coordinate form as $R(\cos\theta + i\sin\theta)$. Then

$$\left|\sum z_k\right| = R = \mathrm{Re}\left\{(\cos\theta - i\sin\theta)\sum z_k\right\}$$

[because $(\cos\theta - i\sin\theta)(\cos\theta + i\sin\theta) = \cos^2\theta + \sin^2\theta = 1$]. Then

$$\left|\sum z_k\right| = \sum \mathrm{Re}\{(\cos\theta - i\sin\theta)z_k\},$$

so by Exercise 1.5 (last part),

$$\left|\sum z_k\right| < \sum |(\cos\theta - i\sin\theta)z_k| = \sum |z_k|,$$

unless $(\cos\theta - i\sin\theta)z_k$ (with θ independent of k) is real for every k, that is, unless all z_k are on the same ray from 0.

Because we often write complex numbers in polar coordinate form, you will need to recall a number of formulas from elementary trigonometry. (Conversely, later on (Sec. **4E**) we shall see how complex algebra can be used to prove the formulas of elementary trigonometry.) You will need the addition formulas for sines and cosines in the following exercise.

EXERCISE 1.11 Show that $\arg(zw) = \arg z + \arg w$, in the sense that the right-hand side is a possible value of the left-hand side. That is, if $z = r(\cos\theta + i\sin\theta)$ and $w = s(\cos\phi + i\sin\phi)$, then $zw = rs[\cos(\theta + \phi) + i\sin(\theta + \phi)]$.

In words, *complex numbers can be multiplied by multiplying their absolute values and adding their arguments.*

Notice that in Exercise 1.11 we do not necessarily get the principal value of the argument of the product even if we started with the principal arguments of the factors; but we do get a unique point in the plane. For example,

$-1 = \cos\pi + i\sin\pi$ and $i = \cos(\pi/2) + i\sin(\pi/2)$, so that $-i = \cos(3\pi/2) + i\sin(3\pi/2)$; however, the principal argument of $-i$ is $-\pi/2$.

The procedure in Exercise 1.11 is often simpler than using the original definition of multiplication.

EXERCISE 1.12 (a) Square $(1 + i\sqrt{3})$ by using the definition of multiplication and also by writing $1 + i\sqrt{3}$ in polar coordinate form. (b) Find the product $(1 + i\sqrt{3})(\sqrt{3} - i)$ by both methods. (c) Calculate $(1 + i\sqrt{3})^{99}$.

Supplementary exercises

1. Find the absolute values and principal angles of the following complex numbers:

(*a*) $2 - 5i$ (*b*) $2.718 - 3.010i$ (*c*) $\dfrac{3 + 2i}{5i - 4}$

(*d*) $\cos 250° + i\sin 250°$ (*e*) $3\left(\cos\dfrac{4\pi}{3} + i\sin\dfrac{4\pi}{3}\right)$

2. Answer the following questions both if angles are principal angles, and if angles are between 0 and 2π.

(*a*) $\arg z + \arg \bar{z} = ?$
(*b*) $\arg(z \pm \bar{z}) = ?$
(*c*) $\arg(z/\bar{z}) = ?$
(*d*) $\arg\overline{(z - \bar{z})} = ?$

1F. Powers and roots of complex numbers

If we apply the trigonometric form of multiplication to the product zz, where $z = r(\cos\theta + i\sin\theta)$, we get $z^2 = r^2(\cos 2\theta + i\sin 2\theta)$. Continuing by induction, we see that if n is a positive integer, we have $z^n = r^n(\cos n\theta + i\sin n\theta)$. This representation of z^n is **De Moivre's formula**.

EXERCISE 1.13 Show that De Moivre's formula is also valid if n is a negative integer and $z \neq 0$.

The next exercise illustrates the advantage of using De Moivre's formula instead of straightforward computation.

EXERCISE 1.14 Find the modulus and argument of

$$z_0 = \frac{i + \sqrt{3}}{-1 - i}.$$

Then compute z_0^{123}.

Although a positive real number x has two square roots, the symbol $\sqrt{x}$ always means the positive square root. Thus $\sqrt{(-2)^2}$ is 2 and not -2. Also, $\sqrt{-1}$ always means the complex number i; the other square root of -1 is $-i$. Except when z is 0 or a positive real number or -1, the symbol $\sqrt{z}$ is

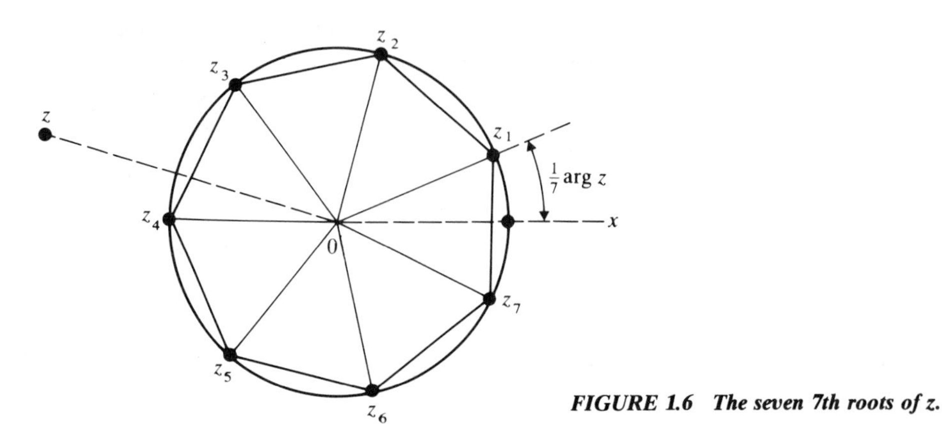

FIGURE 1.6 The seven 7th roots of z.

ambiguous, and we must be careful to specify which square root we are talking about.

De Moivre's formula suggests how nth roots are to be defined when n is a positive integer (greater than 1): *An nth root of the complex number z is any complex number whose nth power is z,* so $r^{1/n}[\cos(\theta/n) + i \sin(\theta/n)]$ certainly qualifies. But which value of θ are we to use, and how many different nth roots will we get?

If θ is some value of $\arg z$ $(z \neq 0)$, then $\theta \pm 2\pi$, $\theta \pm 4\pi$, and so on, are also possible values of $\arg z$; these give rise to values $(\theta/n) \pm (2\pi/n)$, $(\theta/n) \pm (4\pi/n), \ldots$ of $\arg(z^{1/n})$. However, the complex number of modulus $r^{1/n}$ and angle $(\theta/n) + (n \cdot 2\pi/n)$ is the same complex number as $r^{1/n}[\cos(\theta/n) + i \sin(\theta/n)]$. In other words, there are only n values of $z^{1/n}$ (if $z \neq 0$); they all have the same modulus but are located on n rays, each pair separated by the angle $2\pi/n$. Thus, *the nth roots of z* ($\neq 0$) *are located at the vertices of a regular n-gon* (see Fig. 1.6).

EXERCISE 1.15 Show, without using De Moivre's theorem, that every complex number (except 0) has two square roots. [In $(u + iv)^2 = x + iy$, you can find the real numbers x and y by solving a quadratic equation.] Find the square roots of $4 + 10i$ by this method.

EXERCISE 1.16 Find (by using De Moivre's theorem) the square roots of i, $1 + i$, $4 + 10i$, $1 + i\sqrt{3}$; find the fourth roots of -1, the cube roots of $8i$, and the sixth roots of $64(\cos 60° + i \sin 60°)$.

In problems about roots of complex numbers, especially if you want to plot them, it is often more convenient to use degrees instead of radians to measure angles. In the first place, radian protractors are not common; in the second place, I, at least, cannot quickly form a mental picture of what an angle of

$3\pi/5$ looks like, even after decimalizing it as 1.88496; on the other hand, an angle of 108° is easy to visualize.

EXERCISE 1.17 Show that *the sum of the n nth roots of every nonzero complex number w is zero.* (The n nth roots of w are the roots of the equation $z^n - w = 0$.)

*EXERCISE 1.18. Show that a, b, c are the vertices of an equilateral triangle if and only if $a^2 + b^2 + c^2 = ab + ac + bc$.

Supplementary exercises

1. Find (in the form $a + bi$)

(*a*) $(1 - i\sqrt{3})^{85}$

(*b*) $\left(\dfrac{\cos 60° + i \sin 60°}{\cos 45° + i \sin 45°}\right)^{12}$

2. Find the following roots:

(*a*) $(-2i)^{1/2}$ (*b*) $[4(1 - i)]^{1/2}$ (*c*) $(16i + 4)^{1/2}$

(*d*) The 4th roots of $-i$

(*e*) The cube roots of 27

(*f*) The 8th roots of $256(\cos 80° + i \sin 80°)$

3. Find the sum of the squares of the 4th roots of $625[\cos(4\tan^{-1}\frac{4}{3}) + i\sin(4\tan^{-1}\frac{4}{3})]$. (This is easier than it looks at first sight.)

4. Find all the real numbers x and y that satisfy each of the following equations:

(*a*) $x + iy = 3i - 4$

(*b*) $2(x + iy) = (x + iy)^2$

(*c*) $|2 - (x - iy)| = x + iy$

(*d*) $|x + iy| = 2(x + iy)$

(*e*) $\dfrac{x - iy}{x + iy} = i$

(*f*) $|x + iy - 2i| = i(x - iy - 4)$

1G. Limits and continuity

The **distance** between two complex points z_1 and z_2 is $|z_1 - z_2|$, and the distance between two real points x_1 and x_2 is $|x_1 - x_2|$. Consequently, all properties of sets that are expressed exclusively in terms of distances, the definition of the limit of a sequence of numbers, and theorems about convergence, limits, and so on, are formally the same in the complex plane and on the real line. For example, $\lim_{n\to\infty} z_n = z_0$ means that, given $\varepsilon > 0$, there is an $N > 0$ such that $|z - z_0| < \varepsilon$, provided that $n > N$. Notice that, although for real numbers x_n we can express the limit relation by writing $x_0 - \varepsilon < x_n < x_0 + \varepsilon$, we cannot write anything similar for complex numbers, because the complex numbers are not ordered.[4]

In other words, everything is the same as in ordinary calculus except that we may use inequalities only between real numbers. I shall therefore take theorems about limits and convergence for granted and emphasize only properties that are different in the complex plane and on the real line. Notice particularly that convergence in the plane is based on approach in the two-dimensional sense; that is, it is not approach in any particular direction. This is possibly one of the most difficult concepts in complex analysis. It is easy to think of a limit as being independent of the *direction* in which $z \to z_0$, but the definition also covers approach *without* any direction. Think of yourself as being at z_0, and z as a fly buzzing erratically around you; if the fly is to approach you as a limit, it must eventually come and remain very close to you.

EXERCISE 1.19 Show that $\lim_{n\to\infty} z_n = z_0$ if and only if $\lim_{n\to\infty} x_n = x_0$ and $\lim_{n\to\infty} y_n = y_0$ (where $z_n = x_n + iy_n$, $z_0 = x_0 + iy_0$).

EXERCISE 1.20 Show that if $\lim_{n\to\infty} |z_n| = |z_0|$ and if $\lim_{n\to\infty} \arg z_n = \arg z_0$ (with some choice of each arg), then $z_n \to z_0$. (We are tacitly assuming that $z_0 \neq 0$, since $\arg 0$ is not defined.) Conversely, show that if $z_n \to z_0 \neq 0$, then $|z_n| \to |z_0|$, and that $\arg z_n$ and $\arg z_0$ can be specified so that $\arg z_n \to \arg z_0$.

1H. Miscellaneous exercises on complex numbers

EXERCISE 1.21 Notice that a complex number is zero only if its real and imaginary parts are both zero. Using this principle:

(*a*) Find all complex numbers z that satisfy $|z - i| = 2z + i$.

(*b*) Find all real numbers x and y that satisfy $1 + x + iy = (2 + 3i)(x - iy)$.

(*c*) Find all real numbers x and y that satisfy $2ix + 3 = y - i$.

EXERCISE 1.22 Show that if $|w| < 1$, then

$$\left|\frac{z - w}{\overline{w}z - 1}\right| < 1$$

if $|z| < 1$, and

$$\left|\frac{z - w}{\overline{w}z - 1}\right| = 1$$

if $|z| = 1$.

EXERCISE 1.23 Let $P_n(z) = a_n z^n + \cdots + a_0$, $a_n \neq 0$, $n \geq 1$, be a polynomial. Show that when $|z| = R$, and R is sufficiently large, we have $|P_n(z)| \geq \frac{1}{2}|a_n| R^n$. In other words, $|P_n(z)| \to \infty$ as $|z| \to \infty$, just as fast in any one direction as in any other direction.

EXERCISE 1.24 Show that if we define the complex number $z = x + iy$ to be the matrix

$$z = \begin{pmatrix} x & y \\ -y & x \end{pmatrix}$$

and $w = u + iv$ similarly, then the matrix product

$$\begin{pmatrix} x & y \\ -y & x \end{pmatrix}\begin{pmatrix} u & v \\ -v & u \end{pmatrix}$$

corresponds to zw.

EXERCISE 1.25 Verify that the matrix representation (Exercise 1.24) of complex numbers is the representation of the composition of a magnification (or minification) by the factor r and a rotation about the origin through the angle θ. Show also that if $w = s(\cos\varphi + i\sin\varphi)$, then wz corresponds to a magnification factor rs and a rotation about the origin through the angle $\theta + \varphi$.

EXERCISE 1.26 Another method of defining complex numbers is to identify $z = x + iy$ with the vector $\mathbf{z} = x\mathbf{i} + y\mathbf{j}$, where $\mathbf{i}$ and $\mathbf{j}$ are the unit basis vectors in the plane; don't confuse $\mathbf{i}$ with i. Verify that the scalar product $\mathbf{z} \cdot \mathbf{w}$ is the real part of $\bar{z}w$ and that the $\mathbf{k}$ component of the vector product $\mathbf{z} \times \mathbf{w}$ is the imaginary part of $\bar{z}w$.

Notes

1. The complex plane is sometimes called the **Argand diagram**. Argand's geometric interpretation of complex numbers dates from 1806, but he was preceded by Bessel in 1799; the idea was popularized by Gauss in 1831.
2. Another notation, used mostly by engineers, is $r \operatorname{cis} \theta = r(\cos\theta + i\sin\theta)$.
3. The terms "angle" and "phase" are also used. The dual use of "argument" for the angle θ and as in the phrase, "the argument of a function" is unfortunate, but it is conventional.
4. There is no useful way to order the complex numbers—that is, there is no way to order them and at the same time preserve the algebraic properties of the ordering of the real numbers. For real numbers, if $x > 0$ and $y > 0$, then $xy > 0$. If this were to hold for complex numbers, then if $i > 0$ we would have $-1 = i \cdot i > 0$; and if $i < 0$ we would have $-i > 0$ and therefore $(-i) \cdot (-i) = -1 > 0$. For a more elaborate discussion of ordering complex numbers, see R. C. Weimer, "Can the Complex Numbers Be Ordered?," *Two-Year College Math. J.* 7, no. 4 (1976): 10–12.

2 *Functions*

2A. Functions from the complex plane to the complex plane

It is usually important to specify the domain of a function that we are studying. When we do not do this, we are probably assuming that the domain is either the whole plane or the plane minus some isolated points where a formula defining the function breaks down.

A point of notation: In general, we shall be careful to distinguish between f, the name of a function, and $f(z)$, the value of f at the point z. However, especially when the function is defined by a complicated formula, it seems to me to be unnecessarily pedantic not to allow the formula to serve as the name of the function, by way of abbreviation.

Now, a function from the complex plane to the complex plane is simply an assignment of a complex number to each point of the domain of the function, and is formally identical to a pair of "functions of two real variables." In fact, every function f from the complex plane to the complex plane can be thought of as defined by $f(x+iy)=u(x, y)+iv(x, y)$, where u and v are real-valued functions. However, as on the real line, there is not very much of interest to say about perfectly general functions,[1] and we shall progressively narrow the class of functions that we study.

The definitions of **continuity** and of **derivatives** are formally exactly the same as for functions from the real line to the real line. However, you must always remember that limits are taken in the two-dimensional sense. Notice that to say that a function is "differentiable" means different things here and in real two-dimensional calculus. In ordinary calculus we do not usually try to define "the derivative" of a function in more than one dimension, but only its partial derivatives or its differential. Here, however, the derivative is a single number,

$$\lim_{z\to z_0}\frac{f(z)-f(z_0)}{z-z_0},$$

where the limit is taken in the two-dimensional sense.

In the complex plane, the functions that have interesting properties turn out to be the ones that have derivatives at every point of an open connected set. Here "connected" means that every two points of the set can be joined by a polygonal line, without self-intersections, that is in the set; a nonempty connected open set is called a **region**.[2] A function that has a derivative at every point of a region is said to be **analytic** there (the synonym "holomorphic" is common). Functions that have derivatives only at isolated points are not very interesting. By convention, *to say that f is* ***analytic at a point*** *means that f is analytic in a neighborhood of the point. In addition, if S is any set, to say that f is analytic on S means that f is analytic on an open set containing S.*

2B. Landau's o and O notation

Landau's notation is very convenient. We write

$$f(z)=o(g(z))$$

(with the understanding that z is near some point z_0, possibly ∞, that we are interested in) to mean

$$\lim_{z\to z_0}\frac{f(z)}{g(z)}=0;$$

and

$$f(z) = O(g(z))$$

to mean that $|f(z)/g(z)|$ is bounded in a neighbourhood of z_0. We can think of "$f(z) = o(g(z))$" as saying "f is of smaller order than g as $z \to z_0$" and "$f(z) = O(g(z))$" as saying "f is not of larger order than g as $z \to z_0$." For example, $f(z) = o(1)$ means that $f(z) \to 0$; $f(z) = o(z^2)$ means that $f(z)/z^2 \to 0$. Thus, as $x \to 0$, $x^2 = o(x)$ and $1 + x + x^2 = O(1)$, whereas as $x \to \infty$, $x = o(x^2)$ and $1 + x + x^2 = O(x^2)$.

We also write $o(1)$ to stand for an unspecified function whose limit is 0, and $O(1)$ to stand for an unspecified bounded function.

Notice that "$f(z) = o(g(z))$" involves a one-way equality sign: It cannot be converted into $o(g(z)) = f(z)$.

The reason for introducing this notation right now is that it is convenient for writing the definition of a derivative. Since

$$\frac{f(z) - f(z_0)}{z - z_0} \to f'(z_0) \quad \text{as} \quad z \to z_0,$$

we have

$$\frac{f(z) - f(z_0)}{z - z_0} = f'(z_0) + o(1)$$

or

$$f(z) - f(z_0) = (z - z_0)f'(z_0) + o(z - z_0).$$

Supplementary exercises

1. Write in O, o notation as much information as you can express in this way of the content of the following statements:
   (*a*) $f(x)/g(x) \to 0$ as $x \to 0$.
   (*b*) $-2x^3 < f(x) \leq 5x^3$ for $x > 500$.
   (*c*) $x^3 f(x) \to 0$.
   (*d*) $|x^3 f(x)| \to \infty$.
2. Does the statement $f(x) = o(g(x))$ imply anything about $g(x)$? If so, what?
3. If $f(x) \neq o(g(x))$, can we say that $g(x) = o(f(x))$? If not, what *can* we say?
4. If $f(x) = o(x)$ as $x \to \infty$, is it correct to say that $f(x) = O(x)$? $f(x) = O(x^2)$? $f(x) = O(x^{1/2})$? Answer the same questions if $x \to 0$.

2C. Derivatives

Since the definitions of limits and derivatives are formally the same in the real and complex cases, we are going to have the same rules as in ordinary calculus for differentiating sums, products, and quotients. Furthermore, polynomials and rational functions (quotients of polynomials) are analytic, and their derivatives obey the familiar formulas, such as $(d/dz)z^n = nz^{n-1}$ when n is a positive or negative integer. For example, when $n > 0$, this formula results from noticing that it is correct when $n = 1$ and then applying it repeatedly. We cannot yet, however, apply the rule to z^n when n is not an integer, since we have not yet defined even $z^{1/2}$, much less z^{π}. [If you object that we defined the square root of z in Sec. **1F**, you must remember that we defined *two* square roots of z, so that $z^{1/2}$ is not yet defined as a (single-valued!) function.]

At this point we begin to lose the guidance of the real case. On the real line, a function can have a derivative everywhere, yet have no second derivative (consider, for example, the integral of a continuous nowhere differentiable function). In contrast, we shall see in Sec. **7C** that in the complex plane a function that is analytic on an open set has derivatives of all orders on that set. After this section we shall confine our attention to analytic functions.

Notice, in particular, that the **chain rule**,

$$\frac{d}{dz} f(g(z)) = f'(g(z))g'(z),$$

is just as correct for functions in the plane as for functions on the real line. I write the proof in the notation of Sec. **2B**.

Since

$$f(z) - f(z_0) = (z - z_0)[f'(z_0) + o(1)],$$

we have

$$\begin{aligned} f(g(z)) - f(g(z_0)) &= [g(z) - g(z_0)][f'(g(z_0)) + o(1)] \\ &= (z - z_0)[g'(z_0) + o(1)][f'(g(z_0)) + o(1)] \\ &= (z - z_0)g'(z_0)f'(g(z_0)) + o(1). \end{aligned}$$

Here we have tacitly appealed to the principle that $o(1)$ multiplied by a bounded function is still $o(1)$.

I now ask you to work through the following exercises, on the principle that you will understand analytic functions better after you have tried to prove the theorems yourself. If you are too impatient, you can read the proofs at the back of the book.

EXERCISE 2.1 Show that if $u(x, y)$ and $v(x, y)$ are continuous at (x_0, y_0), and if the function f, defined by $f(z) = u(x, y) + iv(x, y)$, has a derivative at $z_0 = (x_0, y_0) = x_0 + iy_0$, then the first partial derivatives of u and v exist at z_0, and

$$\frac{\partial u}{\partial x} = \frac{\partial v}{\partial y} \quad \text{and} \quad \frac{\partial u}{\partial y} = -\frac{\partial v}{\partial x}$$

at z_0. These are the **Cauchy-Riemann equations**. Show that they can also be written as the single complex equation

$$\frac{\partial f}{\partial x} = -i\frac{\partial f}{\partial y}.$$

(The limit defining the derivative is a limit in the two-dimensional sense. If it exists, it can be evaluated by letting $z \to z_0$ along any curve we like—in particular, along a line parallel to the positive x axis and along a line parallel to the positive y axis. Write out the definition of the derivative for each of these cases and equate the results.) You have shown that the Cauchy-Riemann equations provide a *necessary* condition for f to have a derivative at a point. They are, however, not sufficient; see the next exercise.

EXERCISE 2.2 (a) Define a function f by $f(z) = z^4/|z|^3$ for $z \neq 0$, $f(0) = 0$. Show that f is not differentiable at $z = 0$, although the Cauchy-Riemann conditions are satisfied there. [Suggestion: Write $f(z)$ in terms of polar coordinates.]

(b) Let $f(z) = |z|^{3/2}$. Show that f has a derivative at 0 but not anywhere else, and consequently f is not analytic at 0.

In practice, one rarely meets cases in which the Cauchy-Riemann conditions are satisfied but the function is not analytic.

For the next exercise you need to recall the concept of the **differential** (or total differential) from calculus in two dimensions. If g and its first partial derivatives are continuous in a neighborhood of (x_0, y_0), then in that neighborhood, with the notation of Sec. **2B**,

$$u(x, y) - u(x_0, y_0) = (x - x_0)\frac{\partial u}{\partial x}(x_0, y_0) + (y - y_0)\frac{\partial u}{\partial y}(x_0, y_0)$$
$$+ o(x - x_0) + o(y - y_0),$$

or in the condensed notation for partial derivatives (which we will use a good deal),

$$u(x, y) - u(x_0, y_0) = (x - x_0)u_1(x_0, y_0) + (y - y_0)u_2(x_0, y_0)$$
$$+ o(x - x_0) + o(y - y_0).$$

Here is the proof, in case you do not remember it.

We have

$$\begin{aligned} u(x_0 + h, y_0 + k) - u(x_0, y_0) &= u(x_0 + h, y_0 + k) - u(x_0 + h, y_0) \\ &\quad + u(x_0 + h, y_0) - u(x_0, y_0) \\ &= hu_1(x_0 + h', y_0) + ku_2(x_0 + h, y_0 + k'), \end{aligned}$$

where h' is between x_0 and $x_0 + h$ and k' is between y_0 and $y_0 + k$, by the ordinary mean value theorem. Since u_1 and u_2 are continuous functions, we can (by the definition of continuity) write $u_1(x_0 + h', y_0) = u_1(x_0, y_0) + o(1)$, and similarly for u_2.

EXERCISE 2.3 Let $f(z) = u(x, y) + iv(x, y)$ in a neighborhood N of z_0, let u and v be continuous in this neighborhood, and let u and v also have continuous first partial derivatives. Show that if the Cauchy-Riemann equations are satisfied in N, the function f has a derivative at z_0 (and of course throughout N).[3] Outline:

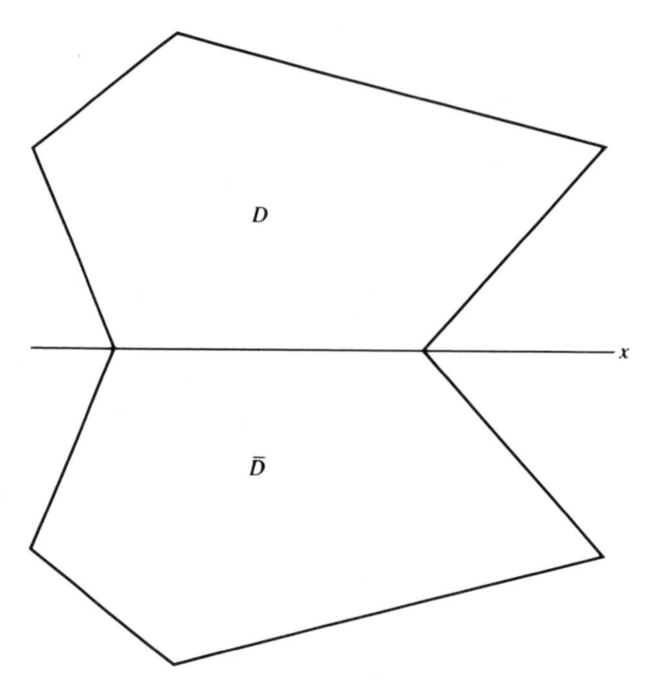

FIGURE 2.1

u and v have total differentials in N; write the difference quotient of f in terms of u and v and apply the Cauchy-Riemann equations.

EXERCISE 2.4 We can also write $f(z) = U(r, \theta) + iV(r, \theta)$, where (r, θ) are plane polar coordinates. Show that if f is differentiable at $z_0 = r_0 \cos \theta_0 + i \sin \theta_0 \neq 0$, then

$$\frac{\partial U}{\partial r} = \frac{1}{r}\frac{\partial V}{\partial \theta} \quad \text{and} \quad \frac{\partial V}{\partial r} = -\frac{1}{r}\frac{\partial U}{\partial \theta}$$

at this point. [These are the Cauchy-Riemann equations for z in polar coordinates and $f(z)$ in rectangular coordinates.]

EXERCISE 2.5 Let $f(z) = u(x, y) + iv(x, y)$ and let f satisfy the Cauchy-Riemann equations in a set D. Let $\bar{D}$ be the set obtained by reflecting D with respect to the real axis; that is, (x, y) is in $\bar{D}$ if $(x, -y)$ is in D. (See Fig. 2.1.) Define the function g in $\bar{D}$ by $g(z) = \overline{f(\bar{z})}$. Show that g satisfies the Cauchy-Riemann conditions in $\bar{D}$.

EXERCISE 2.6 Let f satisfy the Cauchy-Riemann equations in the disk $|z| < R$ and define $g(z) = \overline{f(R^2/\bar{z})}$ for z outside the same disk. Show that g satisfies the Cauchy-Riemann equations. (Use polar coordinates.)

The point of Exercise 2.6 is that $R^2/\bar{z}$ is called the point *symmetric* to z with respect to the circle $|z| = R$, and plays a role similar to that of the reflection of a point with respect to a straight line.

There is another way of looking at the Cauchy-Riemann equations that is interesting and sometimes useful. We have $x = \frac{1}{2}(z + \bar{z})$, $y = -\frac{1}{2}i(z - \bar{z})$. If we make an entirely formal calculation (that is, without letting ourselves think what, if anything, it means) to find $\partial f/\partial z$, we get

$$\frac{\partial f}{\partial z} = \frac{\partial f}{\partial x}\frac{\partial x}{\partial z} + \frac{\partial f}{\partial y}\frac{\partial y}{\partial z} = \frac{1}{2}\frac{\partial f}{\partial x} - \frac{i}{2}\frac{\partial f}{\partial y} = \frac{1}{2}\left(\frac{\partial u}{\partial x} + i\frac{\partial v}{\partial x} - i\frac{\partial u}{\partial y} + \frac{\partial v}{\partial y}\right).$$

If f satisfies the Cauchy-Riemann equations, this is indeed

$$\frac{\partial f}{\partial z} = \frac{\partial u}{\partial x} + i\frac{\partial v}{\partial x} = f'(z).$$

Similarly, we have (still formally)

$$\frac{\partial f}{\partial \bar{z}} = \frac{1}{2}\left(\frac{\partial f}{\partial x} + i\frac{\partial f}{\partial y}\right) = \frac{1}{2}\left(\frac{\partial u}{\partial x} - \frac{\partial v}{\partial y} + i\frac{\partial v}{\partial x} + i\frac{\partial u}{\partial y}\right).$$

If f satisfies the Cauchy-Riemann equations, it follows that

$$\frac{\partial f}{\partial \bar{z}} = 0;$$

conversely, this equation leads back to the Cauchy-Riemann equations. It was calculations of this sort that led mathematicians of a considerably older generation to say that an analytic function is "independent of $\bar{z}$." Taken literally, this last statement makes no sense, of course; but we can *define* the operator $\partial/\partial\bar{z}$ to mean

$$\frac{1}{2}\left(\frac{\partial}{\partial x} + i\frac{\partial}{\partial y}\right),$$

and then it is true that $(\partial/\partial\bar{z})f(z) = 0$ when f is analytic. The operator $\partial/\partial\bar{z}$ is useful in various contexts. For example, Green's theorem in the plane is usually stated in the form

$$\int u\,dx + v\,dy = \iint\left(\frac{\partial v}{\partial x} - \frac{\partial u}{\partial y}\right)dx\,dy,$$

but it can be written compactly, with $f(z) = u(x, y) + iv(x, y)$, as

$$\int f(z)\,dz = 2i\iint\frac{\partial f}{\partial \bar{z}}\,dx\,dy.$$

In fact,

$$\int_C f(z)\,dz = \int_C (u+iv)(dx+i\,dy)$$
$$= \int (u\,dx - v\,dy) + i\int (v\,dx + u\,dy),$$

and if we apply the ordinary form of Green's theorem to each integral, we get

$$\int_C f(z)\,dz = \iint\left(-\frac{\partial v}{\partial x}-\frac{\partial u}{\partial y}\right) + i\iint\left(\frac{\partial u}{\partial x}-\frac{\partial v}{\partial y}\right)$$
$$= 2i\iint \frac{\partial f}{d\bar{z}}.$$

EXERCISE 2.7 Show that the mean value relation

$$f(w)-f(z)=(w-z)f'(t), \qquad t \text{ between } w \text{ and } z,$$

fails to hold in general for analytic functions on the complex plane. Suggestion: Try $f(w)=w^4$, $w=i$, $z=1$; or try to find your own example.

EXERCISE 2.8 Show that if f is analytic and f' is continuous in a region D, and $|f|$ is constant in D, then f is constant in D.

EXERCISE 2.9 Show that if f is analytic and f' is continuous in a region D, and the range of f is real, then f is constant in D. Hence show that if $f \equiv 0$ in a disk in D, then $f \equiv 0$ throughout D.

The only analytic functions that we have seen so far are polynomials and rational functions. In ordinary calculus one goes on to consider trigonometric functions, logarithms and exponentials, and so on, the **elementary functions**. In order to do this in the complex case, we would have to know what symbols like $\sin i$, $\log(2+3i)$, or e^{5+i} mean. Since symbols do not mean anything until we define them, it is better to ask what are useful definitions of such symbols. Rather than making special definitions in each case, we shall define the principal elementary functions as limits of polynomials; this is a reasonable procedure, since we already know about both polynomials and limits. We shall actually use a special kind of limit of polynomials—namely, power series—which we discuss in the next section.

Notes

1. See, for example, H. Blumberg, "New Properties of All Real Functions," *Trans. Amer. Math. Soc. 24* (1922): 113–128.
2. The word "domain" is often used instead of "region," and then "region" is used to mean a region in our sense together with some or all of its boundary points. This use of "domain" has the disadvantage that a statement like "the domain of f is a domain" may sound peculiar.

3. There are weaker conditions than the continuity of the partial derivatives that make $f = u + iv$ analytic if the Cauchy-Riemann conditions are satisfied. One of the most simply stated conditions is that f is analytic in a region D if $|f|$ is locally (Lebesgue) integrable, and the first partial derivatives of u and v exist, are finite, and satisfy the Cauchy-Riemann conditions almost everywhere in D. This was proved recently by G. H. Sindalovskiĭ, "The Cauchy-Riemann Conditions in the Class of Functions with Summable Modulus, and Some Boundary Properties of Analytic Functions," *Mat. Sb. 128*(170), no. 3 (1985): 364–382; an English translation will appear in *Math. USSR Sb. 57* (1987), no. 2. For a summary, see *Soviet Math. Dokl. 29* (1984), 504–505. A special case is the classical Looman-Menshov theorem, in which f is supposed continuous and the partial derivatives are assumed to exist except on a countable set. For references to this theorem and its earlier generalizations, see Burckel, p. 51.

3 *Power Series*

3A. Terminology

An **infinite series** of complex numbers is defined just as an infinite series of real numbers is defined, as the **sequence of its partial sums**. The **sum** of the series (if the series converges) is defined as the limit of the sequence of partial sums. Consequently, almost everything you know about infinite series carries over to the complex case. The exceptions are the properties that depend on the ordering of the real numbers; for example, we do not have an alternating series test, because "alternation of signs" is not a meaningful concept for complex numbers.[1] (See Note 4, Sec. **1**.)

We can think of a **power series** informally as being a polynomial of infinite degree, that is, as

$$a_0 + a_1 z + a_2 z^2 + \cdots \quad \text{or} \quad \sum_{k=1}^{\infty} a_k z^k.$$

The sum of the series is then, as it is in the real case, defined to be

$$\lim_{n\to\infty} \sum_{k=0}^{n} a_k z^k.$$

For a given sequence of coefficients and a given z, this limit may or may not exist; if it does exist for z in a set S, it defines a function whose domain is S. A series of the form

$$\sum_{k=0}^{\infty} a_k (z - z_0)^k$$

is also called a power series.

EXERCISE 3.1 Show that the series $\sum_{k=0}^{\infty} (z+\frac{1}{2})^k$ converges in the set where $|z+\frac{1}{2}|<1$ but that if the powers of $(z+\frac{1}{2})$ are multiplied out and the result arranged as a series in powers of z, this power series does not converge at $z=-1$, even though the original series converges at this point.

A series $\sum_{k=0}^{\infty} a_k z^{-k}$ is also sometimes called a power series. However, in general a series

$$\sum_{k=0}^{\infty} a_k [f(z)]^k$$

is not called a power series unless $f(z) = z - z_0$, even though it is a series of powers of f.

3B. Domain of convergence of a power series

You probably recall that a real power series $\sum_{k=0}^{\infty} a_k x^k$ has an interval of convergence, which may reduce to a point or be the whole real line. The corresponding theorem for complex power series is that *the series either converges only at a single point, or converges in a disk, or converges in the whole plane.* It may seem surprising that there are so few possibilities, but it is perhaps less surprising if you think of an open interval as the one-dimensional analog of an open disk (the set of points distant less than r from a given point).

I give the proof for series $\sum_{k=0}^{\infty} a_k z^k$; the proof is the same for $\sum_{k=0}^{\infty} a_k (z - z_0)^k$, but it takes up more space. To begin with, notice that $\sum_{k=0}^{\infty} a_k z^k$ always converges at the point $z=0$, since when $z=0$, all the terms except the first are zero. (Because $\sum_{k=0}^{\infty} a_k z^k$ is an abbreviation for $a_0 + a_1 z + a_2 z^2 + \cdots$, the beginning term is obtained by putting $k=0$ first, and then $z=0$.) It is possible that the series does not converge for any other value of z.

EXERCISE 3.2 Show that $\sum_{n=1}^{\infty} n^n z^n$ diverges except for $z=0$.

Now suppose that $\sum_{k=1}^{\infty} a_k z^k$ does converge for some nonzero value b of z. Since (just as in the real case) the terms of a convergent series tend to 0, we have $a_k b^k \to 0$. That is, if ε is any (presumably small) positive number, we have $|a_k b^k| < \varepsilon$ if k is large enough, say for $k \geq N$. The first N terms of the series do not affect its convergence, so we need to look only at the terms with $k \geq N$. For these terms, $|a_k z^k| = |a_k|\,|z/b|^k\,|b|^k \leq \varepsilon\,|z/b|^k$. That is, each term of our series (after the Nth) has its absolute value less than or equal to the corresponding term of the series $\sum |z/b|^k$. We express this by saying that the second series **dominates** the first. If $|z| < |b|$, the dominating series is a geometric series with ratio less than 1, and therefore converges. Hence $\sum_{k=0}^{\infty} a_k z^k$ converges by the comparison test (next exercise).

EXERCISE 3.3 Prove the **comparison test** for series of complex numbers: If c_k are nonnegative numbers and $\sum c_k$ converges and dominates $\sum z_k$, then $\sum z_k$ converges.

Exercise 3.3 shows, incidentally, that, just as for series of real numbers, an absolutely convergent series converges. It also shows that if $\sum |z_k|$ converges, so do $\sum \operatorname{Re} z_k$ and $\sum \operatorname{Im} z_k$, because $|\operatorname{Re} z| \leq |z|$ and $|\operatorname{Im} z| \leq |z|$. However, a stronger result follows from Exercise 1.19, namely, that *$\sum \operatorname{Re} z_k$ and $\sum \operatorname{Im} z_k$ converge if $\sum z_k$ converges.*

We have now shown that if $\sum a_k z^k$ converges at $z = b$, then it converges in the disk $|z| < |b|$. It may, of course, converge for some points of larger absolute value, and possibly for all values of z.

EXERCISE 3.4 Show that $\sum_{n=1}^{\infty} z^n/n^n$ converges for every z. [You have to consider only real positive z (why?).]

If our power series does not converge for every z, there must be some z for which it diverges. It then diverges for every z of larger absolute value.

EXERCISE 3.5 Prove the preceding statement.

Thus a power series that converges for some $z \neq 0$ but not for every z must converge inside some disk (center at 0) and diverge outside some larger disk. It follows that *there is a disk (the* **disk of convergence***) inside which the series converges and outside which it diverges.*

EXERCISE 3.6 Prove the preceding statement by considering the least upper bound r of numbers s such that the series converges for $|z| < s$.

Notice that we do not say anything about whether or not the power series converges *on* the boundary of the disk of convergence.

EXERCISE 3.7 Find power series (a) that converge at all points of the boundary of the disk of convergence, or (b) that converge at no points of the boundary. See also Exercise 4.18.

The radius of the disk of convergence is called the **radius of convergence**. We shall refer to the boundary of the disk of convergence as the **circle of convergence**. (You will find this term used in some books to mean the disk of convergence, because older books did not have separate terms for a disk and its boundary.)

If we know the coefficients of a power series $\sum_{k=0}^{\infty} a_k (z - z_0)^k$, *we can calculate its radius of convergence R by the formula*

$$\frac{1}{R} = \limsup_{n \to \infty} |a_n|^{1/n}.$$

I remind you that

$$\limsup_{n \to \infty} x_n = L$$

means, when L is not ∞, that, given $\varepsilon > 0$, we have both $x_n < L + \varepsilon$ for all sufficiently large n, and $x_n > L - \varepsilon$ for an infinite sequence of values of n.

If the formula produces the value 0 for $1/R$, this means that $R=\infty$, that is, the series converges for all z; if $1/R=\infty$, this means that the series converges only at z_0.

EXERCISE 3.8 Formulate a definition for

$$\limsup_{n\to\infty} x_n = \infty.$$

EXERCISE 3.9 Prove the formula for $1/R$, including the cases 0 and ∞.

EXERCISE 3.10 Find the radii of convergence of the following power series. Using the formula for $1/R$ is not necessarily the most efficient method.

(*a*) $\sum_{n=0}^{\infty} (-1)^n (z+i)^n$ (*b*) $\sum_{n=0}^{\infty} (-1)^n 2^n z^{2n+2}$

(c) $1+3z+2z^2+9z^3+\cdots+2^n z^{2n}+3^{n+1}z^{2n+1}+\cdots$

(*d*) $\sum_{m=1}^{\infty} \frac{z^m}{\sqrt{m}}$ (*e*) $\sum_{p=1}^{\infty} \frac{2^p z^p}{(2p)!}$

(*f*) $\sum_{s=1}^{\infty} \frac{z^{2s}(s!)^2}{(2s)!}$ (*g*) $\sum_{k=1}^{\infty} \frac{k!}{k^k}(z+2)^k$

(*h*) $\sum_{n=0}^{\infty} 2^n z^{2^n}$ (*i*) $\sum_{n=2}^{\infty} 2^{\log n} z^n$

(*j*) $\sum_{n=0}^{\infty} \frac{z^{2n}\sqrt{(2n)!}}{n!}$ (*k*) $\sum_{n=1}^{\infty} z^n \left\{\frac{n^2}{(n+1)(n+2)}\right\}^n$

EXERCISE 3.11 Let $\sum_{k=0}^{\infty} a_k z^k$ have radius of convergence R and let m be a positive integer. Show that the series $\sum_{k=0}^{\infty} k^m a_k z^k$ and $\sum_{k=0}^{\infty} k(k-1)\cdots(k-m)a_k z^k$ both have radius of convergence R.

Supplementary exercises

1. Find the domains of convergence of the following power series:

(*a*) $\sum_{n=1}^{\infty} (-1)^n (2i-z)^n$ (*b*) $\sum_{n=1}^{\infty} \frac{n^{100}(z-3)^n}{n!}$

(c) $\sum_{n=1}^{\infty} (-i)^n (2z-3i)^{2n}$ (*d*) $\sum_{n=3}^{\infty} \frac{n+1}{n-2} z^n$

(*e*) $\sum_{n=1}^{\infty} n^{-1} z^{3^n}$ (*f*) $\sum_{n=1}^{\infty} 2^n (2-z)^{2n}$

(*g*) $\sum_{n=1}^{\infty} \frac{(n!)^2}{(2n)!} z^n$ (*h*) $\sum_{n=1}^{\infty} n^{-1/n} z^{n^2}$

2. Show that the sum $\sum (m+ni)^{-3}$, extended over all nonzero integers m and n, converges absolutely.

3C. Differentiation of functions defined by power series

EXERCISE 3.12 Use Exercise 3.11 to show that if we differentiate a power series formally, m times, then the differentiated series and the original series have the same radius of convergence.

We have just seen that the formal derivative of a power series has the same radius of convergence as the original series. This does not in itself say that the function defined by the power series has a derivative, or that if this series has a derivative the differentiated series converges to the derivative of the function. Both of these statements are actually valid, but they have to be proved. Later (Sec. **7D**) we shall be able to give a very short proof by using integration. It is possible to give a straightforward proof now, by operating directly with power series, but it is rather tedious.[2] You may recall from advanced calculus (or elementary real analysis) that an infinite series of functions on a real interval can be differentiated term by term if the differentiated series is uniformly convergent, but it is not easy to apply this theorem to the complex case at this point. We shall therefore *assume* for the present that *power series admit differentiation term by term*; you will be able to see later that we have not actually used the theorem as part of its own proof.

Supplementary exercise

1. Write power series for:

(*a*) The derivative of

$$\sum_{n=0}^{\infty} \frac{(-1)^n z^{2n}}{(2n)!}.$$

(*b*) The derivative of

$$\sum_{n=0}^{\infty} \frac{n!\, z^{2n}}{(2n)!}.$$

(*c*) The kth derivative of

$$\sum_{n=0}^{\infty} (-1)^n z^n \qquad (k = 1, 2, 3, \ldots).$$

3D. Uniform convergence of power series

We can show that *a power series* $\sum_{n=0}^{\infty} a_n (z - z_0)^n$ *converges uniformly in every disk* $|z - z_0| < R_1 < R_2$, *that is, in every disk that is strictly inside its disk of convergence.* (*Caution*: this does *not* mean that the convergence has to be uniform in the *open* disk of convergence. Of course the convergence is not necessarily uniform in the closed disk, because the series does not even have to converge in the closed disk; Exercise 3.7.) Let me remind you what **uniform convergence** means: To say that a series $\sum \phi_n(z)$ *converges uniformly* on the set S means that the Cauchy criterion for convergence is satisfied independently of where z is in S; that is, for every $\varepsilon > 0$ there is a number T, independent of z in S, such that we have

$$\left|\sum_{P}^{Q} \phi_n(z)\right| < \varepsilon$$

whenever P and Q are greater than T. Equivalently, we can say that the condition for uniform convergence is that

$$\sup_{z \in S} \left|\sum_{n=P}^{Q} \phi_n(z)\right| \to 0$$

as P and $Q \to \infty$.

It should be obvious from this definition that a *sufficient* condition for $\sum \phi_n(z)$ to converge uniformly in S is that there is a convergent series of numbers $\sum M_n$ such that $|\phi_n(z)| \le M_n$ for all sufficiently large n and all z in S. (This is the **Weierstrass *M* test** for uniform convergence; M stands for "majorant.") Another way of phrasing the test is to say that $\sum \phi_n(z)$ is *dominated* by a convergent series of numbers M_n that are independent of z when z is in S.

EXERCISE 3.13 Show that the M test is not a necessary condition for uniform convergence: That is, produce an example of a uniformly convergent series for which no suitable series $\sum M_n$ exists.

EXERCISE 3.14 Verify explicitly that $\sum_{n=0}^{\infty} z^n$ converges uniformly for $|z| < 1 - \delta$ (where $0 < \delta < 1$), but not uniformly for $|z| < 1$.

To save space, I give the proof of the uniform convergence of power series for $z_0 = 0$; for the general proof, just replace z by $z - z_0$ everywhere.

We proved in Sec. **3B** that the convergence of the power series for $z = b$ implies convergence for $|z| < |b|$. We actually showed that $|a_k z^k| \le \varepsilon\, |z/b|^k$ when $|z| < |b|$, so that if $|z| < c < |b|$, the series $\sum a_k z^k$ is dominated by the convergent geometric series $\sum \varepsilon\, |c/b|^k$. Hence the series converges uniformly, by the M test, for $|z| < c$.

The uniformity of convergence of power series will turn out to have important consequences later on.

Supplementary exercise

1. Do the following series converge uniformly in the sets indicated?

(a) $\displaystyle\sum_{n=1}^{\infty} \frac{z^n}{n^2} \qquad (|z| < 1)$

(b) $\displaystyle\sum_{n=1}^{\infty} [nz^n - (n+1)z^{n+1}] \qquad (|z| < 1)$

(*c*) $\sum_{n=1}^{\infty} 2^n(z-2)^n \qquad (|z-2|<\frac{1}{3})$

(*d*) $\sum_{n=1}^{\infty} n^{-1}(\sin x)^n \qquad \left(0<x<\frac{\pi}{2}\right)$

Notes

1. However, if we look at the right proof of the alternating series test, we can find a somewhat similar test for complex series. See Exercise 18.1.
2. To show directly that a convergent power series can be differentiated term by term, we would form the different quotient $h^{-1}[f(z+h)-f(z)] = \sum_{n=0}^{\infty} h^{-1}a_n[(z+h)^n - z^n]$. Since it is usually simpler to show that a limit is 0 than that it is something else, we would subtract the formally differentiated series and try to show that the limit of the difference is 0. The natural thing is to expand $(z+h)^n$ by the binomial theorem and then estimate what is left. The rest of the proof is a matter of calculation and occupies about a page. You may want to look ahead to Sec. **7D** and compare the very compact proof that can be given after we have developed enough apparatus.

4 *Some Elementary Functions*

4A. Trigonometric and exponential functions

We are now going to use power series to define many of the familiar functions of elementary calculus at points of the complex plane instead of just on the real line. The idea is straightforward: The sine, cosine, and exponential functions are shown in calculus to be represented on the whole real line by convergent Maclaurin series. Since we now know that a power series $\sum a_n z^n$ that converges for all real z must also converge for all complex z, we simply define these functions by the same power series. (The inverses of these functions are more difficult to deal with, and we shall discuss them in Sec. **10D**.) For example,

$$e^z = 1 + z + \frac{z^2}{2!} + \frac{z^3}{3!} + \dots,$$

$$\sin z = z - \frac{z^3}{3!} + \frac{z^5}{5!} - \dots,$$

$$\cos z = 1 - \frac{z^2}{2!} + \frac{z^4}{4!} - \dots.$$

We sometimes write $\exp(z)$ instead of e^z, especially when z is complicated. These series are going to be used so frequently that you should memorize them. Notice that since the coefficients in the series are real, each of these functions takes conjugate values at conjugate points.

These definitions may seem perfectly reasonable, but you really ought to wonder whether the new functions will behave as their names lead us to expect. For example: Is $e^z e^w = e^{z+w}$? Is $\cos z$ the derivative of $\sin z$? Is $\sin(z+w) = \sin z \cos w + \cos z \sin w$? Is $|\sin z| \le 1$? We will show in Sec. **4D** that the answers to the first three questions are "yes"; but we can see directly that the answer to the fourth is "no."

EXERCISE 4.1 Show by using the series for $\sin z$ that $|\sin i| > 1$. (Exercise 4.14 gives a much stronger result.)

We shall see later (Sec. **7G**) that there is a general principle that says, roughly, that formulas that involve only addition, subtraction, and multiplication remain true in the complex domain, whereas those that involve absolute values usually do not.

4B. The algebra of power series

We shall often want to add two (or more) power series, say $\sum a_n(z-z_0)^n$ and $\sum b_n(z-z_0)^n$ (with the same z_0), as if they were polynomials, that is, by adding them term by term. As long as z is in their common disk of convergence, this is legitimate because any two convergent infinite series can correctly be added term by term. We can also multiply power series just as we multiply polynomials, collecting terms of the same degree in the product. For example, to multiply $\sum_{n=0}^{\infty} z^n$ by $\sum_{m=0}^{\infty} (m+1)z^m$, we write

$$
\begin{array}{r}
1 + z + z^2 + z^3 + \cdots \\
1 + 2z + 3z^2 + 4z^3 + \cdots \\
\hline
1 + z + z^2 + z^3 + \cdots \\
2z + 2z^2 + 2z^3 + \cdots \\
3z^2 + 3z^3 + \cdots \\
4z^3 + \cdots \\
\hline
1 + 3z + 6z^2 + 10z^3 + \cdots
\end{array}
\qquad \text{(product)}
$$

It is often easier to use the formula for the coefficients of the product, which is (see the discussion below)

$$c_n = \sum_{k=0}^{n} a_k b_{n-k}.$$

In our example, $a_k = 1$, $b_k = k + 1$, and

$$\begin{aligned} c_n &= \sum_{k=0}^{n} (n - k + 1) \\ &= \sum_{j=1}^{n+1} j = \frac{(n+1)(n+2)}{2}. \end{aligned}$$

This procedure seems so natural that it would be surprising if it didn't work, but if we are conscientious we ought to verify that it does. The formal computation looks like this:

$$\begin{aligned} \sum_{k=0}^{\infty} a_k z^k \sum_{m=0}^{\infty} b_m z^m &= \sum_{k=0}^{\infty} a_k z^k \sum_{n=k}^{\infty} b_{n-k} z^{n-k} \\ &= \sum_{n=0}^{\infty} \sum_{k=0}^{n} a_k z^k b_{n-k} z^{n-k} \\ &= \sum_{n=0}^{\infty} \left(\sum_{k=0}^{n} a_k b_{n-k} \right) z^n. \end{aligned}$$

This looks somewhat forbidding; you may be happier to think of the second step as an analog of changing one of the iterated integrals $\int_0^\infty dx \int_x^\infty f(x, y)\, dy$ and $\int_0^\infty dy \int_0^y f(x, y)\, dx$ into the other—a conventional mathematical education seems to make people more comfortable with integrals than with sums. To give a rigorous proof that the product series computed this way does converge to the product of the functions represented by the two series requires a knowledge of the rearrangement theorem for absolutely convergent double series (which is tedious to prove), or else some results from later in this book (Sec. **15B**) (which, as in Sec. **3C**, do not depend on what we are doing right here).

Later (Secs. **15D** and **15E**) we shall see that, under natural restrictions, the power series for the quotient of one power series by another, and even the power series of a composition of two power series, can be obtained correctly by formal computation.

*4C. Multiplication of numerical series

It is interesting to notice that the multiplication of power series suggests a method of forming the product of two series $\sum_0^\infty a_n$ and $\sum_0^\infty b_n$ of numbers: Form the associated power series $\sum a_n z^n$ and $\sum b_n z^n$, multiply them, and then let $z = 1$. The result of this process, known as the **Cauchy product**, provides a product series $\sum c_n$ with

$$c_n = \sum_{k=0}^{n} a_k b_{n-k}.$$

It can be shown that this always gives the correct product for absolutely convergent series, or even (*Mertens's theorem*) for convergent series one of which is absolutely convergent. If both series are conditionally convergent, the product may diverge.

*EXERCISE 4.2 Show that the Cauchy product of the convergent series $\sum_{n=1}^{\infty}(-1)^n n^{-1/2}$ by itself is divergent.

It is, however, reassuring to know that *if all three series converge, the sum of the product series is sure to be the product of the sums of the factors* (Exercise 18.2).

4D. Some properties of the elementary functions

EXERCISE 4.3 Find the power series of $\cosh z = \frac{1}{2}(e^z + e^{-z})$.

EXERCISE 4.4 By combining the corresponding power series, show that $e^z e^w = e^{z+w}$ and that $\sin(z+w) = \sin z \cos w + \cos z \sin w$.

EXERCISE 4.5 Show that e^z is never 0.

EXERCISE 4.6 Show that $(d/dz)\sin z = \cos z$, and that $(d/dz)\, e^z = e^z$.

EXERCISE 4.7 Find several terms of the Maclaurin series of $(\cos z)^2$ by squaring the series for $\cos z$.

EXERCISE 4.8 Find the Maclaurin series of $e^z/(1-z)$ by multiplying the series for e^z and $(1-z)^{-1}$.

EXERCISE 4.9 Show that if f is analytic and f' is continuous in a region D, and $\operatorname{Re} f$ is constant in D, then f is constant in D. (Consider e^f.)

Supplementary exercises

1. Find the Maclaurin series of $\sinh z = \frac{1}{2}(e^z - e^{-z})$.
2. Verify from the power series that $\sinh 2z = 2 \sinh z \cosh z$.
3. Express $\sinh z$ in terms of the sine function.
4. Show that $\sinh z$ has infinitely many zeros.
5. Show that $e^{nz} = \cosh nz + \sinh nz$ $(n = 1, 2, 3, \ldots)$.
6. If $a_n \geq 0$ and $\sum_{n=1}^{\infty} n a_n x^{n-1}$ converges for $0 \leq x \leq 1$, show that $\sum_{n=0}^{\infty} a_n x^n$ converges in the same interval.

4E. The connection between trigonometric and exponential functions

Let θ be a real number and consider what $e^{i\theta}$ means. From the exponential series we have

$$e^{i\theta} = 1 + i\theta - \frac{\theta^2}{2!} - i\frac{\theta^3}{3!} + \frac{\theta^4}{4!} + \cdots$$

$$= 1 - \frac{\theta^2}{2!} + \frac{\theta^4}{4!} + \cdots + i\left(\theta - \frac{\theta^3}{3!} + \frac{\theta^5}{5!} + \cdots\right).$$

The two series on the right are, respectively, $\cos\theta$ and $\sin\theta$, so we have

$$e^{i\theta} = \cos\theta + i\sin\theta,$$

$$re^{i\theta} = r\cos\theta + ir\sin\theta.$$

But in terms of polar coordinates in the plane, we have $z = r\cos\theta + ir\sin\theta$, so that we can now write $z = re^{i\theta}$. Here $r = |z|$, since $|e^{i\theta}| = (\sin^2\theta + \cos^2\theta)^{1/2} = 1$.

We can now write De Moivre's formula more compactly as $(re^{i\theta})^n = r^n e^{in\theta}$. We can also write the nth roots of z as $z^{1/n} = r^{1/n}e^{i(\theta+2k\pi)/n}$, $k = 0, 1, 2, \ldots, n-1$.

EXERCISE 4.10 Show that

$$e^{iz} = \cos z + i\sin z,$$

$$\sin z = \frac{e^{iz} - e^{-iz}}{2i}, \quad \text{and} \quad \cos z = \frac{e^{iz} + e^{-iz}}{2}.$$

(The last two formulas are **Euler's formulas** for the sine and cosine.)

The following formulas will be used frequently:

$$e^{i\theta} = \cos\theta + i\sin\theta \quad \text{and} \quad |e^{i\theta}| = 1.$$

Many of the exercises in Sec. **1** can now be done more simply by using formulas like $e^{im\theta}e^{in\theta} = e^{i(m+n)\theta}$.

If we start from Euler's formulas and consider them as definitions of the trigonometric functions, the proofs of the formulas of trigonometry are much simpler. For example, since

$$e^{i\theta}e^{i\phi} = e^{i(\theta+\phi)},$$

we have

$$(\cos\theta + i\sin\theta)(\cos\phi + i\sin\phi) = \cos(\theta+\phi) + i\sin(\theta+\phi),$$

and if we multiply out and equate real and imaginary parts, we get the addition formulas for both the sine and cosine.

EXERCISE 4.11 Show directly (without using the addition formulas) that $\sin 2\theta = 2\sin\theta\cos\theta$ and $\cos 2\theta = \cos^2\theta - \sin^2\theta$.

EXERCISE 4.12 Derive formulas for $\cos 3\theta$ and $\sin 3\theta$.

Of course, all trigonometric identities can now be established by expressing all the functions in terms of $e^{i\theta}$.

Observe that if $w = ze^{i\phi}$ and $\arg z = \theta$, then $\arg w = \theta + \phi$; in words, multiplying a complex number by $e^{i\phi}$ rotates all rays Oz through the angle ϕ.

That is, multiplication by $e^{i\phi}$ is a rotation of the plane. Let us see what this says in rectangular coordinates. If $z = re^{i\theta} = x + iy$, $x = r\cos\theta$, $y = r\sin\theta$, and

$$e^{i\phi}z = re^{i(\theta+\phi)} = r[\cos(\theta+\phi) + i\sin(\theta+\phi)].$$

So the coordinates X, Y of the rotated point are

$$X = x\cos\phi - y\sin\phi,$$
$$Y = y\cos\phi + x\sin\phi,$$

the usual formulas for a rotation.

EXERCISE 4.13 Show that $e^z = 1$ only when z is an integral multiple of $2\pi i$.

EXERCISE 4.14 Show that

$$|\sin z|^2 = \cosh^2 y - \cos^2 x = \sinh^2 y + \sin^2 x;$$

find the corresponding formula for $|\cos z|^2$. Use the first formula to find $\lim_{y\to\infty} |\sin(x+iy)|\, e^{-|y|}$.

Supplementary exercises

1. Show that e^z has period $2\pi i$. What is the period of $|e^z|$?
2. Can an analytic function (not a constant) have arbitrarily small periods? That is, can $f(z+p_n) \equiv f(z)$ with $\lim_{n\to\infty} p_n = 0$, $p_n \neq 0$?
3. Is $e^z + e^{2z}$ periodic? Is $\exp[(1+\sqrt{2})z]$ periodic? Is $\exp(z) + \exp(z\sqrt{2})$ periodic?
4. If $z = x + iy$, find $\operatorname{Im}[\sin(3i + 2z)]$.
5. Is $\tan z$ ever equal to i?
6. Verify the following identities by using the complex forms of the trigonometric functions:
   (a) $\cos^4\theta + \sin^4\theta = \frac{1}{4}\cos 4\theta + \frac{3}{4}$.
   (b) $\cot 2\theta + \csc 2\theta = \cot\theta$.

4F. The uniqueness of power series

First we notice that a convergent power series $\sum_{n=0}^{\infty} a_n(z-z_0)^n$ is the Taylor series of its sum (about the point z_0).

EXERCISE 4.15 Prove this.

This means, in particular, that *a given function can have only one power series*. Equivalently, we can say that if two functions are the same, their Taylor series about the same point have the same coefficients; or, if $\sum a_n z^n = \sum b_n z^n$ throughout a neighborhood of $z = 0$, then $a_n = b_n$ for every n. Indeed, we can prove a much stronger result, that if $f(z) = g(z)$ at a set of points that has 0 as a limit point, then (assuming that both f and g are represented by their Maclaurin series in a neighborhood of 0), f and g are identical in a neighborhood of 0. It amounts to the same thing to say that *if* $f(z_k)=0$ $(z_k \neq 0)$ *and* $z_k \to 0$, *then all the Maclaurin coefficients of* f *are* 0. This is so because, unless $f(z)\equiv 0$, there must be a first nonzero Maclaurin coefficient, so that $a_0=a_1=\cdots=a_{p-1}=0$, $a_p\neq 0$. Then, in the disk of convergence of the Maclaurin series, $f(z)=a_pz^p+a_{p+1}z^{p+1}+\cdots=z^p(a_p+za_{p+1}+\ldots)=z^pg(z)$. Now $g(0)=a_p\neq 0$ and so, since g is continuous, $g(z)\neq 0$ in some neighborhood of 0. Hence $f(z)\neq 0$, except at 0, in some neighborhood of 0. This contradicts the assumed existence of a set of zeros of f with 0 as a limit point.

In particular, *the only power series that represents the constant function* 0 *is the series all of whose coefficients are zero.* This is really a rather remarkable property, which we would appreciate more if it were less familiar. If we had chosen some other set of polynomials, instead of $\{z^n\}$, to use for expanding functions, we might well have series that represent zero although not all their coefficients are zero. For example, let

$$p_0(z)=-1, \quad p_1(z)=1-z, \quad p_n(z)=\frac{z^n}{n!}-\frac{z^{n+1}}{(n+1)!} \qquad (n=2,3,\ldots).$$

Evidently, $\sum_{n=0}^{\infty} p_n(z)=0$ but the coefficients are not zero. Consequently, expansions in terms of $\{p_n(z)\}$ are not unique; that is, a given function (in this case, the identically zero function) can have more than one expansion. Indeed, 0 has the expansion $0p_0+0p_1+\ldots$ in addition to the expansion given just above. Similarly,

$$e^z=-2p_0-p_1+p_2+2p_3+3p_4+\cdots=-p_0+2p_2+3p_3+4p_4+\cdots.$$

*4G. Miscellaneous exercises on power series

EXERCISE 4.16 Show that the function f, defined by $f(z)=\exp(-1/z^4)$ for $z\neq 0$, $f(0)=0$, satisfies the Cauchy-Riemann equations at every point of the finite plane, but is not represented by its Maclaurin series.

EXERCISE 4.17 Sum the geometric progression $1+e^{i\theta}+e^{2i\theta}+\cdots+e^{in\theta}$, take real and imaginary parts, and so obtain compact formulas for

$$\sin\theta+\sin 2\theta+\cdots+\sin n\theta$$

and for

$$\tfrac{1}{2}+\cos\theta+\cos 2\theta+\cdots+\cos n\theta.$$

(These formulas are useful in connection with Fourier series.[1])

EXERCISE 4.18 Show that $\sum_{n=1}^{\infty} z^n n$ converges at every point, except 1, of the unit circle $(z=e^{i\theta})$. (This is most easily done by Exercise 18.1.)

EXERCISE 4.19[2] Find the sum of the distances from the point 1 to the other nth roots of 1. Divide the result by n and let $n\to\infty$ to conclude that the average distance from 1 to a point on $|z|=1$ is $4/\pi$.

EXERCISE 4.20 Find the sum of the squares of the distances in Exercise 4.19.

EXERCISE 4.21 Find the real part of $\cos(z + 3i)$.

EXERCISE 4.22 Let $1, \omega, \omega^2$ be the three cube roots of 1. Let $g(z) = (\cos z)(\cos(\omega z))(\cos(\omega^2 z))$. Show that the Maclaurin series of g has nonzero coefficients a_n only when n is a multiple of 3.

Notes

1. The sum of cosines is the **Dirichlet kernel** used for studying the convergence of Fourier series. It has been remarked that Dirichlet was able to prove the convergence theorem that had eluded Fourier because Dirichlet knew more trigonometry.
2. I learned about this problem from S. L. Zabell.

5 *Curves and Integrals*

5A. Terminology

Many of the most interesting parts of our subject come out of Cauchy's theorem, one version of which says that if a function is analytic inside and on a closed curve C, then $\int_C f(z)\,dz = 0$. This may or may not seem very plausible: It has been claimed[1] that it is so surprising that only a proof can make it credible. It will in any case seem more reasonable if we define our terms: What is a curve, what is its inside, and what kind of integral is being used? As usual, we have to begin with some more terminology.

In Sec. **2A** we defined a **region** to be a nonempty open connected set. Let f be a function that is continuous in a region D. In ordinary calculus we integrate functions over intervals. In more advanced calculus we integrate functions over curves, to get what are called *line integrals* ("line" being a vestigial remnant of the obsolete usage of "line" to mean "curve," whereas nowadays we usually think of a line as a special case of a curve). Our definition of the integral of a complex-valued function over a curve will reduce the integral to the integral of a complex-valued function over a real interval. What we want is to define something that will be denoted by $\int_C f(z)\,dz$, where C is a curve. We had better begin by saying what a curve is.

By a **curve** we are going to mean a parametric curve in the sense of ordinary calculus, that is, a pair of continuous functions $x = x(t)$, $y = y(t)$, with a real interval as common domain, and with the additional requirement that the functions have continuous derivatives, not both zero at the same time, except possibly at a finite number of points. We can abbreviate this latest statement by saying that the functions are *piecewise continuously differentiable*, or that the curve has a continuously turning tangent except at a finite number of

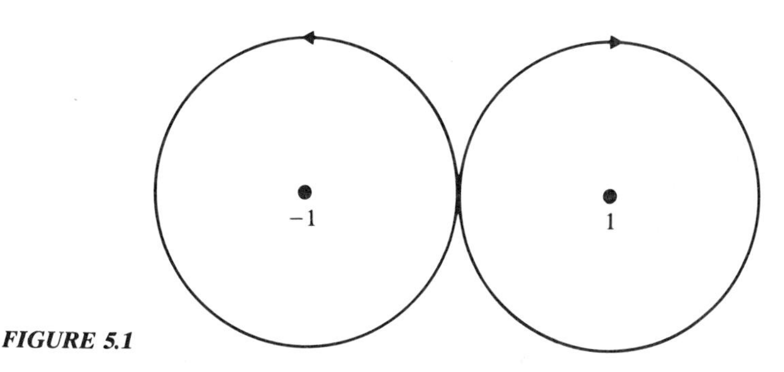

FIGURE 5.1

points. We can write the equations of a curve more concisely in complex form as $z = z(t) = x(t) + iy(t)$. This is a complex-valued function on a real domain, and differentiation is in the one-dimensional sense. The range of the function z is a set of points, which we call the **trace** of the curve. The trace is what we are likely to think of when we see the word "curve." However, a curve can cross itself, touch itself, or even overlap itself, whereas its trace is a set and therefore does not have multiple points.

For example, $z = e^{i\theta}$, $0 \le \theta \le 2\pi$, is a curve whose trace is the unit circle; $z = e^{i\theta}$, $0 \le \theta \le 4\pi$, is a curve with the same trace, but it is a different curve, which we would usually call "the unit circle described twice." The curve $z = -1 + e^{i\theta}$, $0 \le \theta \le 2\pi$; $z = 1 + e^{-i\theta}$, $2\pi \le \theta \le 4\pi$, has a trace consisting of two circles; it represents the first circle described counterclockwise and the second clockwise (Fig. 5.1).

The points $z(a)$ and $z(b)$, if different, are the **initial** and **terminal** points of the curve (and called **endpoints**). If $z(a) = z(b)$, the curve is **closed**. (This is different from the topological meaning of "closed.") If $z(a) \ne z(b)$, we usually call the curve an **arc**; a **contour** is either a curve or an arc. An arc for which $z(c) = z(d)$, where $a \le c < d \le b$, "crosses or touches itself"; otherwise the arc is one-to-one, and is said to be **simple**. Similarly, a closed curve that does not cross or touch itself is simple; simple curves and arcs are also called **Jordan curves** and **arcs**.

EXERCISE 5.1 Show that the trace of a simple closed curve is a closed point set.

We say, for short, that *a curve C lies in (or is in) a region D if the trace of C is a subset of D*; and that *a point z_0 is on C if z_0 is a point of the trace of C*. The trace of a curve will usually look like a geometric curve, possibly with corners, self-intersections, and overlaps; it is to be thought of as being oriented "from $z(a)$ to $z(b)$." The curve with the opposite orientation is defined by $z(a + b - t)$. For example, Fig. 5.2 (p. 36) indicates different curves with the same trace.

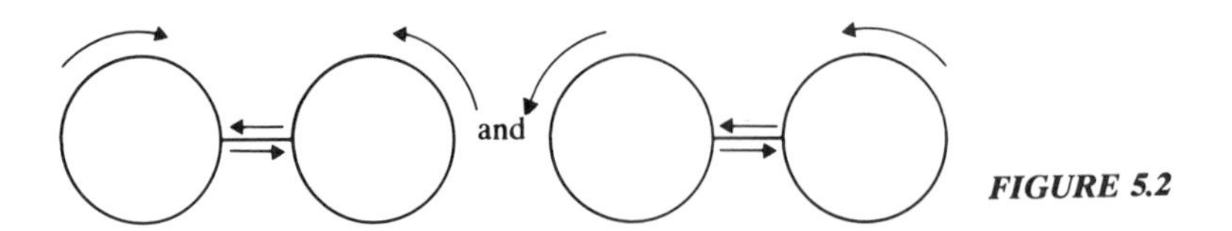

FIGURE 5.2

5B. Integrals

We can now (at last!) define $\int_C f(z)\,dz$. This notation is an abbreviation for the complex number

$$\int_a^b f[z(t)]z'(t)\,dt,$$

which is what is suggested by the chain rule (Sec. **2C**).

If $f(z) = u(x, y) + iv(x, y)$, we can write the integral more explicitly as

$$\int_a^b f[x(t) + iy(t)][x'(t) + iy'(t)]\,dt = \int_a^b (ux' - vy')\,dt + i\int_a^b (vx' + uy')\,dt.$$

The line integrals of ordinary calculus are interpreted in the same way:

$$\int_C u\,dx + v\,dy = \int_a^b u[x(t), y(t)]x'(t)\,dt + \int_a^b v[x(t), y(t)]y'(t)\,dt.$$

In principle, f can be any function that is continuous on the trace of C (or, in informal language, continuous on C).

By way of illustration, suppose that C is defined by $z = t$, $0 \le t \le 1$; $z = 1 + i(t-1)$, $1 \le t \le 2$. The trace of this curve is shown in Fig. 5.3. This conforms to the definition, but usually it would be more convenient to parametrize the same curve by writing

$$z = x, \quad 0 \le x \le 1; \quad z = 1 + iy, \quad 0 \le y \le 1.$$

If we want to integrate $f(z) = \bar{z}$ over this curve, we can write

$$\int_0^1 x\,dx + \int_0^1 (1 - iy)\,dy = \frac{1}{2} + \left(1 - \frac{i}{2}\right) = 1.5 - 0.5i.$$

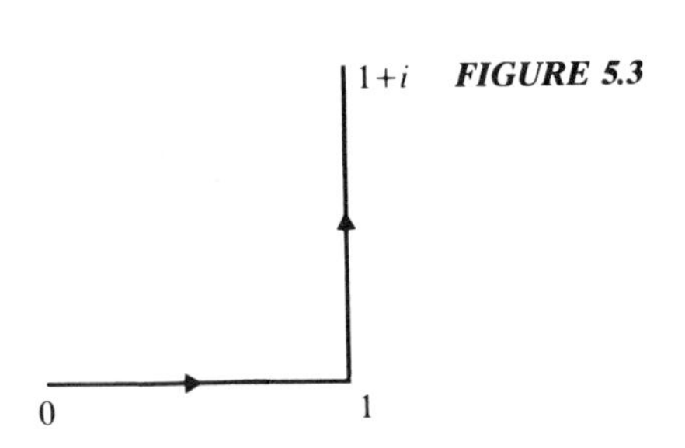

FIGURE 5.3

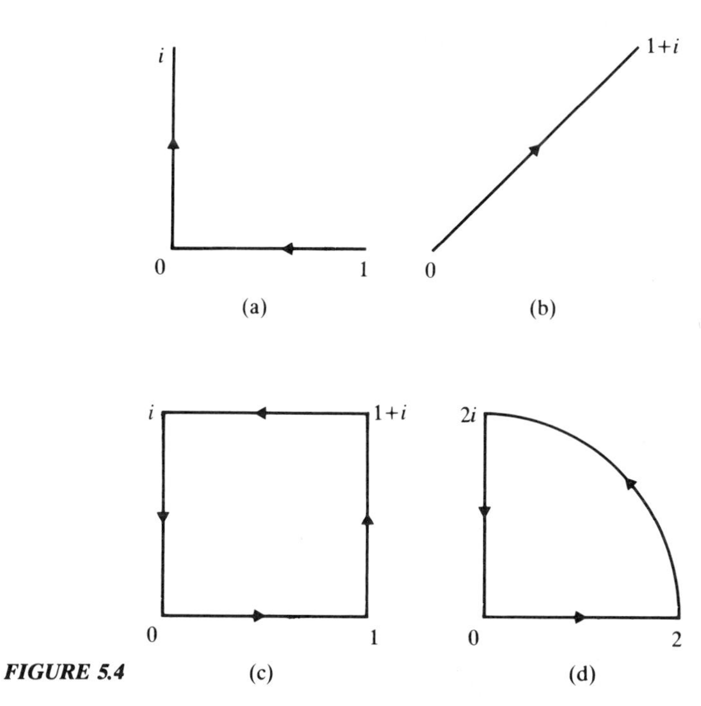

FIGURE 5.4

EXERCISE 5.2 Parametrize the indicated curves and calculate the integrals directly from the definition.

(*a*) $\int_{C_1} \bar{z}\, dz$, where C_1 is shown in Fig. 5.4*a*.

(*b*) $\int_{C_2} \bar{z}\, dz$, where C_2 is shown in Fig. 5.4*b*.

(*c*) $\int_{C_3} z\, dz$, where C_3 is shown in Fig. 5.4*c*.

(*d*) $\int_C z^{-1}\, dz$, where C is the unit circle ($|z| = 1$), described once in the positive direction.

(*e*) $\int_C \bar{z}\, dz$, where C is shown in Fig. 5.4*d*.

EXERCISE 5.3 Parametrize the curves in Fig. 5.5 (p. 38). Problems like these should be interpreted as saying: "assuming that the given figure is the trace of a curve, construct a parametric curve of which it is the trace." Lacking more explicit instructions, you should find the most economical curve, one such that no part of the trace is covered more often than necessary; there is not always a unique way of doing this.

Notice (compare Exercise 5.2b with p. 36) that $\int_C f(z)\, dz$ depends on C as well as on f, not (generally speaking) just on the trace of C and on f. For

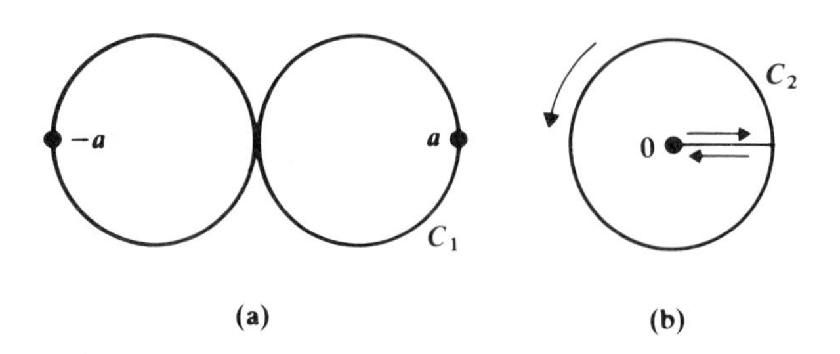

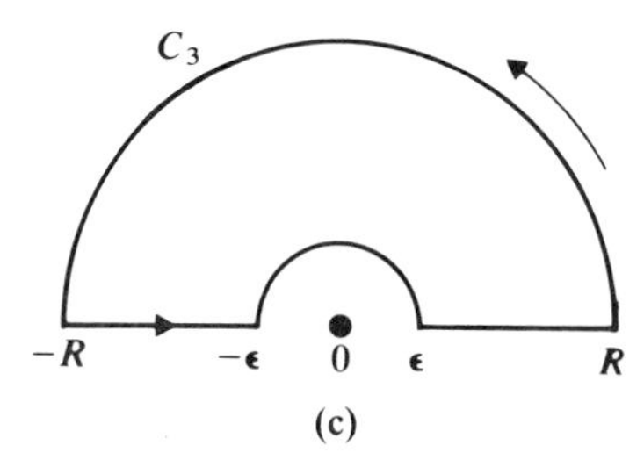

FIGURE 5.5

example, let C_1 be $z_1 = e^{it}$, $0 \le t \le 2\pi$, and let C_2 be $z_2 = e^{it}$, $0 \le t \le 4\pi$. These curves have the same trace (the unit circle) but are different curves.

EXERCISE 5.4 If C_1 and C_2 are these curves, find a function f such that

$$\int_{C_1} f(z)\, dz \neq \int_{C_2} f(z)\, dz.$$

EXERCISE 5.5 If C is a circle of radius 1 with center at $z = 1$ (described once counterclockwise), calculate

$$\int_C \frac{1+z}{1-z}\, dz.$$

According to our definition, changing the parametrization of a curve produces a different curve. However, if $z(t)$, $a \le t \le b$, defines a curve, then $z[t(u)]$, $r \le u \le s$, with $t(r) = a$, $t(s) = b$, defines a curve whose trace covers the same points in the same order, provided that t is an increasing function with piecewise continuous derivative. We can therefore "identify" the two curves; that is, we consider them to be the same curve. In more formal language, we can say that the word "curve" now really stands for an **equivalence class of curves** as originally defined.

EXERCISE 5.6 Show that equivalent curves C give the same value to $\int_C f(z)\, dz$.

If C is defined by $z = z(t) = x(t) + iy(t)$, then $z'(t) = x'(t) + iy'(t)$, and we see that the real and imaginary parts of $z'(t)$ are (except perhaps for a finite number of points) the components of the tangent vector to C. Similarly,

$|z'(t)|\,dt$ is the differential ds of arc length, as defined in calculus, so that the length of C is given by $\int_C ds = \int_a^b |z'(t)|\,dt$ (this is the length of the curve, not necessarily the length of its trace). We often abbreviate the length integral to $\int_C |dz|$, and $\int_a^b f(z)\,|z'(t)|\,dt$ to $\int_C f(z)\,|dz|$.

For example, if C is $z = re^{i\theta}$, $0 \le \theta \le 2\pi$, then the length of C is $\int_0^{2\pi} r\,d\theta = 2\pi r$.

Supplementary exercises

Parametrize the indicated curves and compute the integrals from the definition.

1. $\displaystyle\int_C z\,dz$, C as shown in Fig. 5.6*a*.

2. $\displaystyle\int_C (1/\bar{z})\,dz$, C as shown in Fig. 5.6*b*.

3. $\displaystyle\int_C \bar{z}\,dz$, C as shown in Fig. 5.6*c*.

4. Evaluate

$$\int_C \frac{2z+3}{z-4i}\,dz,$$

where C is a circle of radius 4, described counterclockwise, centered at $z = 4i$.

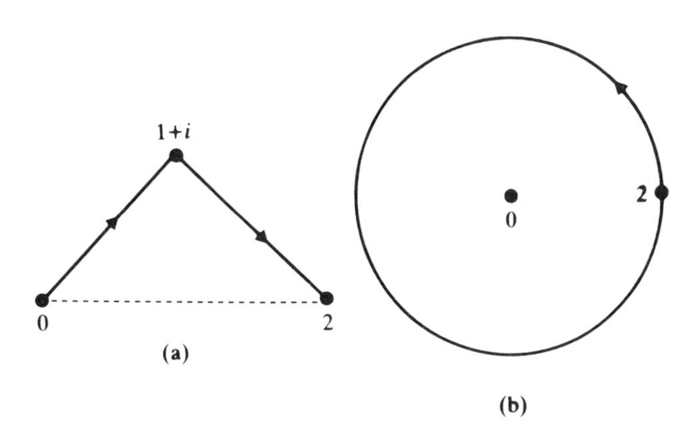

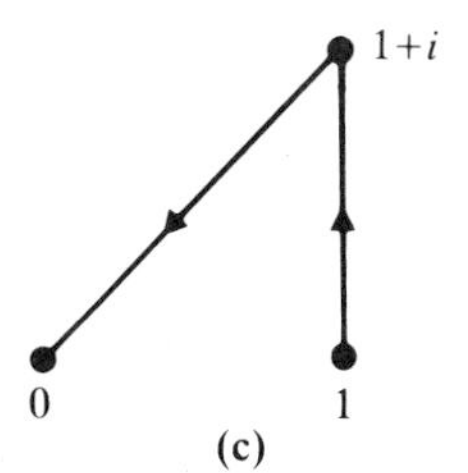

FIGURE 5.6

5C. Inequalities for integrals

We shall often need the inequalities

$$\left|\int_C f(z)\,dz\right| \le \max|f(z)| \cdot (\text{length of } C)$$

and, more generally,

$$\left|\int_C f(z)\,dz\right| \le \int_a^b |f[z(t)]|\,|z'(t)|\,dt = \int_C |f(z)|\,|dz|\,.$$

These are complex analogs of the familiar elementary inequalities

$$\left|\int_a^b f(t)\,dt\right| \le (b-a) \max_{a\le t\le b} |f(t)|$$

and

$$\left|\int_a^b f(t)\,dt\right| \le \int_a^b |f(t)|\,dt$$

for real functions on a real interval; they are also continuous analogs of the inequality for sums in Exercise 1.10.

The proofs are not quite obvious. Let $\int_C f(z)\,dz = Ae^{i\phi}$; then

$$\begin{aligned}\left|\int_C f(z)\,dz\right| = A &= \mathrm{Re}\left[e^{-i\phi}\int_C f(z)\,dz\right] = \mathrm{Re}\left\{e^{-i\phi}\int_a^b f[z(t)]z'(t)\,dt\right\}\\ &= \mathrm{Re}\left\{\int_a^b e^{-i\phi}f[z(t)]z'(t)\,dt\right\}\\ &= \int_a^b \mathrm{Re}\{e^{-i\phi}f[z(t)]z'(t)\}\,dt\\ &\le \int_a^b |e^{-i\phi}f[z(t)]z'(t)|\,dt\\ &= \int_a^b |f[z(t)]|\,|z'(t)|\,dt\\ &= \int_C |f(z)|\,|dz|\,.\end{aligned}$$

The step in this argument that may not be obvious is going from $\mathrm{Re}\int\{\ldots\}$ to $\int \mathrm{Re}\{\ldots\}$. This is a correct statement because $\int(p+iq)\,dt = \int p\,dt + i\int q\,dt$.

In "real" calculus we have the mean value theorem,

$$\int_a^b f(t)\,dt = (b-a)f(c), \qquad \text{where } a<c<b.$$

This is not usually correct for complex contour integrals.

EXERCISE 5.7 Show that the mean value theorem fails for $\int_0^{1+i} z\,dz$, where integration is along the curve in Fig. 5.3.

5D. Some easy integrations

There are some quite extensive classes of integrals that we can evaluate with little or no calculation.

We can always evaluate $\int_C f(z)\,dz$ when f is continuous and is the derivative of a function F in a region that contains C (more precisely, the trace of C). In particular, we can do this whenever f is a polynomial, or a linear combination of functions such as e^z or $\sin z$ that we know are derivatives, or when f is defined by a power series whose disk of convergence contains C.

In fact, if $f(z)=F'(z)$ and C is defined by $z=z(t)$, $a \le t \le b$, we have

$$\int_C f(z)\,dz = \int_a^b F'[z(t)]z'(t)\,dt.$$

The integral on the right is (I remind you) an ordinary integral of a complex-valued function with real domain $[a, b]$, and the ordinary rules of calculus apply to it. The integrand is the derivative of $F[z(t)]$ with respect to t (by the chain rule). Hence the integral equals

$$F[z(b)] - F[z(a)],$$

and this (for a given f) depends only on the endpoints of C. That is, if C_1 and C_2 are two curves with the same endpoints P and Q (described from P to Q), then

$$\int_{C_1} f(z)\,dz = F(Q) - F(P) = \int_{C_2} f(z)\,dz.$$

In particular, if C is a closed curve (simple or not), then (under our current hypotheses) $\int_C f(z)\,dz = 0$. Notice that we do not know (yet) that f is differentiable if it is itself a derivative. (See Sec. **7C**.)

For example, $\int_C P(z)\,dz = 0$ for every polynomial and every closed curve C; also, $\int_C z^{-2}\,dz = 0$ for every closed curve that does not go through the origin.

Consequently, problems like Exercise 5.2c become trivial because z is a polynomial and hence

$$\int_P^Q z\,dz = z^2/2\Big|_P^Q = \frac{Q^2 - P^2}{2},$$

independently of the curve along which we integrate.

Similarly, $\int_C f(z)\,dz = 0$ if C is inside the disk of convergence of a power series whose sum is f, because the power series converges uniformly on C by Sec. **3D**, and so can be integrated term by term; but all the integrals of individual terms are 0. Thus, for example, we see that $\int_C \sin z\,dz = 0$ for every closed curve C.

On the other hand, there are functions for which the integral does depend on the curve C; compare the illustrative example just before Exercise 5.2 with Exercise 5.2b.

We might have expected z^{-1} to be a derivative, indeed the derivative of $\log z$ (which we have not yet defined), but we now see that there cannot be a (single-valued) function $\log z$, with the properties that we would expect a logarithm to have, in any region that contains a disk with center at 0. Indeed, if there were such a function, the integral of its derivative around the circumference of such a disk would be 0; but this is just what it is not: See Exercise 5.2d. Logarithms will be discussed in Sec. **10**.

Note

1. Jeffreys and Jeffreys, p. v.

6 *Cauchy's Theorem*

6A. Simple closed curves

I remind you that Cauchy's theorem deals with points inside simple closed curves and with integrals around the curves. If we are going to give a precise statement of Cauchy's theorem we need to know both what points are inside a simple closed curve and in which direction the curve is to be traced. The simplest way (if not the most elementary way) to answer questions like these is to appeal to the **Jordan curve theorem**, which states that *the trace of a simple closed curve* (without any hypotheses about differentiability) *separates the extended plane into two regions, one of which is bounded whereas the other contains* ∞. The bounded region is called the **inside** of the curve. The Jordan curve theorem is far from easy to prove;[1] the proof belongs to topology rather than to analysis, so I shall simply assume the result. At the same time, you should realize that the theorem is rather intuitive for curves that are not too complicated. It is easily proved for convex curves, or more generally for **star-shaped curves**. A simple curve C is star-shaped if there is a point P, not on C, such that each line segment from P to a point of C intersects C in one and only one point. Obvious examples of star-shaped curves are circles, triangles,

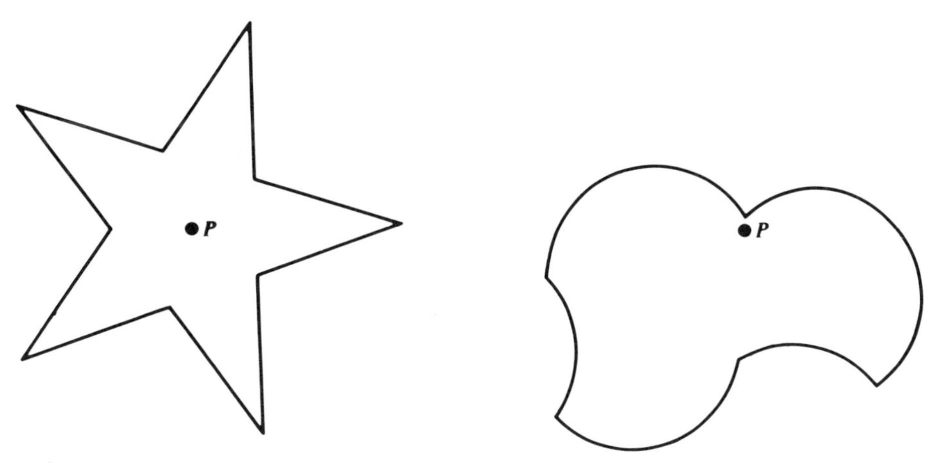

FIGURE 6.1

rectangles, and curves like those in Fig. 6.1. Here the inside of C consists of the union of all the open line segments from P to C (together with P).

However, if C is a complicated curve, it is not always easy to see which points are inside C. Consider, for example, Fig. 6.2. Is the point P inside or outside? A convenient way to decide is to draw a ray from P to ∞; if the ray cuts C an odd number of times, P is inside C; we shall not prove this.

FIGURE 6.2

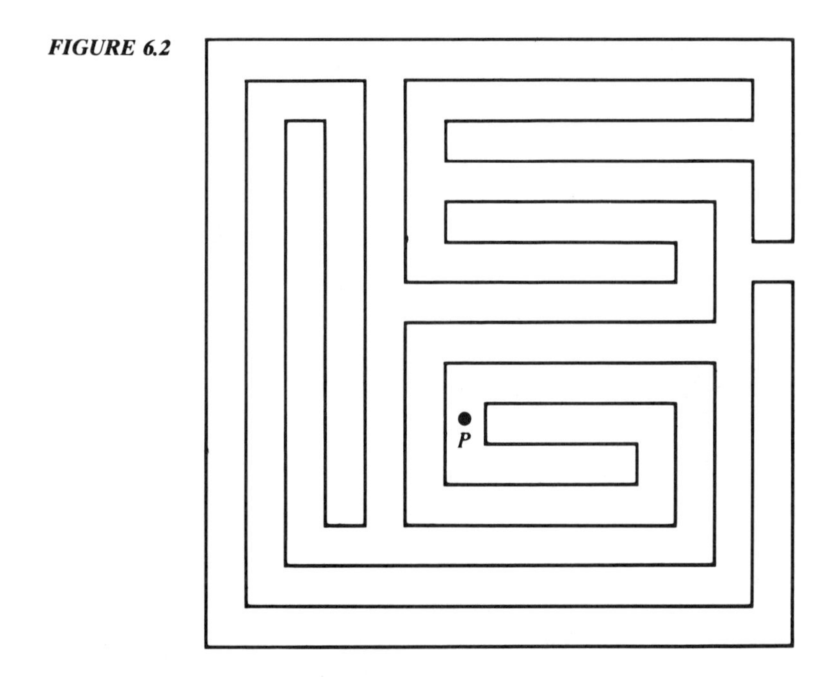

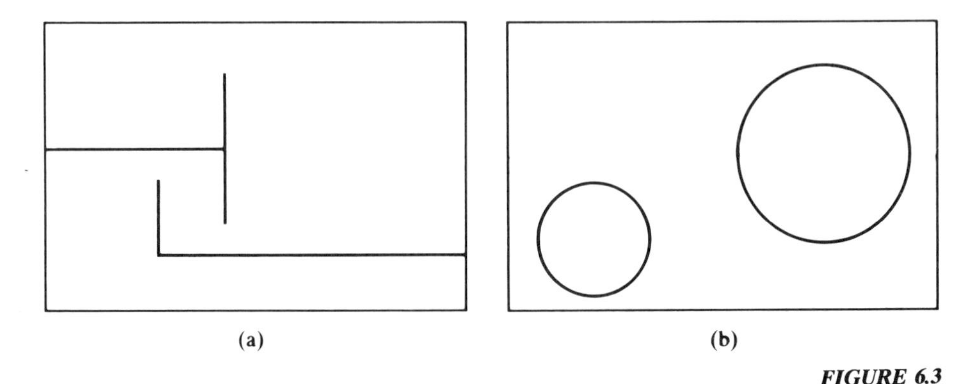

FIGURE 6.3

6B. Simple connectedness

The region D inside a simple closed curve C has the property of being **simply connected**, which can be defined in several ways.[2] One definition is that *every closed curve in D can be shrunk continuously to a point in D without bumping into the boundary C of D.* Regions more general than the interiors of simple closed curves can be simply connected; for example, the region indicated in Fig. 6.3a is simply connected, but that in Fig. 6.3b is not. If a region is not very complicated, it is usually easy to tell by inspection whether or not it is simply connected.

When D is unbounded and has ∞ as an interior point, we call D simply connected if it is simply connected as a subset of the Riemann sphere, that is, as a subset of the extended plane. For example, the outside of a circle (the exterior of a disk) is simply connected in the extended plane, but not in the finite plane.

A curve C has an **orientation** imposed by the orientation of the parameter interval. That is, if $z = z(t)$, $a \le t \le b$, the points of C are taken in the order of increasing values of t. For a *simple* closed curve C, it is usual to say that the curve is **positively oriented** if the parametrization is chosen so that the curve is traversed counterclockwise, that is, so that if you walked around the curve in the positive sense, the inside of the curve would be on your left. However, if ∞ is an interior point, the positive direction is clockwise, so that ∞ would be on your left.

6C. Cauchy's theorem

As we said in Sec. **2A**, function f is **analytic in a region** D if f has a derivative at every point of D; f is **analytic at a point** P if f is analytic in some neighborhood of P, that is, in some open set that contains P. If S is any set, f is **analytic on** S if f is analytic in some open set that contains S. For example, if f

is analytic in the closed disk $|z| \leq 1$, then f is analytic in some open disk $|z| < 1 + \varepsilon$, where $\varepsilon > 0$. Recall that examples of analytic functions are functions defined by power series, and (in particular) polynomials and the elementary functions discussed in Sec. **4**.

We can now give a precise statement of **Cauchy's theorem** (not the most general statement possible, but one that will be adequate for what we are going to do later).

Let C be a simple closed curve whose derivative is continuous except at a finite number of points. Suppose that C is inside a simply connected region D in which f is analytic. Then $\int_C f(z)\, dz = 0$.

A more general version of Cauchy's theorem,[3] but one that is much harder to prove, is that *if f is analytic in the region D inside a simple closed rectifiable curve C, and continuous in the closure of D, then* $\int_C f(z)\, dz = 0$. (Here we are not assuming that C is piecewise differentiable; a rectifiable curve is any continuous curve that has finite length; that is, the lengths of polygons inscribed in C have a finite least upper bound.)

We first prove Cauchy's theorem[4] when $C = C_1$ is the boundary of a rectangle R_1. The key to the proof is the realization that if C, of length P, bounds a very small rectangle R, then we have

$$\int_C f(w)\, dw = o(P^2)$$

as $P \to 0$. This is true because if w is on C and z inside, the definition of the derivative (see Sec. **2B**) implies that

$$f(w) = f(z) + f'(z)(w - z) + \varepsilon(w)(w - z),$$

where $\varepsilon(w) \to 0$ as $P \to 0$. If we integrate this equation over C with respect to w, the integrals of the first two terms are zero because Cauchy's theorem is true for polynomials. Hence, by the inequalities in Sec. **5C**, we have, as $P \to 0$,

$$\left| \int_C f(w)\, dw \right| \leq P \cdot \max |w - z| \cdot \max |\varepsilon(w)| = o(P^2),$$

because $|w - z| < P$.

Notice that without using Cauchy's theorem for polynomials we would have had only

$$\left| \int_C f(w)\, dw \right| \leq \max |f(w)| \cdot P,$$

that is,

$$\int_C f(w)\, dw = O(P).$$

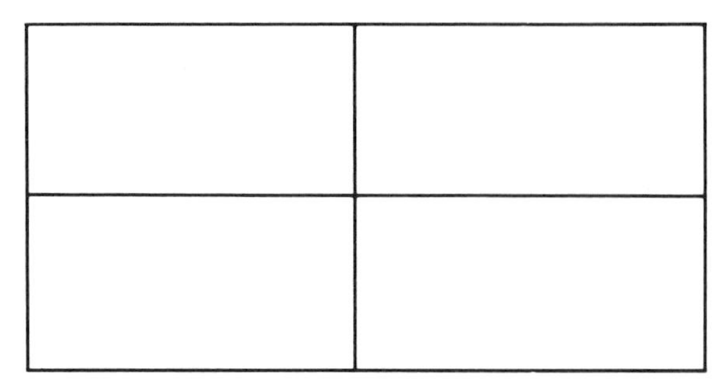

FIGURE 6.4

Thus $\int f(w)\,dw$ around a very small rectangle is much smaller for an analytic f than it would be for a general f. To obtain Cauchy's theorem for a large rectangle we have, in effect, to inflate C to the boundary of a large rectangle.

It is simpler to start from the large rectangle R_1 with boundary C_1 of length L, assume that

$$\int_{C_1} f(w)\,dw = A \neq 0,$$

and deduce a contradiction by cutting R_1 into small pieces.

Divide R_1 into four congruent similar rectangles as in Fig. 6.4. Notice that since R_1 is in the simply connected region D, the interior of R_1 is in D. Each of the small rectangles has perimeter $L/2$. The sum of the integrals of f around the four rectangles is the integral around the original rectangle because the integrals back and forth along the interior sides cancel. Hence at least one of the integrals around a small rectangle (call it R_2) has absolute value at least $A/4$. At the next step we divide R_2 into four rectangles, each of perimeter $L/2^2$, and obtain a rectangle R_3 for which the integral has absolute value at least $A/4^2$. Continuing in this way, at the nth step we have a rectangle R_n of perimeter $L/2^n$ for which the absolute value of the integral is at least $A/4^n$. By Cantor's nested set theorem,[5] there is a point z_0 that is in all the rectangles R_k.

For a rectangle R_n containing z_0, we have

$$\left|\int_{R_n} f(w)\,dw\right| \geq A/4^n.$$

The perimeter of R_n is $P_n = 2^{-n}L$, so

$$\left|\int_{R_n} f(w)\,dw\right| \geq AL^{-2}P_n^2.$$

For large n, this contradicts what we just proved, namely, that

$$\int_{R_n} f(w)\,dw = o(P_n^2) \qquad \text{as } P_n \to 0.$$

Therefore A must be 0.

Hence we have proved Cauchy's theorem when C is a rectangle. We now have to prove the theorem for an arbitrary curve.

We do this indirectly, by showing that our function f is a derivative.[6] Let D be any simply connected region in the finite plane. Choose a convenient point z_0 in D and consider $F(z) = \int_C f(w)\, dw$, where C is a simple arc from z_0 to z, in D, composed of a finite number of line segments that are parallel either to the real or the imaginary axis. We can find such an arc because there is (by hypothesis) a polygon connecting z_0 and z in D, and we can approximate it by another polygon of the required kind. Now I claim that F is a function, that is, that if Γ is another arc of the same kind then $\int_\Gamma f(w)\, dw = \int_C f(w)\, dw$.

It will be enough to consider the case when $C - \Gamma$ is a *simple* closed curve. In fact, if we start at z_0 and follow C, there will be a first point z_1 at which C meets Γ; that is, the arc (z_0, z_1) of C followed by the arc (z_1, z_0) of Γ is a simple curve. Then start from z_1 and repeat the process. In this way we can represent $C - \Gamma$ as a finite union of simple closed curves together with some line segments traced in opposite directions [along which $\int f(w)\, dw = 0$].

Now if $C - \Gamma$ is a simple closed curve, we can represent it as a union of boundaries of rectangles by drawing lines parallel to the axes, as in Fig. 6.5. The integrals around the rectangles are zero by Cauchy's theorem for rectangles, and the integrals back and forth along the line segments cancel. The simple connectivity of D enters because it prevents any of the rectangles from containing points that are not in D.

FIGURE 6.5

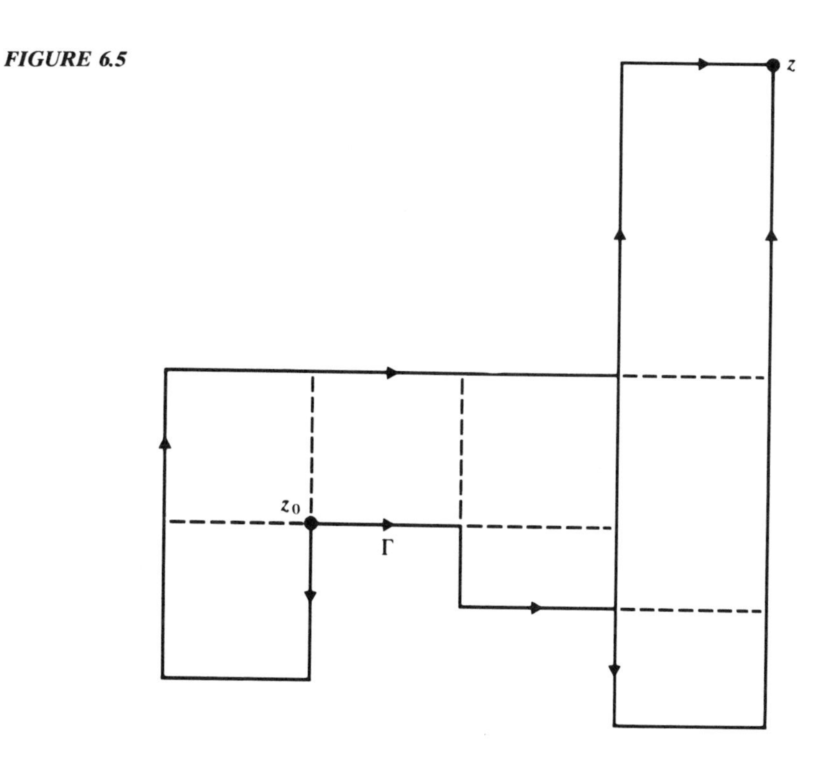

EXERCISE 6.1 Show that $F'(z) = f(z)$ by calculating $F'(z)$ from the definition of a derivative.

This completes the proof of Cauchy's theorem.

EXERCISE 6.2 Although Cauchy's theorem says that if f is analytic in a simply connected region that contains the simple closed curve C, then $\int_C f(z)\,dz = 0$, you must not assume that $\int_C f(z)\,dz$ is necessarily different from 0 if f is not analytic or if the region is not simply connected. Consider, as examples,

$$\int_{|z|=1} (\bar{z})^2\,dz$$

and

$$\int_{|z|=1} z^{-2}\,dz, \qquad \text{the region being the annulus } \frac{1}{2} < |z| < 2.$$

6D. The fundamental theorem of algebra

As a first application of Cauchy's theorem (which we need here only for a circle), let us prove that *every polynomial of positive degree has at least one zero.*[7]

Assume the contrary. Since every polynomial of degree 1 has a zero, there is a polynomial $P(z)$ of degree $n > 1$ with no zeros. We may assume that P is real on the real axis.

EXERCISE 6.3 Show that if $P(z)$ is not real on the real axis, then $P(z)\bar{P}(z)$ (where $\bar{P}$ is P with all its coefficients conjugated) is a polynomial that is real on the real axis and has no zeros if P has no zeros.

Since $P(z) \neq 0$, $P(2\cos\theta) \neq 0$. Consider the integral

$$I = \int_{-\pi}^{\pi} \frac{d\theta}{P(2\cos\theta)}.$$

Since $P(z)$ is real for real z and never 0, we see that $P(2\cos\theta)$ is always of the same sign, and hence $I \neq 0$. Now interpret I as an integral around the unit circle, parametrized by $z = e^{i\theta}$, $0 \leq \theta \leq 2\pi$:

$$I = -i\int_{|z|=1} \frac{dz}{zP(z + z^{-1})}.$$

If $P(z) = a_0 + a_1 z + \cdots + a_n z^n$, then

$$P\left(z + \frac{1}{z}\right) = a_0 + a_1\left(z + \frac{1}{z}\right) + \cdots + a_n\left(z + \frac{1}{z}\right)^n$$
$$= z^{-n}Q(z),$$

where $Q(z)$ is a polynomial with $Q(0) = a_n \neq 0$. Thus the integrand in I is $z^{n-1}/Q(z)$ with $1/Q(z)$ analytic, $n \geq 1$, and therefore $I = 0$ by Cauchy's theorem. This contradicts our earlier conclusion, so P must have at least one zero.

Supplementary exercises

1. An exponential polynomial is a finite sum $\sum_{k=1}^{n} a_k \exp(b_k z)$. Must a nonconstant exponential polynomial have at least one zero?

2. The function $f(z) = 2 + \cos z$ has no real zeros. Does it have nonreal zeros? If so, find them.

6E. Green's theorem and Cauchy's theorem

If you know Green's theorem from two-dimensional calculus, you can prove Cauchy's theorem for functions that are not only analytic, but have a *continuous* derivative, and for all curves for which Green's theorem has been proved. At least, you can prove it for rectangles, and then continue as in Sec. **6C**.

For our purposes, Green's theorem can be written most simply in the complex form of Sec. **2C** (p. 19),

$$\int_C f(z)\,dz = 2i \iint_D \frac{\partial f}{\partial \bar{z}}\,dx\,dy,$$

where the left-hand integral is along a simple closed curve C and the right-hand integral is over the region D bounded by C. Now if f is analytic in D, we have $\partial f/\partial \bar{z} = 0$ in D (p. 19), and Cauchy's theorem follows.

6F. A converse of Cauchy's theorem (Morera's theorem)

Whether $f(z)$ is analytic or not, Green's theorem still says that

$$\int_C f(z)\,dz = 2i \iint_D \frac{\partial f}{\partial \bar{z}}\,dx\,dy.$$

If the left-hand side is zero, then the right-hand side is zero. Suppose that the left-hand side is zero for *every* C in some region R, where D is the inside of C.

EXERCISE 6.4 Show that then the integrand on the right is always 0, that is, that f is analytic in the interior of every closed curve in R, and hence analytic in R.

This is **Morera's theorem**: If $\int_C f(z)\,dz = 0$ for a continuous f and every closed curve C in R, then f is analytic in R.

Essentially the same proof shows that we need only have $\int_C f(z)\,dz = 0$ for every circle in R, or every square.[8]

We could also prove Morera's theorem under weaker hypotheses on f, by constructing the function F of Sec. **6C**. This would show that f is a derivative, but we will not know until Sec. **7C** that a derivative is necessarily analytic.

An application of Morera's theorem is given in Sec. **7E**.

Notes

1. There is no really simple proof of the Jordan curve theorem. For discussion and references, see Burckel, p. 116.

2. Burckel (p. 344) gives 15 different (equivalent) definitions of simple connectivity of a region.

3. For references to proofs of the general form of Cauchy's theorem, see Burckel, p. 341.

4. This was first done by Goursat, and consequently the theorem is sometimes called the Cauchy-Goursat theorem.

5. Cantor's nested set theorem says that if S_n are closed, bounded, and nonempty sets such that $S_n \supset S_{n+1}$, and the diameter of $S_n \to 0$, then there is a unique point that is in all S_n.

6. See D. V. Widder, "A Simplified Approach to Cauchy's Integral Formula," *Amer. Math. Monthly 53* (1946): 359–363.

7. There are dozens of proofs of the fundamental theorem of algebra. For references see Burckel, pp. 117–118; also Remmert, pp. 187–189. For the following one, which does not depend on Exercise 1.23, see R. P. Boas, "Yet Another Proof of the Fundamental Theorem of Algebra," *Amer. Math. Monthly 71* (1964): 180.

8. For further developments along these lines, see L. Zalcman, "Real Proofs of Complex Theorems (and Vice Versa)," *Amer. Math. Monthly 81* (1974): 115–137.

CHAPTER 2

Applications of Cauchy's Theorem

7 Cauchy's Integral Formula

We suppose that f is analytic in a simply connected region D.

Let C be a simple closed positively oriented curve in D, and z a point inside C. Then Cauchy's formula says that

$$f(z) = \frac{1}{2\pi i} \int_C \frac{f(w)\,dw}{w - z}, \qquad z \text{ inside } C.$$

This is a remarkable formula, because it shows that the function f can be reconstructed, inside C, just from the values that it takes *on* C. It has many applications, some of them quite unexpected, and occupies a central place in our subject. In this section we prove the formula.

7A. Curves inside curves

For the proof of Cauchy's formula, and for use later, we need a version of Cauchy's theorem that applies to a region like that between the circles C and γ in Fig. 7.1 (p. 52) and to a function f that is analytic inside and on C except at the center of γ. We cannot apply Cauchy's theorem directly, both because the region is not simply connected and because its boundary is not a single curve. However, we can avoid both difficulties by drawing line segments L_1 and L_2 from γ to C and considering $\int f(z)\,dz$ separately along the upper and lower

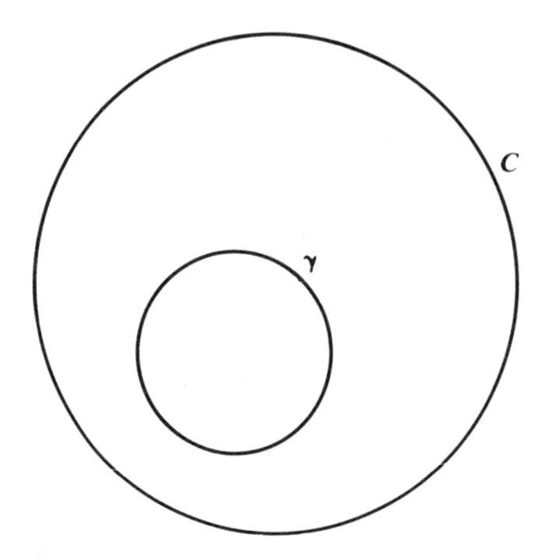

FIGURE 7.1

curves formed by parts of C and γ, and L_1 and L_2. See Fig. 7.2. Each of these curves is in a simply connected region in which f is analytic, so each integral is 0 and the integrals along L_1 and L_2 cancel, so that we have $\int_C f(z)\,dz = \int_\gamma f(z)\,dz$. Also, as long as γ does not meet C, $\int_\gamma f(z)\,dz$ is independent of the radius of γ.

If C is not a circle, but any simple positively oriented curve, such as the one shown in Fig. 7.3, we can proceed in the same way, except that it less obvious how to draw L_1 and L_2. In this case we pick any two points P_1 and P_2 on γ and draw rays from the center of γ through P_1 and P_2. Extend these rays until they meet C for the first time, and let L_1 and L_2 be the parts of the rays between γ and C. [There *is* a first point of L_1 (or L_2) that is on C, because the trace of C is a closed point set.] We can then proceed as before.

Finally, if there are several points inside C where f fails to be analytic, we

FIGURE 7.2

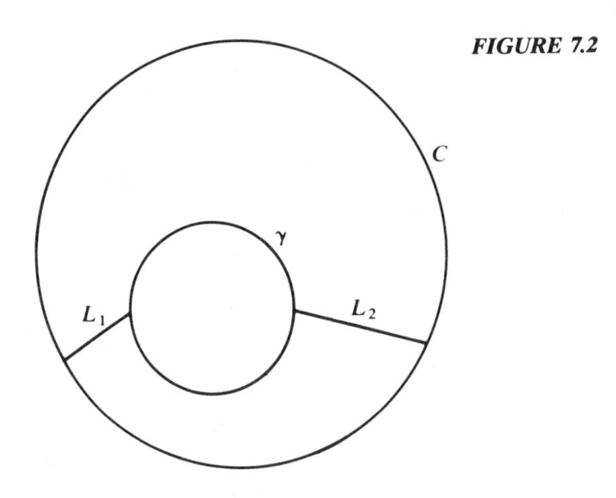

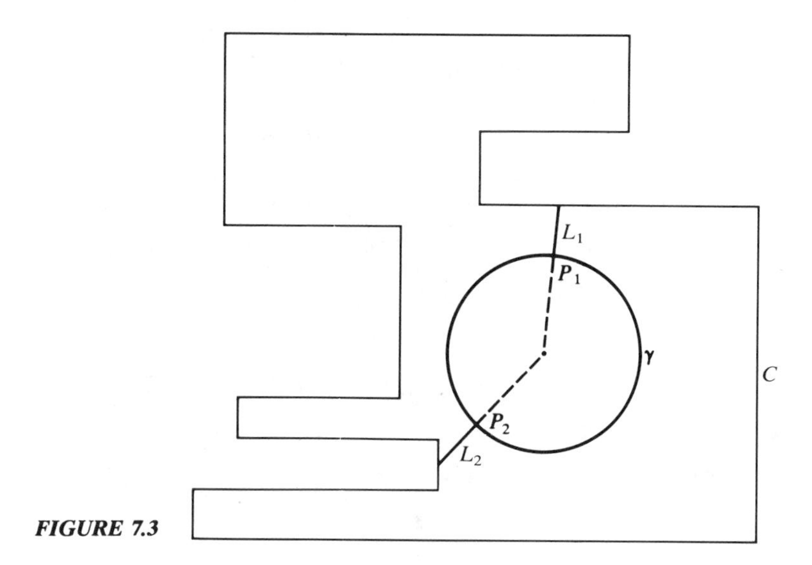

FIGURE 7.3

can draw circles around each one and construct corresponding line segments, provided that we make the circles small enough.

Consequently we have the following lemma.

LEMMA *Let D be a simply connected region, C a simple closed curve in D; let Δ_k $(k = 1, \ldots, n)$ be disjoint disks lying inside C, with boundaries γ_k; and let f be analytic inside C except at the centers of the Δ_k. Then*

$$\int_{\gamma_1} f\,dz + \cdots + \int_{\gamma_n} f\,dz = \int_C f\,dz.$$

7B. Proof of Cauchy's formula

The formula was stated at the beginning of Sec. **7A**. To prove it, we apply the lemma at the end of Sec. **7A** (just above), with $n = 1$, to the function

$$\frac{f(w)}{w - z}$$

(with a given z). This function is analytic on and inside C except at $w = z$. If γ is a small circle around z (so small that it is inside C), we then have

$$\frac{1}{2\pi i}\int_C \frac{f(w)}{w - z}\,dw = \frac{1}{2\pi i}\int_\gamma \frac{f(w)}{w - z}\,dw.$$

We can write the last integral as

$$\frac{1}{2\pi}\int_0^{2\pi} f(z + re^{i\theta})\,d\theta,$$

where r is the radius of γ. Since $f(z+re^{i\theta})\to f(z)$ as $r\to 0$, uniformly with respect to θ (by the definition of limits), we obtain

$$\lim_{r\to 0}\frac{1}{2\pi}\int_0^{2\pi} f(z+re^{i\theta})\,d\theta = f(z)$$

by the following exercise (with $n\to\infty$ replaced by $r\to 0$).

EXERCISE 7.1 (Integration of a Uniformly Convergent Sequence). Use the inequality

$$\left|\int_C f(w)\,dw\right| \le \int_C |f(w)|\,|dw|$$

from Sec. **5C** to show that if $f_n(w)\to f(w)$ uniformly on a bounded curve C, then $\int_C [f_n(w)-f(w)]\,dw\to 0$ as $n\to\infty$, so that $\int_C f_n(w)\,dw \to \int_C f(w)\,dw$.

EXERCISE 7.2 Show that if z is outside C, the integral in Cauchy's formula is equal to 0.

7C. Differentiation of functions represented by Cauchy's formula

We have defined a function f to be analytic if it has a derivative f' (in a region), but we do not yet know that f' is analytic. We are now going to use Cauchy's formula to show that f' is indeed analytic. It is no harder to prove the following result:

Every function g defined by a formula

$$g(z) = \int_C \frac{\varphi(w)\,dw}{w-z},$$

where φ is a continuous function, C is a simple closed curve, and z is inside C, has a derivative in the region inside C, and furthermore has derivatives of all orders, so that g is analytic inside C.

Notice that the integral also defines an analytic function in the exterior of C; in general, there is no obvious connection between the functions defined by the two integrals (but see Sec. **9E**).

It ought to appear plausible, especially if you know Leibniz's rule for differentiating a definite integral with respect to a parameter, that

$$g'(t) = \int_C \frac{\varphi(w)\,dw}{(w-z)^2}.$$

We can prove this as follows. The derivative of the integral for g is, by definition,

$$\lim_{z_1 \to z} \left\{ \int_C \frac{\varphi(w)\,dw}{w - z_1} - \int_C \frac{\varphi(w)\,dw}{w - z} \right\} \frac{1}{z_1 - z}$$

$$= \lim_{z_1 \to z} \frac{1}{z_1 - z} \int_C \varphi(w) \left\{ \frac{1}{w - z_1} - \frac{1}{w - z} \right\} dw$$

$$= \lim_{z_1 \to z} \int_C \frac{\varphi(w)\,dw}{(w - z_1)(w - z)}.$$

Since z is at a positive distance from C, say $|z - w| \geq 2\delta > 0$ for all w on C, then $|z_1 - w| \geq \delta$ eventually and so $|w - z_1|\,|w - z|$ has a positive lower bound. Consequently,

$$\frac{\varphi(w)}{(w - z_1)(w - z)} \to \frac{\varphi(w)}{(w - z)^2},$$

uniformly with respect to w. Therefore (see Exercise 7.1)

$$\lim_{z_1 \to z} \int_C \frac{\varphi(w)\,dw}{(w - z_1)(w - z)} = \int_C \frac{\varphi(w)}{(w - z)^2}\,dw.$$

Here z and z_1 are inside C and w is on the trace of C. Since the trace is a closed set (Exercise 5.1), the distances from z and z_1 to C have a positive lower bound and hence it is indeed true that g has a derivative, and moreover

$$g'(z) = \int_C \frac{\varphi(w)}{(w - z)^2}\,dw, \qquad z \text{ inside } C.$$

We can now show in the same way that

$$g''(z) = 2 \int_C \frac{\varphi(w)}{(w - t)^3}\,dw,$$

and generally,

$$g^{(n)}(z) = n! \int_C \frac{\varphi(w)}{(w - z)^{n+1}}\,dw, \qquad n = 1, 2, 3, \ldots.$$

Notice that we have shown at the same time that *an analytic function has derivatives of all orders.*

This (for first-order derivatives) is what we needed to complete the discussion of Morera's theorem in Sec. **6F**.

Since the previous discussion was carried through without the factor $1/(2\pi i)$ in front of the integrals, you should note particularly that, for an analytic function f, **Cauchy's formula for derivatives** reads

$$f^{(n)}(z) = \frac{n!}{2\pi i} \int_C \frac{f(w)}{(w - z)^{n+1}}\,dw.$$

Since analytic functions have derivatives of all orders, we can begin to see how restrictive it was to make the apparently mild assumption that our functions are differentiable. In addition, see Sec. **7F**.

EXERCISE 7.3. Carry through the proof of the formula for $g''(z)$.

EXERCISE 7.4. A student argues that if

$$\sum_{n=1}^{\infty} |u_n(z)| \le \sum_{n=1}^{\infty} M_n < \infty, \qquad z \in D,$$

M_n independent of z, then $\sum u_n(z)$ converges uniformly in D. Criticize this reasoning.

EXERCISE 7.5. With the hypotheses of Cauchy's formula, evaluate

$$\frac{1}{2\pi i}\int_C \frac{f(w)-f(z)}{w-z}\frac{dw}{w-z}.$$

EXERCISE 7.6. Suppose that $|f(z)| \le M$ on the circumference of a square whose side is L, and let z_0 be the center of the square. If f is analytic in the square, show that $|f'(z_0)| \le 8M/(\pi L)$.

The following two theorems are both famous and extremely useful.

CAUCHY'S INEQUALITIES *Let f be analytic for $|z-z_0| \le r$, and let $M = \max|f(z)|$ for $|z-z_0| \le r$. Let a_n be the coefficients of the Taylor series of f about z_0. Then $|a_n| \le Mr^{-n}$.*

We have

$$a_n = \frac{f^{(n)}(z_0)}{n!} = \frac{1}{2\pi i}\int_{|z-z_0|=r} \frac{f(w)\,dw}{(w-z_0)^{n+1}}.$$

Hence, by the inequalities in **5C**, we have

$$|a_n| \le \frac{1}{2\pi}\cdot 2\pi r \cdot M \cdot \frac{1}{r^{n+1}} = Mr^{-n}.$$

A function that is analytic in the whole finite plane is called **entire**.[1]

LIOUVILLE'S THEOREM *An entire function that is bounded in the whole finite plane must be a constant.*

For the proof, let $|f(z)| \le M$. For each z, take $R > |z|$ and write

$$f(z) - f(0) = \frac{1}{2\pi i}\int_{|w|=R} f(w)\left(\frac{1}{w-z} - \frac{1}{w}\right) dw$$

$$= \frac{z}{2\pi i}\int_{|w|=R} \frac{f(w)}{w(w-z)}\,dw.$$

Then we have

$$|f(z)-f(0)| \leq \frac{|z|MR}{R(R-|z|)}.$$

For each z, the right-hand side approaches 0 as $R \to \infty$, and hence $f(z)-f(0)=0$ for every z.

EXERCISE 7.7. If P is a polynomial of degree at least 1 and has no zeros, apply Liouville's theorem to $1/P(z)$ to obtain a contradiction. (See Exercise 1.23.) This proves the fundamental theorem of algebra again.

EXERCISE 7.8. Let f be entire, and let there be constants A and c such that $|f(z)| \leq |z|^c$ for $|z| \geq A$. Show that f is a polynomial of degree at most c.

EXERCISE 7.9 Show that Liouville's theorem remains true if we replace the boundedness of f by the boundedness of

$$\int_0^{2\pi} |f(re^{i\theta})|\, d\theta.$$

EXERCISE 7.10 Show that if f is analytic in $|z|<1$ and $|f(z)| \leq 1-|z|$, then $f(z) \equiv 0$.

EXERCISE 7.11 Can an analytic function satisfy $|f(z)| \geq 1/(1-|z|)$ for $|z|<1$?

EXERCISE 7.12 (Derivatives of Infinite Order) Suppose that $f(z)$ is analytic in $|z| \leq R$ and that $\lim_{n\to\infty} f^{(n)}(z) = g(z)$ uniformly in $|z| \leq R$. Show that $g(z) = ce^z$, c constant, for $|z| \leq R$. (*Hint*: What is $\lim_{n\to\infty}[f^{(n)}(z)]'$?)

Supplementary exercises

1. If $\sum M_n$ is a convergent series of constants, and $u_n(z) \leq M_n$, does $\sum u_n(z)$ necessarily converge uniformly?

2. If each $u_n(x)$ is continuous on $0 \leq x \leq 1$ and $u_n(x)$ converges uniformly on $0 \leq x < 1$, does $u_n(x)$ necessarily converge uniformly on $0 \leq x \leq 1$?

3. Does the sequence $\{[4x(1-x)]^n\}$ converge uniformly on $0<x<1$?

4. The sequences $\{x^n(1-x^n)\}$ and $\{nx^n(1-x^n)\}$ both converge pointwise to 0 on $0 \leq x \leq 1$. Does either sequence converge uniformly?

5. Deduce Liouville's theorem from Cauchy's inequalities.

6. Explain why $f(z) = z^{1/2}$ does not violate Exercise 7.8 with $c = \frac{1}{2}$.

7. Suppose that f is analytic in $|z| \leq R$. Define $F_1(z) = \int_0^z f(w)\, dw$, $F_n(z) = \int_0^z F_{n-1}(w)\, dw$. What conclusion (if any) can you deduce from the hypothesis that $\lim_{n\to\infty} F_n(z) = g(z)$ uniformly in $|z| \leq R_1 < R$?

7D. Uniform convergence of sequences of analytic functions

I remind you that a sequence $\{f_n\}$ of functions **converges uniformly** on a set S if

$$\sup_{z \in S} |f_n(z) - f_m(z)| \to 0 \qquad \text{as } m \text{ and } n \to \infty;$$

and $\{f_n\}$ converges uniformly to f on S if

$$\sup_{z \in S} |f_n(z) - f(z)| \to 0 \qquad \text{as } n \to \infty.$$

A series of functions converges uniformly if the sequence of its partial sums converges uniformly. (Compare Sec. **3D**.)

You are probably aware that the limit of a uniformly convergent sequence of continuous functions is continuous. For analytic functions, we have a much stronger result. Loosely stated, in a form that is easy to remember, it says that *the limit of a uniformly convergent sequence of analytic functions is analytic.* This is an abbreviation for a more precise theorem.

> ***UNIFORM CONVERGENCE THEOREM*** *If the functions f_n are all analytic (from some value of n onward) in a simply connected region D, and $\{f_n\}$ converges uniformly on a simple closed curve C in D, then*[2] *$\{f_n\}$ converges, in the region S inside C, to an analytic function f. Furthermore, $f_n' \to f'$ uniformly on each compact subset of S, and consequently, $f_n'(z) \to f'(z)$ for each z in S.*

In short, *a uniformly convergent series of analytic functions can be differentiated term by term.*

The uniform convergence theorem tells us that convergence on C propagates inward. We shall see later (and could prove now) that convergence also propagates outward (Sec. **28D**): If f_n are analytic and uniformly bounded, and $\{f_n\}$ converges on a subset of D that has a limit point inside D, then $\{f_n\}$ converges uniformly on each compact subset of D.

In particular, since a power series converges uniformly in each compact subset of its (open) disk of convergence, we conclude that a power series can be differentiated term by term inside its disk of convergence (not necessarily in the closed disk), as claimed in Sec. **3C**.

Notice the difference from the corresponding situation on the real line—there, a uniformly convergent series can be integrated term by term but not necessarily differentiated term by term.

To prove the uniform convergence theorem, we represent $f_n(z)$ for z inside C by Cauchy's formula,

$$f_n(z) = \frac{1}{2\pi i} \int_C \frac{f_n(w)}{w - z}\, dw,$$

so that

$$f_n(z) - f_m(z) = \frac{1}{2\pi i} \int_C \frac{f_n(w) - f_m(w)}{w - z} \, dw.$$

Since the trace of C is a closed set that does not contain z, there is a positive distance δ between z and the trace of C. Hence

$$|f_n(z) - f_m(z)| \le \frac{1}{2\pi\delta} \cdot (\text{length of } C) \cdot \sup_{w \in C} |f_n(w) - f_m(w)|.$$

This says that $\{f_n(z)\}$ converges. Moreover, if z is confined to a compact subset T of S, the distance between T and the trace of C is positive, so that the convergence is uniform on T. If $f_n \to f$, we also have

$$f(z) = \frac{1}{2\pi i} \int_C \frac{f(w)}{w - z} \, dw,$$

so that f is represented by Cauchy's formula and consequently is analytic (Sec. **7C**). We also know that

$$f_n'(z) = \frac{1}{2\pi i} \int_C \frac{f_n(w)}{(w - z)^2} \, dw$$

and

$$f'(z) = \frac{1}{2\pi i} \int_C \frac{f(w)}{(w - z)^2} \, dw.$$

Therefore

$$|f_n'(z) - f'(z)| \le \frac{1}{2\pi\delta^2} (\text{length of } C) \cdot \sup_{w \in C} |f_n(w) - f_n(z)| \to 0.$$

This establishes the uniform convergence of $\{f_n'\}$ to f'.

7E. Functions defined by integrals with a parameter

In Secs. **6F** and **7C** we proved Morera's theorem, that if in a simply connected region D we have f continuous and $\int_C f(w) \, dw = 0$ for every simple closed curve, then f is analytic in D. This may have seemed interesting as a converse of Cauchy's theorem, but somewhat impractical—how can one satisfy the hypotheses? Here I want to exhibit one of the applications of Morera's theorem to results that would otherwise be rather complicated to establish. Rather than presenting a general theorem, I illustrate with some specific examples. We will need a lemma, which you can easily prove.

EXERCISE 7.13 Show that if φ is analytic in the whole finite plane and g is continuous on $[a, b]$, then $\int_a^b g(t)\varphi(zt) \, dt$ is continuous.

Now let g be continuous on $[0, 1]$ and define $F(z) = \int_0^1 g(t) \sin zt\, dt$. Let us show that F is analytic in the whole finite plane.

After Exercise 7.13, all that is needed is to observe that, for every closed curve C,

$$\int_C F(z)\, dz = \int_C \left\{ \int_0^1 g(t) \sin zt\, dt \right\} dz = \int_0^1 g(t) \left\{ \int_C \sin zt\, dz \right\} dt = 0,$$

because $\sin zt$ is an analytic function of z for each t. The change of order of integration is justified because if we write the integrals in terms of the parameter on C, we can apply the elementary theory of changing the order of integration in an iterated integral.

We can evidently replace $[0, 1]$ by any finite interval, and $\sin zt$ by $f(z, t)$, where f is continuous (in the two-dimensional sense) in z and t, and analytic in z for each t.

An integral of the form

$$F(z) = \int_0^\infty e^{-zt} \varphi(t)\, dt$$

is called a **Laplace transform**. We can now show that, at least when φ is continuous and satisfies $|\varphi(t)| \le Ae^{ct}$, *the Laplace transform F is analytic in the half-plane* $\operatorname{Re} z > c$. Here we have, for $x = \operatorname{Re} z > c$, and all positive numbers R and S, with $0 < R < S$,

$$\left| \int_R^S e^{-zt} \varphi(t)\, dt \right| \le A \int_R^S e^{-xt} e^{ct}\, dt.$$

Consequently, if $\operatorname{Re} z > c + \delta$, $\delta > 0$, the integral

$$\int_0^\infty e^{-zt} \varphi(t)\, dt$$

converges uniformly, and hence is an analytic function of z by the uniform convergence theorem (Sec. **7D**). Strictly speaking, the uniform convergence theorem was proved only for sequences, but we can consider the sequence

$$\left\{ \int_0^n e^{-zt} \varphi(t)\, dt \right\}_{n=1}^\infty .$$

Supplementary exercises

1. Let g be analytic in the unit disk. Is

$$F(z) = \int_0^z g(t) \sin(z + t)\, dt$$

analytic, and (if so) in what domain?

2. Answer the same question for

$$F(z) = \int_0^\infty (\sin zt) g(t)\, dt,$$

if $\int_0^\infty |g(t)|\, dt$ converges.

3. If $F(x) = \int_0^x f(x+t)\, dt$, where f is the restriction to the real axis of an entire function, is F also the restriction of an entire function?

7F. Taylor series

We can use Cauchy's integral formula, with C a circle, to show that *if f is analytic at a point z_0* (remember that this means that f is analytic in some neighborhood of z_0), *then f is always represented by its Taylor series in a disk centered at z_0.*

To do this, we start again from

$$f(z) = \frac{1}{2\pi i} \int_C \frac{f(w)\, dw}{w - z},$$

where C is the circumference of a disk centered at z_0 and contained in a simply connected region D in which f is analytic, and z is inside C.

The discussion depends on the idea that if we expand the "kernel" $(w-z)^{-1}$ in a series of functions that converges uniformly on C, we can integrate term by term and deduce an expansion of $f(z)$. This is a powerful method for obtaining various kinds of expansions of functions. Since in the present case we want a series of powers of $(z - z_0)$, we look for an expansion of $(w-z)^{-1}$ in powers of $(z - z_0)$. It is probably not immediately obvious what this expansion should be. However, if we first think about the simplest case, when $z_0 = 0$, we have

$$f(z) = \frac{1}{2\pi i} \int_C \frac{f(w)\, dw}{w - z}$$

and

$$\frac{1}{w-z} = \frac{1}{w} \frac{1}{1 - (z/w)} = \frac{1}{w} \sum_{n=0}^{\infty} \left(\frac{z}{w}\right)^n = \sum_{n=0}^{\infty} \frac{z^n}{w^{n+1}},$$

where the series converges uniformly for w on C, and uniformly in w and z if z is in any closed disk centered at 0 and completely inside C, since then $|z/w|$ is less than some number less than 1. This suggests that in general we should write

$$\frac{1}{w-z} = \frac{1}{(w-z_0) - (z-z_0)} = \frac{1}{w-z_0} \sum_{n=0}^{\infty} \frac{(z-z_0)^n}{(w-z_0)^n} = \sum_{n=0}^{\infty} \frac{(z-z_0)^n}{(w-z_0)^{n+1}}.$$

Here $|z - z_0|/|w - z_0| < 1$ and the series converges uniformly if z is confined to a disk completely inside C. Now we can substitute this series for $(w-z)^{-1}$ in

Cauchy's formula and integrate term by term to get

$$f(z) = \frac{1}{2\pi i} \sum_{n=0}^{\infty} (z - z_0)^n \int_C \frac{f(w)\, dw}{(w - z_0)^{n+1}}, \qquad z \text{ inside } C.$$

By Cauchy's formula for derivatives in Sec. **7C**, we have

$$\frac{1}{2\pi i} \int_C \frac{f(w)\, dw}{(w - z_0)^{n+1}} = \frac{f^{(n)}(z_0)}{n!}.$$

On the right we have the usual formula for the coefficients of the Taylor series of f. We now also have a way of expressing the coefficients as integrals.

We now have proved that every analytic function is represented locally by its Taylor expansion about any point where it is analytic; in fact, *the Taylor series will converge in the largest disk, centered at the point, in which the function is analytic.*

In addition, we have an alternative way of finding the coefficients in a power series. Looking at it from another point of view, we can regard it as a method of evaluating integrals of a special kind if we know enough about their integrands.

EXERCISE 7.14 Evaluate

$$\int_{|w|=1} \frac{\sin w\, dw}{[w - (\pi/4)]^3}.$$

EXERCISE 7.15 Evaluate

$$\int_{|w|=2} \frac{e^{3w}\, dw}{(w - 1)^4}.$$

7G. Zeros of analytic functions

We now know that if f is analytic at z_0 then $f(z)$ is represented by its Taylor series in a neighborhood of z_0. If it happens that $f(z_0) = 0$, we say that f has a **zero** at z_0, and then the Taylor series has the form

$$f(z) = a_1(z - z_0) + a_2(z - z_0)^2 + \cdots$$

Unless $f(z) \equiv 0$ in a neighborhood of z_0, some a_k must not be zero; let a_n be the first nonzero coefficient. Then

$$f(z) = a_n(z - z_0)^n + a_{n+1}(z - z_0)^{n+1} + \cdots,$$

that is,

$$\begin{aligned} f(z) &= (z - z_0)^n(a_n + a_{n+1}(z - z_0) + \cdots) \\ &= (z - z_0)^n \varphi(z), \end{aligned}$$

where φ is analytic at z_0 (as the sum of a power series) and $\varphi(z_0) = a_n \neq 0$. Since φ is continuous, $\varphi(z) \neq 0$ in some neighborhood of z_0. Consequently,

$f(z) \neq 0$ in some neighborhood of z_0 (except at z_0). The number n is the **order** of the zero of f at z_0. Notice that the order is necessarily an integer.

Now z_0 could have been any point at which f is analytic; consequently, we say that the zeros of an analytic function are isolated (if the function is not identically zero in a neighborhood of one of them).

This statement does not, by itself, assure us that there is not a limit point of isolated zeros of f. However, if f had a set of zeros with limit point z_0 inside the region where f is analytic, it would follow that $f(z_0) = 0$, by continuity.

We assumed that $f(z)$ is not identically zero in a neighborhood of z_0. We now show that if $f(z) \equiv 0$ in a neighborhood of z_0, then $f(z) \equiv 0$ throughout any region D, containing z_0, in which f is analytic. If not, there would be a point z_1 of D at which $f(z_1) \neq 0$. Let S be the largest connected subset of D that contains z_0 and in which $f(z) = 0$. If S is not all of D, S has a boundary point z_2 in D. Since S is connected, z_2 is not an isolated point of S; consequently $f(z_2) = 0$. Since z_2 is a limit point of S, we have $f(z) \equiv 0$ in a neighborhood of z_2; but z_2, being a boundary point, has points arbitrarily close to it but not in S. This is a contradiction, so S must be all of D, and $f(z) \equiv 0$ in D.

We can therefore say that:

The zeros of an analytic function are isolated unless the function is the constant 0.

It follows at once that *if f and g are both analytic in a region D, and $f(z) = g(z)$ for $z = z_n$, where $\{z_n\}$ has a limit point* **in** *D, then f and g are the same function in D.*

Notice that if the limit point is on the boundary of D, we cannot assert that $f \equiv g$. For example, the functions $\sin \pi z$ and 0 have the same values at the integers; 0 and $\sin(\pi/z)$ have the same values at $1/n$ ($n = 1, 2, \ldots$).

The **coincidence principle**, that *two analytic functions are the same if they coincide on a set with a limit point,* has many applications. Here is a simple example. It is obvious that a polynomial P with real coefficients takes conjugate values at conjugate points, that is, that $P(\bar{z}) = \overline{P(z)}$. Similarly, if $f(z)$ is represented by a power series $\sum_{k=0}^{\infty} a_k(z - c)^k$ with both c and the a_k real, then $f(\bar{z}) = \overline{f(z)}$ in the disk of convergence. We can now prove something that is less obvious, but extremely plausible, namely, that *if f is analytic in a region D that contains an interval I of the real axis and is symmetric about the real axis* (see Fig. 7.4, p. 64), *and if $f(z)$ is real on I, then f takes conjugate values at conjugate points in D* (even when D is not a disk). Recall from Exercise 2.5 that if f satisfies the Cauchy-Riemann equations in a region S, then the function g defined in the reflection of S with respect to the real axis by the equation $g(z) = \overline{f(\bar{z})}$ satisfies the Cauchy-Riemann equations. In our case S and its reflection are both D. Thus f and g are analytic in D and coincide on I; hence they coincide in D.

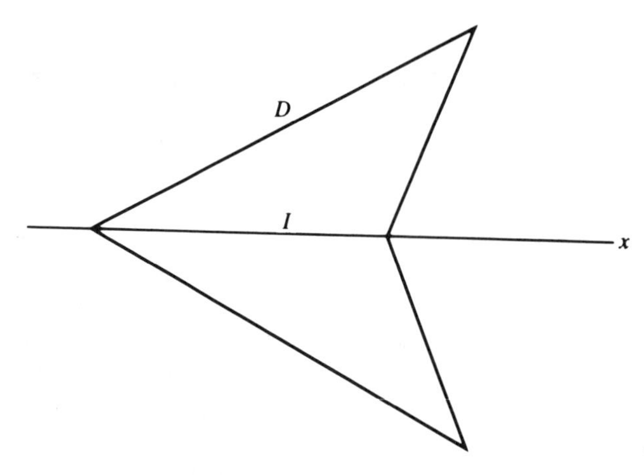

FIGURE 7.4

As another application, suppose that we want to verify that an identity like $\sin 2x = 2\sin x \cos x$ can be extended to all complex z. We could, of course, multiply out the corresponding power series to verify the identity. However, it is much easier to consider the analytic function f defined by $f(z) = \sin 2z - 2\sin z \cos z$. We have f analytic in the whole finite plane and zero on the real axis (zero for $0 < x < \pi$ would be more than sufficient), and hence by the coincidence principle, $f(z) \equiv 0$.

EXERCISE 7.16 Show (without going back to Euler's formulas or power series) that

$$\sin(z + w) = \sin z \cos w + \cos z \sin w.$$

Supplementary exercises

1. Show that the identities of Sec. **4E**, Supplementary Exercise 6 on p. 32, are valid if θ is replaced by any z for which the formulas make sense.

2. Show that

$$\lim_{k\to\infty}\left(1 + \frac{z}{k}\right)^k = e^z \qquad \text{for all complex } z.$$

3. Suppose that for $n = 1, 2, \ldots,$ we have $f_n(z) \to L_n$ when $z \to z_0$, and that $|f_n(z)| \le M_n$ for all z, with $\sum M_n < \infty$. Show that if $p = p(z) \to \infty$ as $z \to z_0$, then

$$\lim_{z\to z_0}\{f_1(z) + f_2(z) + \cdots + f_p(z)\} = \sum L_n.$$

You may assume that all $f_n(z)$ are analytic in a common region D.

7H. The gamma function

The **gamma function** is defined initially by

$$\Gamma(z) = \int_0^\infty t^{z-1}e^{-t}\,dt, \qquad \operatorname{Re} z > 0;$$

this is our first explicit nonelementary function. To see that Γ is analytic for $\operatorname{Re} z > 0$, we need to know that the integral converges uniformly (at both ends, when $0 < \operatorname{Re} z < 1$) on each compact subset of the right-hand half-plane. The proof is much like the one that we gave in Sec. **7E** for Laplace transforms, but the details are more complicated and not very interesting, so I omit them.

It is easy to see that when n is 0 or a positive integer we have $\Gamma(n+1) = n!$. In fact, the Γ function seems to have been introduced originally as a continuous function that interpolates between the factorials. For this reason, $x!$ is sometimes written instead of $\Gamma(x+1)$. It is no harder to show that $\Gamma(z+1) = z\Gamma(z)$ for $\operatorname{Re} z > 0$. To do this, consider

$$\int_0^R t^{z-1} e^{-t}\, dt,$$

and integrate by parts, differentiating e^{-t} and integrating t^{z-1}, to get

$$z^{-1} e^{-t} t^z \,\Big|_0^R + \frac{1}{z} \int_0^R t^z e^{-t}\, dt.$$

When $R \to \infty$, the integrated terms drop out because $\lim_{R\to\infty} e^{-R} R^z = 0$, and the integral becomes $\Gamma(z+1)$. Thus we have $\Gamma(z+1) = z\Gamma(z)$, the functional equation for the gamma function, when $\operatorname{Re} z > 0$. We can now define $\Gamma(z)$ for $-1 < \operatorname{Re} z < 0$ by the equation $\Gamma(z) = z^{-1}\Gamma(z+1)$. We can repeat this process, and extend $\Gamma(z)$ to successive strips[3] $-n < \operatorname{Re} z < -n+1$. See Fig. 7.5 for a graph[3] of real values of z.

FIGURE 7.5 The gamma function for real x.

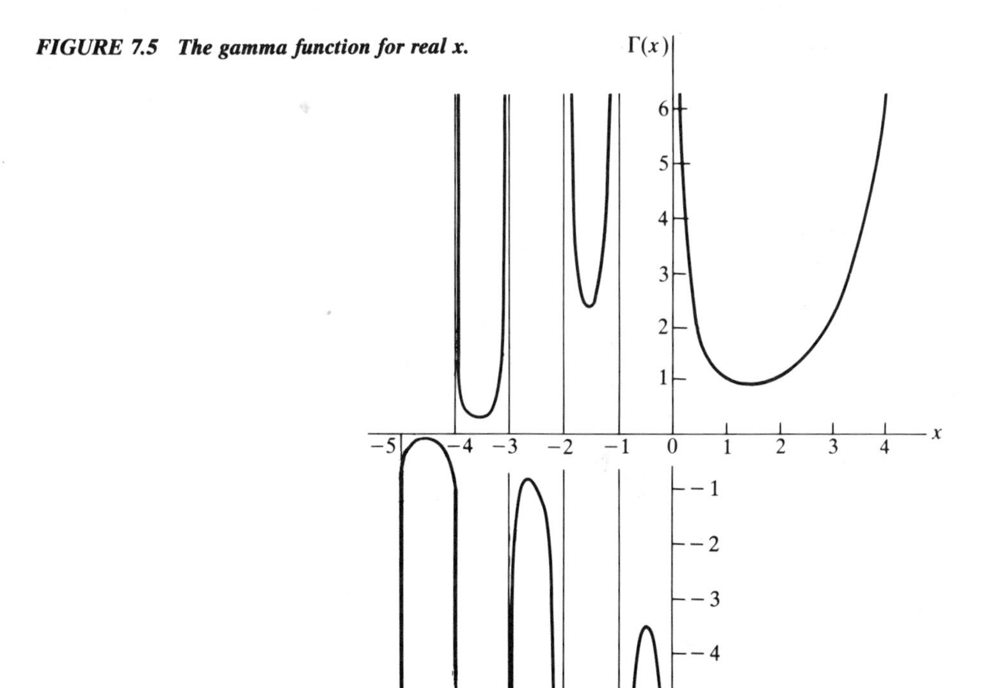

We shall see below that the gamma function is analytic in the finite plane, except at 0 and the negative integers; also that $\Gamma(z)$ is never zero, and consequently $1/\Gamma$ is an entire function.

We are now going to obtain a useful formula for the gamma function by some ingenious eighteenth-century manipulation of integrals. We start from

$$\Gamma(p) = \int_0^\infty t^{p-1} e^{-t}\, dt, \qquad p > 0,$$

and obtain, with $t = y^2$,

$$\Gamma(p) = 2\int_0^\infty y^{2p-1} e^{-y^2}\, dy,$$

and similarly,

$$\Gamma(q) = 2\int_0^\infty x^{2q-1} e^{-x^2}\, dx,$$

so that

$$\Gamma(p)\Gamma(q) = 4\iint_0^\infty x^{2q-1} y^{2p-1} e^{-(x^2+y^2)}\, dx\, dy.$$

Now we rewrite the double integral in polar coordinates:

$$\begin{aligned}\Gamma(p)\Gamma(q) &= 4\int_0^\infty e^{-r^2} r\, dr \int_0^{\pi/2} (r\cos\theta)^{2q-1}(r\sin\theta)^{2p-1}\, d\theta \\ &= 4\int_0^\infty r^{2p+2q-1} e^{-r^2}\, dr \int_0^{\pi/2} (\cos\theta)^{2q-1}(\sin\theta)^{2p-1}\, d\theta \\ &= 2\Gamma(p+q)\int_0^{\pi/2} (\cos\theta)^{2q-1}(\sin\theta)^{2p-1}\, d\theta.\end{aligned}$$

Incidentally, if we take $p = q = \frac{1}{2}$ at this point, we get

$$\int_0^\infty e^{-x^2}\, dx = \tfrac{1}{2}\pi^{1/2}.$$

Next let $x = \sin^2\theta,\ 1 - x = \cos^2\theta,\ dx = 2\sin\theta\cos\theta\, d\theta$:

$$2\int_0^{\pi/2} (\cos\theta)^{2q-1}(\sin\theta)^{2p-1}\, d\theta = \int_0^1 x^{p-1}(1-x)^{q-1}\, dx.$$

The last integral is the **beta function** $\mathrm{B}(p, q)$. Now let $x = y/(1+y)$; then we have

$$\mathrm{B}(p, q) = \int_0^\infty \frac{y^{p-1}\, dy}{(1+y)^{p+q}}.$$

We have therefore proved incidentally that

$$B(p, q) = \frac{\Gamma(p)\Gamma(q)}{\Gamma(p+q)}.$$

When $0 < p < 1$ and $q = 1 - p$, we now have

$$\int_0^\infty \frac{y^{p-1}}{1+y}\, dy = \Gamma(p)\Gamma(1-p).$$

The integral on the left will be evaluated directly in Sec. **11A**: it is equal to $\pi/\sin \pi p$ when $0 < p < 1$. Hence we have

$$\Gamma(z)\Gamma(1-z) = \frac{\pi}{\sin \pi z}, \qquad 0 < z < 1.$$

This equation holds for all z for which all the functions concerned are analytic, by the coincidence principle.

In particular, we can take $z = \frac{1}{2}$ to obtain $\Gamma(\frac{1}{2}) = \pi^{1/2}$. It also follows that $\Gamma(z)$ is never 0, since $\Gamma(1-z) = \infty$ only when $z = 0, 1, 2, \ldots$. Moreover,

$$\Gamma(z) \sin \pi z = \frac{\pi}{\Gamma(1-z)} \pi$$

as $z \to 0$, or

$$z\Gamma(z) \frac{\sin \pi z}{\pi z} \to 1 \qquad \text{as } z \to 0.$$

Since $(\pi z)^{-1} \sin \pi z \to 1$, we have $z\Gamma(z) \to 1$ as $z \to 0$.

EXERCISE 7.17 Show that Γ is analytic on the line $\operatorname{Re} z = 0$ except at $z = 0$.

Supplementary exercises

1. Compute $\Gamma(3/2)$ and $\Gamma(-5/2)$.
2. Given that $\Gamma(1.3) = 0.89747$, find $\Gamma(5.3)$ and $\Gamma(-3.7)$.
3. Show that $\Gamma''(x) > 0$ for $x > 0$, and hence that, for positive x, the graph of $\Gamma(x)$ has a single minimum point.

Notes

1. Entire functions are also called integral functions, especially in British usage.

2. Actually, as we shall see in Sec. **28D**, the convergence of $\{f_n\}$ is uniform, not only in S, but in each compact subset of D.

3. More accurate graphs of the gamma function are available in a number of places: E. P. R. Duval, "Graphs of the Functions Π and Ψ," *Ann. Math.* (2) 5 (1904): 64–65, reproduced in Widder, p. 370; also in Sibagaki, p. 87; M. L. Boas, *Solutions*, p. 347.

Nevertheless, the graphs in many reference books are rather incorrect for negative x. The gamma function is tabulated for $1 \leq x \leq 2$, and the values for other real arguments are easily calculated from the functional equation.

8 *Isolated Singular Points*

8A. Examples of singular points

A **singular point** (or **singularity**) of a function f is a point at which f is not analytic, for example, 0 for $f(z) = 1/z$ or $f(z) = e^{1/z}$. Other examples are 0 for $f(z) = z/z$ and $f(z) = z^{-1} \sin z$. There are obvious differences between the first two examples and the second two. Although all four functions fail to be analytic at 0 because they are not defined there, the second two approach limits as $z \to 0$, and become analytic at 0 if we define them at 0 by their limits: $z/z = 1$, and $z^{-1} \sin z$ by the sum of the series

$$1 - \frac{z^2}{3!} + \frac{z^4}{5!} - \cdots .$$

On the other hand, $1/z$ does not have a finite limit as $z \to 0$; and $e^{1/z}$ fails altogether to have a limit as $z \to 0$.

EXERCISE 8.1 Show that $e^{1/z}$ does not approach a limit as $z \to 0$. (Consider its behavior as $z \to 0$ through real and through pure imaginary values.)

To say that z/z has a singular point at 0 may seem rather artificial, or even pedantic: The singular point is there only because we defined the function by an inappropriate formula. The singularity of $z^{-1} \sin z$ is not much more complicated; we would not have noticed it if we had thought of $\sin z$ as defined by its Maclaurin series.

In these examples, we are dealing with functions that are analytic in a "punctured" neighborhood of the singular point, that is, in some disk centered at the point, with the point itself removed. In these circumstances we say that the point is an **isolated** singular point of the function. The singular points of all four of our examples are isolated, but it is easy to find analytic functions that have nonisolated singular points.

EXERCISE 8.2 Show that the singular point of $\tan(1/z)$ at 0 is not isolated.

8B. Classification of isolated singular points

It turns out to be useful to distinguish three kinds of isolated singular point. Let the singular point be z_0.

1. $f(z)$ is bounded in some neighborhood of z_0.

2. $|f(z)| \to \infty$ as $z \to z_0$.

3. Everything else. (That is, $|f(z)|$ is unbounded but $|f(z)| \not\to \infty$.)

In case 1, the singularity is said to be **removable**. The reason for the terminology is that (as we are about to show) in this case $f(z)$ necessarily approaches a (finite) limit as $z \to z_0$, and that the singularity disappears if $f(z_0)$ is defined to be this limit.

8C. Removing a removable singular point

The following theorem tells us not only that this can be done, but how to do it.

RIEMANN'S THEOREM ON REMOVABLE SINGULAR POINTS
If f is analytic and bounded in some punctured neighborhood of z_0, then $\lim_{z \to z_0} f(z)$ exists; and if $f(z_0)$ is defined as this limit, then f becomes analytic at z_0.

To prove the theorem, consider[1] the function g, defined by

$$g(z) = (z - z_0)^2 f(z), \qquad g(z_0) = 0,$$

which we can show is analytic at z_0. Indeed, it is evident that g' exists for $z \neq z_0$. We have $g'(z_0) = 0$ because

$$\frac{g(z) - g(z_0)}{z - z_0} \to 0$$

as $z \to z_0$. Since g is differentiable in a neighborhood of z_0, it is analytic at z_0.

If we now expand g in a Taylor series about z_0, the first two terms are zero, so

$$g(z) = \sum_{n=2}^{\infty} \frac{g^{(n)}(z_0)}{n!} (z - z_0)^n = (z - z_0)^2 h(z),$$

where h is a function that is analytic at z_0 (because all convergent power series represent analytic functions). Hence $(z - z_0)^2 h(z) = (z - z_0)^2 f(z)$, so that $h(z) = f(z)$ for $z \neq z_0$. Moreover, $\lim_{z \to z_0} h(z)$ exists, so $\lim_{z \to z_0} f(z)$ exists. If we redefine $f(z_0)$ as this limit, we have $f(z) \equiv h(z)$ and therefore f is analytic at the point z_0.

It is worth noticing that the boundedness of $|f(z)|$ can be replaced by $f(z) = O(|z - z_0|)^{-c}$, where $0 \leq c < 1$.

When a singularity is removable, we often think of it as having been removed, even when we do not say so explicitly.

EXERCISE 8.3 Decide which of the functions have removable singularities at the indicated points:

(a) $\dfrac{\sin z}{z^2 - \pi^2}$ at π $\qquad$ (b) $\dfrac{\sin z}{(\pi - z)^2}$ at π

(c) $\dfrac{1-\cos z}{z}$ at 0 (d) $z \cot z$ at 0

(e) $\dfrac{6-z-z^2}{2-z}$ at 2 (f) $\dfrac{\sin z}{z^2+z}$ at 0

(g) $\dfrac{\cos z}{1-\sin z}$ at $\pi/2$

For additional exercises on the material in this section, see Exercise 8.4.

8D. Poles

If $f(z) \to \infty$ as $z \to z_0$, there will be a disk centered at z_0 in which $|f(z)| > c > 0$. Consider $g(z) = 1/f(z)$ in this disk. In the punctured disk, g is analytic and bounded, so g must have a removable singular point at z_0; and the singularity is removed by setting $g(z_0) = 0$. We can then write $g(z) = (z - z_0)^n \varphi(z)$, where φ is analytic at z_0 and $\varphi(z_0) \neq 0$ (see Sec. **7G**). We say that f has a **pole of order** n at z_0; notice that n has to be a positive integer. If $n = 1$, the pole is **simple**. If $n > 1$, the **order** of the pole is the order of the zero of g, or, alternatively, the smallest integer n such that $(z - z_0)^n f(z)$ is bounded in a neighborhood of z_0. The term "pole" is used because a graph of $|f(z)|$ over the (x, y) plane would have a sharp spike, or pole, at z_0. If we think of the values of f as being on the Riemann sphere, we can say that $f(z_0) = \infty$, the north pole of the sphere (in a different sense of the word "pole").

A function that is analytic in a region except for poles is called **meromorphic** in that region.

We now see that the gamma function (Sec. **7H**) is a meromorphic function with simple poles at the points $0, -1, -2, \ldots$.

EXERCISE 8.4 Find the orders of the poles of the following functions at the indicated points:

(a) $\dfrac{e^{z^2}-1}{z^3}$ at 0 (b) $\dfrac{\cos z - 1}{z^2}$ at 0

(c) $\dfrac{z}{1-\sin z}$ at 0 (d) $\tan^2 z$ at $\dfrac{\pi}{2}$

(e) $\sum_{n=0}^{\infty} n(-1)^n z^n$ at -1 (f) $\dfrac{\cos z}{[(\pi/2)-z]^7}$ at $\dfrac{\pi}{2}$

Supplementary exercises

Classify as removable or poles the singularities of the following functions at the indicated points. State the orders of any poles.

1. $\dfrac{z}{(\sin z)^2}$ at 0

2. $\dfrac{1}{e^{z^2}-1}$ at 0

3. $\dfrac{(\sin z)^2}{(z-\pi)^2}$ at π

4. $\dfrac{z\cos z - \sin z}{z^3}$ at 0

5. $\dfrac{1}{z}\displaystyle\int_0^z e^{-t^2}\,dt$ at 0

6. $\dfrac{\pi^2 - z^2}{(\sin z)^3}$ at π

7. $\dfrac{z^2-4z+4}{z-2}$ at 2

8. $f(z) = \sum_{n=0}^{\infty} (\sin z)^n$ at $\dfrac{\pi}{2}$

9. $\dfrac{e^z}{z^2-z-6}$ at 3

10. $\dfrac{(e^z-1)^3}{z^6}$ at 0

8E. Essential singular points

The essential singular points are the isolated singular points that are neither poles nor removable singular points. For an essential singular point z_0, the function $f(z)$ approaches different limits as $z \to z_0$ along different sequences tending to z_0. The following theorem makes this statement even stronger.

CASORATI-WEIERSTRASS THEOREM[2] *In every neighborhood of an essential singular point of a function f, $f(z)$ takes values arbitrarily close to every complex number.*

In fact, suppose that there were a number A such that the values of f are (in some neighborhood of z_0) all at least some $\delta > 0$ distant from A. That is, in some neighborhood of z_0 we have $|f(z) - A| > \delta > 0$. Then $1/[f(z) - A]$ would be bounded in a neighborhood of z_0, so z_0 would be a removable singular point and $1/[f(z) - A]$ would have a limit L as $z \to z_0$. If $L = 0$, $f(z) - A$ would become infinite and f would have a pole, which we have excluded. If $L \neq 0$, $f(z) - A \to 1/L$ and f would have a removable singularity at z_0, a possibility that we also excluded.

For example, $e^{1/z}$ must take values arbitrarily close to every complex number in every neighborhood of 0. This means that we can find a sequence of points z_n with $z_n \to 0$ on which $\exp(1/z_n) \to 1$, another on which $\exp(1/z_n) \to -100$, another on which $\exp(1/z_n) \to \infty$, and so on.

However, the Casorati-Weierstrass theorem is by no means the full story of how chaotic the behavior of an analytic function is in the neighborhood of an essential singular point. Picard's theorem, which is beyond the scope of this book, says that in every neighborhood of z_0 there are at most two points in the extended plane that the function does not actually assume infinitely often. For $e^{1/z}$ these are 0 and ∞.

8F. Residues

Let f be a function with an isolated singular point, of any kind, at z_0. Consider the integral

$$\frac{1}{2\pi i}\int_C f(z)\,dz,$$

where C is a simple positively oriented curve surrounding z_0 and no other singular points of f. The value of the integral is called the **residue** of f at z_0. Because of the lemma in Sec. **7A**, the residue is independent of the choice of the curve C, since C can be replaced by any circle that is inside C and centered at z_0. We shall therefore usually suppose that C has been taken to be such a circle.

It also follows from Sec. **7A** that:

If C surrounds a finite number of isolated singular points, the integral yields the sum of the residues at the singular points inside C. This is the **residue theorem**.

EXERCISE 8.5 (*a*) Show that if f has a *simple* pole at z_0 with residue R, then

$$f(z)-\frac{R}{z-z_0}$$

has a removable singular point at z_0.

(*b*) Show that the residue is always 0 at a finite removable singularity.

Residues at ∞ will be defined and discussed in Sec. **8G**.

It is not obvious why residues are going to be useful; indeed, if we had to rely only on the definition, they would not be very convenient to use. I shall therefore take up techniques for calculating residues (without integration) before showing you some of their applications. We shall usually want to find residues at poles.

Simple poles Let f have a simple pole at z_0. By Sec. **8D**, we have

$$f(z)=\frac{1}{z-z_0}\varphi(z),$$

where $\varphi(z_0)\neq 0$ and φ is analytic at z_0. Then the residue of f at z_0 is, by definition,

$$\frac{1}{2\pi i}\int_C f(z)\,dz = \frac{1}{2\pi i}\int_C \frac{\varphi(z)}{z-z_0}\,dz,$$

and the integral on the right equals $\varphi(z_0)$ by Cauchy's formula. On the other hand, by the way $\varphi(z)$ was defined, we have $\varphi(z) = \lim_{z\to z_0} f(z)(z-z_0)$, which we can usually calculate easily. In summary:

The residue of f at a pole z_0 is $\lim_{z\to z_0} f(z)(z-z_0)$ if the limit exists (and then the pole is simple).

This rule is especially convenient when we happen to have a function like

$$\frac{e^z}{(z-2)(z-3)}$$

with a factored denominator; then the residue at $z=2$ is evidently $e^2/(2-3) = -e^2$.

On the other hand, we do not always have the poles so explicitly exhibited.

EXERCISE 8.6 Let $f(z) = g(z)/h(z)$, where h has a simple zero at z_0 and $g(z_0) \neq 0$. Show that the residue of f at z_0 is $g(z_0)/h'(z_0)$.

For example, the residue of $e^{2z}/\sin z$ at $z=\pi$ is $e^{2\pi}/\cos\pi = -e^{2\pi}$.

Multiple poles When f has a pole of multiplicity $n>1$ at z_0, the calculation of the residue is less simple. We have

$$f(z) = (z-z_0)^{-n}\varphi(z), \qquad \varphi(z_0)\neq 0.$$

Then the residue is

$$\frac{1}{2\pi i}\int_C \frac{\varphi(z)\,dz}{(z-z_0)^n},$$

and this is just $1/(n-1)!$ times Cauchy's formula for $\varphi^{(n-1)}(z_0)$ (see Sec. **7C**). Thus

---

The residue at a pole of f of order n is

$$\frac{1}{(n-1)!}\left(\frac{d}{dz}\right)^{n-1}\{f(z)(z-z_0)^n\}.$$

In words, to find the residue, multiply $f(z)$ by $(z-z_0)^n$, differentiate $n-1$ times, divide by $(n-1)!$, and evaluate at z_0.

Since it is easy to overestimate the order of a pole, it is fortunate that we still get the correct residue if we multiply by $(z - z_0)^m$ with $m > n$, differentiate $m - 1$ times, divide by $(m - 1)!$, and evaluate at z_0.

EXERCISE 8.7 Prove the statement above.

EXERCISE 8.8 What happens if we *under*estimate the order of the pole?

In applying this rule, it is useful to be able to differentiate $(z - z_0)^n f(z)$, $n - 1$ times, by applying **Leibniz's rule** for differentiating a product:

$$(fg)^{(k)} = f^{(k)}g + \binom{k}{1} f^{(k-1)}g' + \binom{k}{2} f^{(k-2)}g'' + \cdots + fg^{(k)},$$

where

$$\binom{k}{j} = \frac{k!}{j!(k-j)!}$$

are the binomial coefficients. Leibniz's rule is easily proved by induction.

EXERCISE 8.9 People sometimes make incorrect generalizations. Show that when h has a double zero at z_0, then the residue of $g(z)/h(z)$ at z_0 is not $g(z_0)/h''(z_0)$; and that the residue of

$$\frac{g(z)}{(z - z_0)^2 h(z)}, \qquad h(0) \neq 0,$$

is not

$$\frac{1}{h(z_0)} \left\{ \text{residue of } \frac{g(z)}{(z - z_0)^2} \right\}.$$

EXERCISE 8.10 Show that if you happen to know the Taylor series of $g(z)$ about z_0, then you can find the residue of $g(z)/(z - z_0)^k$ very easily. Find the residue of $z^{-7}e^{-z^2}$ at $z = 0$.

Residues at conjugate points It is often useful to know that when $f(z)$ takes only real values on the real axis, the residue of f at a pole z_0 is the conjugate of the residue at $\overline{z_0}$. More precisely:

Let f be meromorphic in a region D that is symmetric about the real axis, and let f be real on the real axis; then the residues of f at z_0 and $\overline{z_0}$ are conjugates.

Here is the proof for simple poles; for the general case see Exercise 8.11. Since f takes conjugate values at conjugate points (Sec. **7G**), if f has a pole at z_0 with $\operatorname{Im} z_0 > 0$, then f also has a pole of the same order at $\overline{z_0}$. We will suppose that z_0 is a simple pole with residue R, and that the residue at $\overline{z_0}$ is S. Then

$$R = \lim_{z \to z_0} f(z)(z - z_0), \qquad S = \lim_{z \to \bar{z}_0} f(z)(z - \overline{z_0}).$$

We can just as well write

$$S = \lim_{w \to \bar{z}_0} f(w)(w - \overline{z_0}).$$

Now we can call w by the name $\bar{z}$ if we like, so

$$S = \lim_{\bar{z} \to \bar{z}_0} f(\bar{z})(\bar{z} - \overline{z_0}) = \lim_{\bar{z} \to \bar{z}_0} \overline{f(z)}(\overline{z - z_0}),$$

because $f(\bar{z}) = \overline{f(z)}$. But to say that $\bar{z} \to \overline{z_0}$ is to say that $z \to z_0$, so

$$S = \lim_{z \to z_0} \overline{f(z)(z - z_0)} = \bar{R}.$$

EXERCISE 8.11 Prove the same result for poles of any order.

EXERCISE 8.12 For practice, find the residues of the following functions at the indicated points:

(a) $\dfrac{z}{(2-3z)(4z+3)}$ at $\dfrac{2}{3}, -\dfrac{3}{4}$

(b) $\dfrac{e^{z-1}}{e^z - 1}$ at 0

(c) $\dfrac{2z^2+1}{z^2+25}$ at $5i$

(d) $\dfrac{e^{i\pi z}}{16 - z^4}$ at 2

(e) $\dfrac{\sin z}{1 - 2\cos z}$ at $\dfrac{\pi}{3}$

(f) $\dfrac{\sinh z - z}{z^8}$ at 0

(g) $\dfrac{\cos^2 z}{(2\pi - z)^3}$ at 2π

(h) $z \tan z$ at $\dfrac{\pi}{2}$

(i) $\dfrac{z^2}{z^4+1}$ at $\dfrac{1+i}{\sqrt{2}}$

(j) $\dfrac{z+1}{(z^2+4)^2}$ at $2i$

EXERCISE 8.13 (a) Is the residue of $f + g$ at $z = a$ equal to the residue of f plus the residue of g?

(b) Is the residue of fg equal to the product of the residues of f and g?

(c) Compare the residue of f at a simple pole $z = a \neq 0$ with the residue of $zf(z^2)$ at $z = a^{1/2}$.

Supplementary exercises

Find the residues of the following functions at the indicated points:

1. $\dfrac{z^2}{(i-z)(4z+5)}$ at i and at $-\dfrac{5}{4}$

2. $\dfrac{4z}{z^2 - 5z + 6}$ at 2 and 3

3. $\dfrac{z - \frac{1}{6}z^3 - \sin z}{z^8}$ at 0

4. $\dfrac{1 - \cos z}{2\sin z - \sqrt{3}}$ at $\dfrac{\pi}{3}$

5. $\dfrac{z^2+1}{z^4-1}$ at 1 and i

6. $\dfrac{\cot z}{z-\pi}$ at π

7. $\dfrac{2z+3}{(z^2-9)^2}$ at 3

8. $\dfrac{(\sin z)^3}{(z-\pi)^3}$ at π

8G. The point at infinity as an isolated singular point

If a function f is analytic in the part of the finite plane that is outside some disk, it seems reasonable to say that f has an isolated singular point at ∞. We classify isolated singular points at ∞ by saying that $f(z)$ *has the same kind of singular point at* ∞ *that* $f(1/z)$ *has at* 0: Thus there can be removable singular points, poles, or essential singular points at ∞. For example, $1/z$ has a removable singular point at ∞, z has a simple pole at ∞, and e^z has an essential singular point at ∞. If f has a removable singular point at ∞, we usually say that f is **analytic at** ∞ (assuming tacitly that the singularity has been removed).

The residue of f at an isolated singular point at ∞ is defined like the residue at a finite singular point, as $(2\pi i)^{-1}\int_C f(z)\,dz$, where C is a circle (or other simple closed curve) that surrounds all the other singular points (if any) of f. However, we must remember that C has to be described in the positive sense *with respect to* ∞, so that ∞ would be on your left if you walked around C. That is, C is to be considered as oriented clockwise.

Since, by definition, a function f has at ∞ the same kind of singular point as $f(1/z)$ has at 0, it is tempting to guess that the residue of f at ∞ should be the residue of $f(1/z)$ at 0. This is not true!

For example, it follows directly from the definition that the residue of $1/z$ at ∞ is -1; this shows that a function may have a nonzero residue at ∞ even when the function is analytic at ∞. The residue of z at ∞ is 0; thus the residue of a function can be 0 at a simple pole at ∞.

EXERCISE 8.14 If f is a rational function or, more generally, if f is analytic in the extended plane except for a finite number of isolated singular points, then the sum of all the residues of f in the extended plane is 0.

EXERCISE 8.15 Show that the residue at ∞ of z^{-n}, $n \geq 2$, is zero.

EXERCISE 8.16 Show that the residue at ∞ is zero for every rational function $R(z) = P(z)/Q(z)$ with the degree of Q at least 2 more than the degree of P.

Sometimes it is convenient to have a more computational method of finding residues at ∞. Let us show that *the residue of f at* ∞ *is equal to the residue of* $-z^{-2}f(1/z)$ *at* 0. We can see this by formal transformation of the integral that

defines the residue. As we would expect, if we set $w = 1/z$, the integral

$$\frac{1}{2\pi i}\int_C f(z)\,dz,$$

where C is the circle $|z| = R$ described clockwise, becomes

$$\frac{1}{2\pi i}\int f\left(\frac{1}{w}\right)\left(-\frac{1}{w^2}\right)dw,$$

where w describes the circle $|w| = 1/R$ in the opposite sense [because $\arg(1/z) = -\arg z$]. We can justify this formal procedure by parametrizing C. Let the residue of f at ∞ be A. Then

$$A = \frac{R}{2\pi}\int_{2\pi}^{0} f(Re^{i\theta})e^{i\theta}\,d\theta.$$

Here we are dealing with an integral over a real interval, and we can manipulate it by the usual techniques of calculus. Thus

$$A = -\frac{R}{2\pi}\int_0^{2\pi} f(Re^{i\theta})e^{i\theta}\,d\theta.$$

Now put $R = 1/r$ and $\theta = -\varphi$ to get

$$A = \frac{1}{2\pi r}\int_0^{-2\pi} f(r^{-1}e^{-i\varphi})e^{-i\varphi}\,d\varphi.$$

If we let $z = re^{i\varphi}$, we get

$$A = \frac{1}{2\pi}\int_0^{-2\pi} f(z^{-1})z^{-1}\,d\varphi,$$

which we can interpret as the contour integral

$$-\frac{1}{2\pi i}\int_\Gamma f(z^{-1})z^{-2}\,dz,$$

where Γ is the circle $|z| = r$ described counterclockwise; this last integral is indeed the residue of $-z^{-2}f(1/z)$ at 0.

We can now compute residues at ∞ by the methods of Sec. **8F**.

EXERCISE 8.17 Show that if f is analytic at all points of the extended plane, then f is a constant.

EXERCISE 8.18 Show that if f is meromorphic in the extended plane, then f is a rational function.

Supplementary exercises

Find the residues at ∞ of

1. $\dfrac{z^2+3}{5z^4-7z^2+6z}$

2. $\dfrac{2z^2}{z^3+27}$

3. $\dfrac{z^2+5z+6}{z}$

4. $\dfrac{2z-3}{z^2}$

5. Define $f(z)$ in a neighborhood of ∞ by taking $f(x)=\sqrt{(x-1)(x+1)}$ and $f(x)$ positive for large positive values of x. Find the residue of f at ∞.

8H. Residues and partial fractions

We can use residues to give a systematic method (not necessarily the most efficient method in any particular case) for finding the partial fraction expansion of a rational function $P(z)/Q(z)$, in lowest terms, with P of lower degree than Q. It is convenient to suppose that Q has leading coefficient 1.

Let the zeros of Q be $\omega_1, \omega_2, \ldots$, with respective orders $n, m, \ldots$. Then the expansion looks like this:

$$\frac{P(z)}{Q(z)}=\frac{A_1}{(z-\omega_1)^n}+\frac{A_2}{(z-\omega_1)^{n-1}}+\cdots+\frac{A_n}{z-\omega_1}+$$
$$\frac{B_1}{(z-\omega_2)^m}+\cdots+\frac{C_1}{(z-\omega_3)^p}+\cdots.$$

We see that A_1 is the residue at ω_1 of $(z-\omega_1)^{n-1}P(z)/Q(z)$, A_2 is the residue of $(z-\omega_1)^{n-2}P(z)/Q(z)$, and so on; A_n is the residue of $P(z)/Q(z)$ at ω_1; and so on. If we are to find the $A_k, B_k, \ldots$ explicitly, it is usually best to write Q as a product of linear factors, $Q(z)=(z-\omega_1)^n(z-\omega_2)^m\cdots$, so that the residue of $(z-\omega_1)^kP(z)/Q(z)$ is the residue of

$$\frac{P(z)}{(z-\omega_1)^{n-k}(z-\omega_2)^m\cdots}=\frac{P(z)}{(z-\omega_1)^{n-k}S(z)}, \qquad S(z)=\frac{Q(z)}{(z-\omega_1)^n},$$

and therefore equals

$$\frac{1}{(n-k-1)!}\left(\frac{d}{dz}\right)^{n-k-1}\left\{\frac{P(z)}{S(z)}\right\} \quad \text{at } z=\omega_1.$$

In particular,

$$A_1=\frac{P(\omega_1)}{(\omega_1-\omega_2)^m(\omega_1-\omega_3)^p\cdots}.$$

Notes

1. Osgood, pp. 370–371.
2. The Casorati-Weierstrass theorem is often called Sohotsky's theorem in the Russian literature (and appears in translations with various transliterations of the name).

9 *Evaluation of Definite Integrals*

Cauchy's theorem and integral formula have many applications to properties of analytic functions in general. Before going into these, I am going to show you one of the more practical applications, indeed the one for which Cauchy seems to have invented the formula. This is the evaluation of definite integrals.[1]

9A. Trigonometric functions integrated over a period

As a first example, let us consider

$$I = \int_{-\pi}^{\pi} \frac{d\theta}{5 + 3\cos\theta},$$

which can be evaluated by looking up the indefinite integral (which is rather messy) or by ordinary calculus methods, but which illustrates the idea of Cauchy's method in a simple situation. The first step is to write the cosine in terms of exponential functions:

$$I = \int_{-\pi}^{\pi} \frac{d\theta}{5 + \frac{3}{2}(e^{i\theta} + e^{-i\theta})} = 2\int_{-\pi}^{\pi} \frac{d\theta}{10 + 3e^{i\theta} + 3e^{-i\theta}} = 2\int_{-\pi}^{\pi} \frac{e^{i\theta}\,d\theta}{3e^{2i\theta} + 10e^{i\theta} + 3}.$$

We can (once we think of the idea) interpret the last integral as an integral around the unit circle C: $\{z = e^{i\theta}\}$, thus:

$$I = -2i \int_C \frac{dz}{3z^2 + 10z + 3}.$$

The last integral would be zero if the denominator had no zeros in the unit disk, but in fact the denominator is $(3z + 1)(z + 3)$ and the integral becomes

$$I = -\frac{2i}{3} \int_C \frac{dz}{(z + \frac{1}{3})(z + 3)}.$$

This looks like, and in fact is, an instance of Cauchy's integral formula (or of the residue theorem). The curve C surrounds the point $z = -\frac{1}{3}$ but not $z = -3$, and hence

$$I = 2\pi i\left(\frac{-2i}{3}\right)\frac{3}{8} = \frac{\pi}{2}.$$

Since the original integrand was an even function, we also have

$$\int_0^{\pi} \frac{d\theta}{5+3\cos\theta} = \frac{\pi}{4}.$$

The same idea applies, in principle, to any rational function of $\sin\theta$ and $\cos\theta$ (or of $e^{i\theta}$), as long as the denominator does not have a zero for real θ, which means that the corresponding analytic function does not have a pole on the unit circle.

EXERCISE 9.1 Evaluate the following integrals:

(a) $\displaystyle\int_0^{2\pi} \frac{d\theta}{13-5\sin\theta}$ (b) $\displaystyle\int_0^{\pi} \frac{d\theta}{1-2r\cos\theta+r^2}$, $0 \le r < 1$

(c) $\displaystyle\int_0^{2\pi} \frac{\cos^2\theta\, d\theta}{5+3\sin\theta}$ (d) $\displaystyle\int_0^{\pi} \frac{d\theta}{(3+2\cos\theta)^2}$

(e) $\displaystyle\int_{-\pi}^{\pi} \frac{d\theta}{a+b\sin\theta}$, $|a|>|b|$, a and b real

Supplementary exercises

1. $\displaystyle\int_0^{\pi} \frac{\cos 2\theta\, d\theta}{5-3\cos\theta}$

2. $\displaystyle\int_0^{2\pi} \frac{d\theta}{2+\sin\theta}$

3. $\displaystyle\int_0^{2\pi} \frac{(\sin\theta)^2\, d\theta}{5+4\cos\theta}$

4. $\displaystyle\int_0^{2\pi} \frac{d\theta}{(2+\sin\theta)^2}$

5. $\displaystyle\int_0^{2\pi} \frac{e^{i\theta}\, d\theta}{2-e^{-i\theta}}$

9B. Rational functions integrated over $(-\infty, \infty)$

If P and Q are polynomials, Q has no real zeros, and the degree of Q is at least 2 more than the degree of P, then the integral

$$\int_{-\infty}^{\infty} \frac{P(x)}{Q(x)}\, dx$$

converges because the behavior of $P(x)/Q(x)$ is dominated, as $|x| \to \infty$, by the quotient of the terms of highest degree. This integral could always be evaluated by breaking the integrand into partial fractions and using integral tables, or perhaps by using a symbol manipulation program on a computer to evaluate the indefinite integral first. If P/Q is at all complicated, and we are interested in the *definite* integral, it may well be easier to use the residue theorem in a way that will be described below (see, for example, Exercises 9.2, 9.3). The method is also applicable to some definite integrals where the indefinite integrals are not expressible in terms of elementary functions. For example, $\int x^{-1} \sin x \, dx$ is not an elementary function, but $\int_0^\infty x^{-1} \sin x \, dx$ can be calculated exactly by using contour integration, as we shall see in Sec. **9D**. However, the application to integrals of rational functions will show the method at work in a simpler situation.

At first sight, an integral $\int_{-\infty}^{\infty} [P(x)/Q(x)] \, dx$ looks rather different from the integrals that we saw in Sec. **9A**. There, once we interpreted the integral as a contour integral, we could apply the residue theorem directly. Here we appear not to have a closed contour. Nevertheless, the closed contour is implicitly there if we look at the situation on the Riemann sphere, where the real axis becomes a circle through the point at infinity. We still cannot apply the residue theorem directly, because we proved the theorem only for bounded curves. However, we can argue that it ought to be good enough to apply the residue theorem to a closed curve in the finite plane that approximates (on the sphere) the circle through ∞, at least if our integrand is small near the point at infinity. Such a curve could look like the one in Fig. 9.1, where R is large, the poles of P/Q are inside the closed curve made up of C_1 and the interval $(-R, R)$ of the real axis, and C_1 moves off toward ∞ as $R \to \infty$. Now the residue theorem gives us

$$\int_{-R}^{R} \frac{P(x)}{Q(x)} \, dx = 2\pi i \left(\text{sum of residues of } \frac{P}{Q} \right) - \int_{C_1} \frac{P(z)}{Q(z)} \, dz.$$

As $R \to \infty$, the limit of the integral on the left is $\int_{-\infty}^{\infty} [P(x)/Q(x)] \, dx$, the sum of residues does not change, and it is a reasonable guess that (under our hypotheses on P and Q)

$$\int_{C_1} \frac{P(z)}{Q(z)} \, dz \to 0$$

FIGURE 9.1

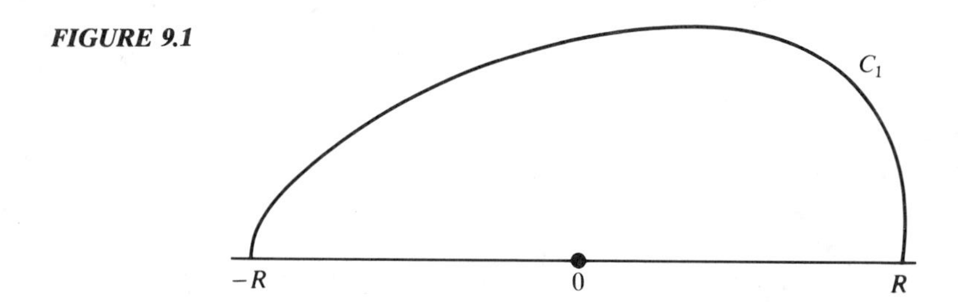

as $R \to \infty$. Hence we expect that:

If Q has no real zeros and the degree of Q is at least 2 more than the degree of P, then

$$\int_{-\infty}^{\infty} \frac{P(x)}{Q(x)}\,dx = 2\pi i \left(\text{sum of residues of } \frac{P}{Q} \text{ in the upper half-plane}\right).$$

This reduces the evaluation of a definite integral to the calculation of a moderate number of residues (half the degree of Q if Q is even and real on the real axis). Notice that if Q is of odd degree and real on the real axis, it has to have at least one real zero, and does not satisfy our hypotheses; but see Sec. **9C**.

What is missing so far is a proof that the integral along C_1 approaches zero as $R \to \infty$. Naturally the proof is simpler if C_1 has a simple equation; the obvious choice for C_1 is a semicircle. Rather than giving a general proof, let us work out the details for a specific example; it will then be easy to see how it generalizes.

Let us consider

$$I = \int_{-\infty}^{\infty} \frac{dx}{x^2 + 2x + 2}.$$

All that we need in our order to evaluate I are, first, the sum of the residues of the integrand in the upper half plane, that is, the residue at $-1 + i$; and a proof that

$$\int_{C_1} \frac{dz}{z^2 + 2z + 2} \to 0, \qquad R \to \infty,$$

where C_1 is the semicircle $|z| = R$, $\operatorname{Im} z \geq 0$, and $R \to \infty$. By Exercise 8.6 the residue is $1/(2i)$, so that our value for I will be $2\pi i/(2i) = \pi$.

To show that

$$\int_{C_1} \frac{dz}{z^2 + 2z + 2} \to 0,$$

we use the inequality $|z^2 + 2z + 2| > \frac{1}{2}R^2$ for $|z| = R$ and R sufficiently large (Exercise 1.23), and the inequality for integrals in Sec. **5C**, to obtain

$$\left|\int_{C_1} \frac{dz}{z^2 + 2z + 2}\right| \leq \frac{2(\text{length of } C_1)}{R^2} = \frac{2\pi}{R}$$

for sufficiently large R. Since the right-hand side tends to 0, we are done.

A similar argument will apply to any rational function P/Q as long as Q is of degree at least 2 more than the degree of P.

The method works just as well when P/Q is not real on the real axis, although in practice one does not see such integrals very often. If P/Q *is* real on the real axis, the value of the definite integral will of course be real; this is a crude but useful check on one's work. Remember also that if the integrand is positive, the integral must be positive.

EXERCISE 9.2 Use the general result above to evaluate the following integrals:

(*a*) $\displaystyle\int_{-\infty}^{\infty} \frac{x+4}{(x^2+2x+2)(x^2+9)}\,dx$

(*b*) $\displaystyle\int_{-\infty}^{\infty} \frac{dx}{(x^2-2ix-2)(x+i)}$

(*c*) $\displaystyle\int_{0}^{\infty} \frac{dx}{x^4+6x^2+8}$ [Note: this is half the integral over $(-\infty, \infty)$.]

EXERCISE 9.3 Evaluate

$$I = \int_{-\infty}^{\infty} \frac{dx}{1+x^4}$$

by the method of this section. Also evaluate I by looking up the indefinite integral in tables; unless you are very careful, you will get the absurd answer $I = 0$. Can you explain the contradiction? Note: We shall see a somewhat easier method for integrals like this in Exercise 11.2.

Supplementary exercises

Evaluate the following integrals:

1. $\displaystyle\int_{-\infty}^{\infty} \frac{dx}{x^2+2x+5}$

2. $\displaystyle\int_{-\infty}^{\infty} \frac{x^2+2x}{(9x^2+4)(x^2+25)}\,dx$

3. $\displaystyle\int_{-\infty}^{\infty} \frac{x\,dx}{(x-2i)(x+3i)(x+4i)}$

4. $\displaystyle\int_{-\infty}^{\infty} \frac{(2x+3)\,dx}{(x^2+2x+5)(x^2+2x+10)}$

5. $\displaystyle\int_{0}^{\infty} \frac{dx}{x^4+2x^2+1}$

9C. Principal value integrals

The integrals in Sec. **9B** were rather obviously convergent in the strict sense that $\int_{-\infty}^{0} [P(x)/Q(x)]\,dx$ and $\int_{0}^{\infty} [P(x)/Q(x)]\,dx$ both converge. In practice we sometimes want to attach a value to integrals that are technically divergent, for example, to

$$\int_{-\infty}^{\infty} \frac{2x+3}{x^2+2x+2}\,dx \quad \text{or} \quad \int_{-\infty}^{\infty} \frac{dx}{(x-1)(x^2+4)}.$$

In the first integral the integrand is not small enough for large $|x|$ to let the integral converge; in the second, the integrand has a pole on the interval of integration. It is often useful to interpret such integrals as symmetric limits, that is, as

$$\lim_{R\to\infty} \int_{-R}^{R} \frac{2x+3}{x^2+2x+2}\,dx$$

and

$$\lim_{\varepsilon\to 0} \left\{ \int_{-\infty}^{1-\varepsilon} + \int_{1+\varepsilon}^{\infty} \right\} \frac{dx}{(x-1)(x^2+4)}.$$

These are called **principal values** (or *Cauchy principal values*), and are indicated by the notation PV $\int \cdots$.

EXERCISE 9.4 To see what happens if we do not take symmetric limits, calculate

$$\lim_{R\to\infty} \int_{-R}^{2R} \frac{x\,dx}{1+x^2}$$

and compare with

$$\lim_{R\to\infty} \int_{-R}^{R} \frac{x\,dx}{1+x^2}.$$

EXERCISE 9.5 Show that if c is real, $Q(z)$ has a simple zero at c, and no other real zeros, and $P(c) \neq 0$, then

$$\text{PV} \int_{-\infty}^{\infty} \frac{P(z)}{Q(z)}\,dz$$

(with degree of Q at least $2+$ degree of P) is equal to $2\pi i$ times (the sum of the residues of P/Q in the upper half-plane plus half the residue at c). (See Fig. 9.2.

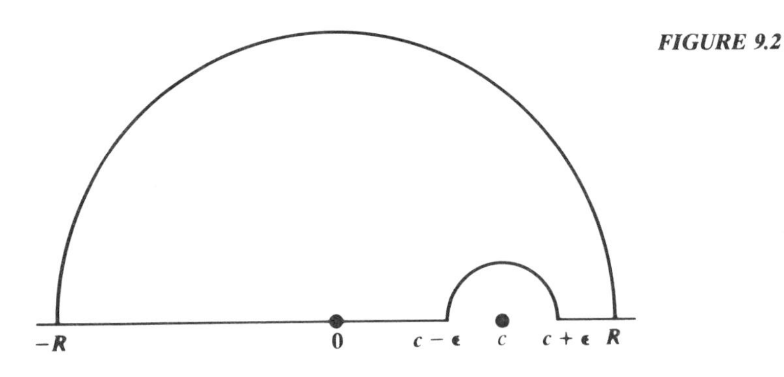

FIGURE 9.2

The effect is that we can think of the contour as passing through the pole at the expense of counting only half the residue.)

The rule in Exercise 9.5 extends in the obvious way to rational functions with several simple poles on the real axis.

EXERCISE 9.6 By considering $P(z)/Q(z) = 1/z^2$, show that a principal value will not exist for a double pole.

EXERCISE 9.7 Evaluate

(*a*) $\mathrm{PV}\displaystyle\int_{-\infty}^{\infty} \frac{dx}{(2-x)(x^2+4)}$

(*b*) $\mathrm{PV}\displaystyle\int_{-\infty}^{\infty} \frac{dx}{x^3+4x+5}$

EXERCISE 9.8 Show that if f is analytic inside and on a closed curve C except for a simple pole at a point z_0, where C has a corner with interior angle α, then

$$\mathrm{PV}\int_C f(z)\,dz$$

is equal to $i\alpha$ times the residue of f at z_0. (Compare Exercise 9.5.)

EXERCISE 9.9 Find

$$\mathrm{PV}\int_{-\infty}^{\infty} \frac{2x+3}{x^2+2x+2}\,dx$$

by using the method of Sec. **9B** with the semicircular contour and showing that the integral over the semicircle approaches a limit, not equal to 0.

Supplementary exercises

1. $\mathrm{PV}\displaystyle\int_{-\infty}^{\infty} \frac{dx}{x^3-x^2+x-1}$
2. $\mathrm{PV}\displaystyle\int_{-\infty}^{\infty} \frac{dx}{x^3+2x^2+5x}$
3. $\mathrm{PV}\displaystyle\int_{-\infty}^{\infty} \frac{(x+2)\,dx}{(3-x)(x^2+3x+3)}$
4. $\mathrm{PV}\displaystyle\int_{-\infty}^{\infty} \frac{dx}{x^2+13x-45}$

   (*Hint*: You can find the sum of the residues of the integrand without calculation.)
5. Let f be analytic in a closed disk bounded by the circle C. If z is a point on C, evaluate the integral

$$\frac{1}{2\pi i}\mathrm{PV}\int_C \frac{f(w)\,dw}{w-z}.$$

9D. Rational functions times trigonometric functions

The integral

$$\int_0^\infty \frac{\sin x}{x}\, dx$$

is typical of a class of integrals whose integrands do not have elementary indefinite integrals but can nevertheless be evaluated.

Your first guess might be to try integrating $z^{-1}\sin z$ (an even function) around a semicircle in the upper half plane, with diameter $(-R, R)$. However, since the integrand has a removable singularity at 0, the contour integral is zero, and what we find is that

$$2\int_0^R \frac{\sin x}{x}\, dx = -\int_0^\pi \frac{\sin(Re^{i\theta})}{Re^{i\theta}} iRe^{i\theta}\, d\theta = -i\int_0^\pi \sin[R(\cos\theta + i\sin\theta)]\, d\theta$$

$$= -i\int_0^\pi [\sin(R\cos\theta)\cosh(R\sin\theta) + i\cos(R\cos\theta)\sinh(R\sin\theta)]\, d\theta.$$

Since the left-hand side is real, so is the right-hand side; therefore its imaginary part is zero and we find that

$$2\int_0^R \frac{\sin x}{x}\, dx = \int_0^\pi \cos(R\cos\theta)\sinh(R\sin\theta)\, d\theta.$$

Since the integral on the right looks even more intractable than the integral on the left, we have not accomplished much in this way.

What does work is to consider the integral

$$\int_C \frac{e^{iz}}{z}\, dz$$

around a contour in the upper half plane. This suggests itself because $|e^{iz}| = e^{-r\sin\theta}$, which is small in the upper half plane except for $\theta = 0$. Then the imaginary part of the integral along the real axis is the integral of $x^{-1}\sin x$. We take[2] C to be the triangle with vertices $-R$, iR, and R. (See Fig. 9.3.) Then we have $\int_C z^{-1}e^{iz}\, dz = \pi i$ (residue of $z^{-1}e^{iz}$ at 0) $= i\pi$. The limit of the integral along $(-R, R)$ is

$$\mathrm{PV}\int_{-\infty}^\infty \frac{\cos x + i\sin x}{x}\, dx = \mathrm{PV}\int_{-\infty}^\infty \frac{\cos x}{x}\, dx + i\mathrm{PV}\int_{-\infty}^\infty \frac{\sin x}{x}\, dx$$

$$= 2i\int_0^\infty \frac{\sin x}{x}\, dx$$

because $x^{-1}\cos x$ is an odd function, and so has a zero integral over a symmetric interval; and $x^{-1}\sin x$ is even, and bounded near $x = 0$, so that the last integral is an ordinary integral. If we can show that the integrals over the

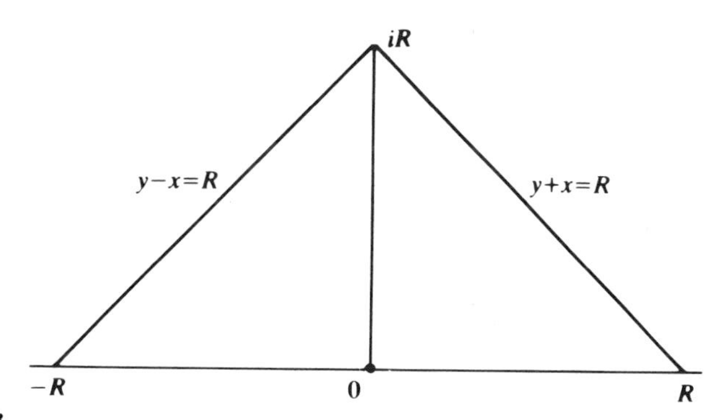

FIGURE 9.3

other two sides of the triangle approach zero, we infer that

$$\int_0^\infty \frac{\sin x}{x}\,dx = \frac{\pi}{2}.$$

The equations of the two slanting lines are $y + x = R$ and $y - x = R$, or in complex form, $z = x + i(R - x)$ and $z = x + i(R + x)$. Then the integral of $z^{-1}e^{iz}$ over the right-hand slanting line is

$$(1 - i)\int_0^R \frac{e^{i[x+i(R-x)]}}{x + i(R - x)}\,dx,$$

and its absolute value does not exceed

$$|1 - i|\int_0^R \frac{e^{-(R-x)}}{|x + i(R - x)|}\,dx \le \frac{2}{R}e^{-R}\int_0^R e^x\,dx = \frac{2}{R}e^{-R}(e^R - 1) \to 0.$$

(I used the inequality $|x + i(R - x)| \ge R/2^{1/2}$, which follows from $|x + i(R - x)|^2 = x^2 + (R - x)^2 \ge R^2/2$.)

The other integral is estimated similarly.

This completes the proof that $\int_0^\infty x^{-1}\sin x\,dx = \pi/2$.

EXERCISE 9.10[3] Evaluate the same integral by using as contour the rectangle with vertices $\pm R$ and $\pm R + iR$.

By a similar proof we obtain the following general result:

Let $R(z) = P(z)/Q(z)$, where P and Q are polynomials with real coefficients and the degree of Q is at least one more than the degree of P. Then PV $\int_{-\infty}^{\infty} R(x)\cos x\,dx$ is the real part, and PV $\int_{-\infty}^{\infty} R(x)\sin x\,dx$ the imaginary part, of πi times [twice the sum of the residues of $R(z)e^{iz}$ in the upper half plane, plus the sum of the residues of $R(z)e^{iz}$ on the real axis].

EXERCISE 9.11 Evaluate the following integrals, using principal values when appropriate:

(*a*) $\displaystyle\int_0^\infty \frac{x \sin x \, dx}{16x^2+9}$ (*b*) $\displaystyle\int_{-\infty}^\infty \frac{\cos x \, dx}{x^3+x^2+4x+4}$

(*c*) $\displaystyle\int_0^\infty \frac{\cos x \, dx}{(x^2+4)^2}$ (*d*) $\displaystyle\int_{-\infty}^\infty \frac{x \sin x}{4-x^2}\, dx$

(*e*) $\displaystyle\int_{-\infty}^\infty \frac{\sin x \, dx}{(x-1)(x+2)}$

EXERCISE 9.12 Without doing any more contour integration, evaluate $\int_0^\infty t^{-1} \sin xt \, dt$ and observe that it is a discontinuous function of x.

Supplementary exercises

Interpret integrals as principal values, if necessary.

1. $\displaystyle\int_{-\infty}^\infty \frac{\sin x}{x+x^2}\, dx$

2. $\displaystyle\int_{-\infty}^\infty \frac{\sin x + \cos x}{x^2+9}\, dx$

3. $\displaystyle\int_{-\infty}^\infty \frac{\cos x \, dx}{(x-2)(x^2+1)}$

4. $\displaystyle\int_0^\infty \frac{\cos x}{x^2-9}\, dx$

5. $\displaystyle\int_{-\infty}^\infty \frac{e^{ix}\, dx}{x-\pi}$

In the following exercises, use a contour consisting of a long rectangle with one side on the real axis and height 2π for Exercise 6, height π for Exercise 7.

6. $\displaystyle\int_{-\infty}^\infty \frac{e^{x/n}}{1-e^x}\, dx, \qquad n = 2, 3, \ldots.$

7. $\displaystyle\int_0^\infty \frac{x \, dx}{\sinh x}$

9E. Integrals of Cauchy type

We saw in Sec. **7C** that the equation

$$\varphi(z) = \frac{1}{2\pi i}\int_C \frac{f(w)\, dw}{w-z},$$

with f continuous on the closed curve C, defines a function that is analytic inside C (and a different function outside C). Later (Sec. **20C**) we are going to need a corresponding result when C is a simple (or Jordan) arc, that is, an arc that does not intersect itself. We can show, just as in Sec. **7C**, that if f

is continuous on the simple arc A, then

$$\varphi(z)=\frac{1}{2\pi i}\int_A\frac{f(w)\,dw}{w-z}dw$$

is analytic in the finite plane except for the points on (the trace of) A.

Integrals of the two kinds just mentioned are known as integrals of **Cauchy type**. When C is a closed curve, an integral of Cauchy type defines two functions φ_+ and φ_-, one inside C and the other outside. (When f is analytic on and inside C, the "outside" function is zero, by Cauchy's theorem.) It is important in some applications[4] to know what the limits $\varphi_+(z_0)$ and $\varphi_-(z_0)$ are when z approaches a point z_0 of C. If z is actually on C, the integral must be interpreted as a principal value $\varphi_p(z)$, the limit of the integral along C with a short symmetric piece around z_0 deleted. When C is an arc, we define $\varphi_+(z_0)$ as the limit of the integral as $z\to z_0$ from the left of C (considering C as described in the positive direction), and $\varphi_-(z_0)$ as the limit from the right. We shall suppose that C has a tangent at z_0; corners, and endpoints in the case of an arc, can be discussed similarly.

The **Sohotsky** (or **Plemelj) formulas** are suggested by our discussion of principal value integrals in Sec. **9C**. They say that

$$\varphi_+(z_0)=\tfrac{1}{2}f(z_0)+\varphi_p(z_0)\quad\text{and}\quad\varphi_-(z_0)=-\tfrac{1}{2}f(z_0)+\varphi_p(z_0)$$

When f is analytic on C (and inside C when C is closed), we essentially proved the formulas in Sec. **9C**. For,

$$\varphi(z)=\frac{1}{2\pi i}\int_C\frac{f(t)\,dt}{t-z}=\begin{cases}0, & z\text{ outside } C,\\ f(z), & z\text{ inside } C,\end{cases}$$

$$\varphi_+(z_0)=\lim_{\substack{z\to z_0\\ \text{inside}}}\varphi(z)=f(z_0)=\varphi_p(z_0)+\frac{1}{2}f(z_0),$$

$$\varphi_p(z_0)=\frac{1}{2\pi i}\,\mathrm{PV}\int_C\frac{f(t)}{t-z_0}\,dt=\frac{1}{2}f(z_0).$$

(Compare Exercise 9.8.)

Now suppose, more generally, that f is analytic only in a neighborhood N of z_0. Then $C=C_1+C_2$, where C_1 is the part of C inside N and C_2 is the rest of C. Let γ be an arc joining the ends of C_1 inside C. (See Fig. 9.4, p. 90.) Then

$$\begin{aligned}\varphi(z)&=\frac{1}{2\pi i}\int_{C_2+C_1}\frac{f(t)}{t-z}\,dt\\&=\frac{1}{2\pi i}\int_{C_2+\gamma}\frac{f(t)}{t-z}\,dt+\frac{1}{2\pi i}\int_{C_1-\gamma}\frac{f(t)}{t-z}\,dt\\&=K(z)+F(z),\end{aligned}$$

where K is analytic in a neighborhood of z_0.

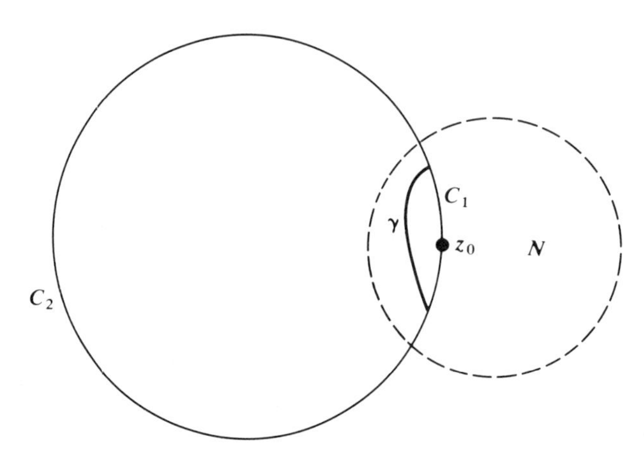

FIGURE 9.4

Then K approaches the same limit whether z approaches z_0 from inside or outside C. Since f is analytic inside and on $C_1 - \gamma$, we have just seen that

$$F_+(z_0) = F_p(z_0) + \tfrac{1}{2}f(z_0).$$

Consequently,

$$\begin{aligned}\varphi_+(z_0) = K_p(z_0) + F_+(z_0) &= K_p(z_0) + F_p(z_0) + \tfrac{1}{2}f(z_0)\\ &= \varphi_p(z_0) + \tfrac{1}{2}f(z_0).\end{aligned}$$

The other formula and the case when C is an arc can be discussed similarly.

It is possible, but much more difficult, to show that the formulas hold when f is merely bounded and integrable on C, and satisfies a Hölder condition, $|f(z) - f(z_0)| \le c\,|z - z_0|^\alpha$, $0 < \alpha \le 1$.

Notes

1. You might suppose that techniques for evaluating definite integrals are now unimportant, since integrals can be evaluated by computers. However, a computer calculation may produce a decimal approximation, whereas an application of Cauchy's theorem produces an expression "in closed form" like $3\pi/8$ or $6e$, which is more informative since numbers like e and π are known with great precision. In addition, the evaluation of integrals by contour integration is an important technique in many branches of mathematics.

2. H. P. Boas and E. Friedman, "A Simplification in Certain Contour Integrals," *Amer. Math. Monthly 84* (1977): 467–468.

3. Ahlfors, pp. 157–158. One can also use the semicircular contour of Sec. **9B**, but the proof that the integral along the semicircle approaches zero causes more difficulty.

4. See Carrier, Krook, and Pearson, pp. 412 ff; Markushevich, *Theory of Functions,* pp. 309 ff.; and especially Henrici, Chapter 14.

10 *Logarithms and General Powers*

It is not completely obvious how to define logarithms of complex numbers. At least, it was not obvious a few centuries ago, when mathematicians argued about whether -1 has a logarithm, and if it does, whether the logarithm is 0 or something else. This is hardly surprising, since the same mathematicians did not even have a good definition of complex numbers.[1]

10A. Definition of $\log z$

There are two rather natural ways of defining logarithms of complex numbers. One is to say that *since* $z = re^{i\theta} = |z|\, e^{i \arg z}$, *we should take*

$$\log z = \ln r + i\theta = \ln |z| + i \arg z,$$

provided that $z \neq 0$. [I use $\ln r$ for the ordinary "natural" logarithm (to base e) of a positive real number, to avoid confusing it with the logarithm of a general complex number.[2]] Since $\arg z$ is not a function (not single-valued), $\log z$ is not a function either, but we should have expected this (see Sec. **5D**). Which value we use for $\arg z$, and so for $\log z$, will depend on what we are going to do with the logarithm.

On the other hand, there is a familiar definition of $\ln x$ as a function on positive real numbers, namely,

$$\ln x = \int_1^x t^{-1}\, dt.$$

We can imitate this definition in the complex plane. Let D be a simply connected region that does not contain 0 or ∞. Let $z_0 \in D$ and choose any admissible value for $\arg z_0$. Then we define

$$\log z = \int_{z_0}^{z} w^{-1}\, dw + \log z_0,$$

with integration along any arc in D that connects z_0 and z. Since $1/w$ is analytic in D, it follows from Cauchy's theorem that the integral is independent of the arc and defines a function $\log z$ in D. As in Sec. **6C**, we see that $\log z$ *is analytic in* D, *its derivative being* $1/z$. However, different values of $\log z_0$ may give us different functions.

Now we have to see how to connect these two definitions of $\log z$. Even when D is simply connected and does not contain 0 (as is the case here), it is not obvious how to define $\arg z$ as a function in D. However, there is no problem in a neighborhood of z_0: In a small disk $\Delta \subset D$, centered at z_0 (hence not containing 0), we can take $\arg z$ to be the principal value, unless z_0 is a negative real number, in which case we take $\arg z$ between $\pi/2$ and $3\pi/2$. In other words, $\arg z$ is definable locally. Furthermore, the function

$$L_1(z) = \ln(x^2 + y^2)^{1/2} + i \tan^{-1}\left(\frac{y}{x}\right) = \ln r + i\theta$$

is analytic locally, as we can see by verifying the Cauchy-Riemann equations for it.

EXERCISE 10.1 Do this.

Since L_1 is differentiable, we can calculate (as in Exercise 6.1) its derivative, which of course turns out to be $1/z$.

The function

$$L_2(z) = \int_{z_0}^{z} w^{-1}\, dw + L_1(z_0)$$

is analytic in D and also has the derivative $1/z$. Hence L_1 and L_2 differ only by a constant in Δ, and $L_2(z_0) = L_1(z_0)$. Consequently, $L_2(z) \equiv L_1(z)$ in Δ, and we have $\operatorname{Re} L_2(z) = \ln r$, $\operatorname{Im} L_2(z) = \theta$ in Δ. Now define $\log z$ (in D) to be $L_2(z)$ and define $\arg z$ (in D) to be $\operatorname{Im} \log z$. We now have, throughout D, an analytic function $\log z$ of the form $\ln r + i\theta$, where θ is determined by the initial choice of $\arg z_0$. In particular, $\arg z$ is now a function in D, and since different values of $\arg z_0$ differ by integral multiples of 2π, different determinations of $\log z$ differ by integral multiples of $2\pi i$. We refer to these different functions $\log z$ as **branches** of $\log z$; a particular branch is determined by the region D and the value of $\arg z$ at some point in D.

The next exercise shows that *the logarithm has the basic property that we expect of a logarithm.*

EXERCISE 10.2 Show that $\log(z_1 z_2) = \log z_1 + \log z_2$ provided that z_1, z_2 and $z_1 z_2$ are all in the same region D (simply connected and not containing 0), and that the initial values for the logarithms (or the branches of the functions) are correctly chosen.

10B. The logarithm of a function

We can of course define (many values of) the number $\log f(z)$, provided that $f(z) \neq 0$. What we need is a definition of $\log f$ as a function. For this, we need to have f analytic and zero-free in a simply connected region D; then we define

$$\log f(z) = \int_{z_0}^{z} \frac{f'(w)}{f(w)}\, dw + \log f(z_0),$$

where $\log f(z_0)$ is any specified one of its possible values. We have

$$\frac{d}{dz} \log f(z) = \frac{f'(z)}{f(z)},$$

and so any two determinations ("branches") differ only by a constant throughout D. We then have

$$\log f(z) = \ln|f(z)| + i \arg f(z), \qquad z \in D,$$

the only ambiguity being in which value of $\arg f(z_0)$ was selected.

Principal value The principal value of a logarithm is usually taken to be $\ln|z| + i \arg z$ with $-\pi < \arg z \leq \pi$; this is an analytic function for $-\pi < \arg z < \pi$, but discontinuous on the negative real axis.

EXERCISE 10.3 Show that $\log(zw) = \log z + \log w$ does not always hold if logarithms are taken as principal values.

EXERCISE 10.4 Show that $e^{\log z} = z$. In what sense is $\log(e^z) = z$?

EXERCISE 10.5 Find all the values of $\log(-1)$, $\log i$, and $\log(\frac{1}{2} + \sqrt{3}i/2)$.

10C. General powers

We define a^b, provided that $a \neq 0$ and $a \neq e$, to mean (all values of) $e^{b \log a}$. The reason for excluding $a = e$ is to avoid an infinite regress—we want e^z to remain single-valued and to mean the power series that we defined it to be in Sec. **4A**. Thus e^z does not mean any other value of $e^{z \log(2.71828\ldots)}$. It would be less confusing if we always wrote $\exp(z)$ for the power series and then treated e like any other complex number, but the notation e^z is traditional. We shall, however, use the notation exp( ) when the expression in ( ) is complicated.

EXERCISE 10.6 Find all values of i^{-2i}, $i^{(i+1)}$, $(1+i)^i$, $i^{-\log i}$, i^{i^i}. Remember that a^{b^c} means $a^{(b^c)}$, not $(a^b)^c$.

EXERCISE 10.7 If n is a positive integer, show that $z^{1/n}$ has only n different values (for $z \neq 0$).

EXERCISE 10.8 By $\sqrt[p]{z}$ we mean $z^{1/p}$. Compute $\sqrt[i]{-1}$.

You will avoid complications in problems dealing with exponents if you will accept that the "laws of exponents" do not hold without qualification except for positive real numbers.

EXERCISE 10.9 Consider the calculation

$$e^z = e^{2\pi i(z/2\pi i)} = 1^{z/2\pi i} = 1.$$

Show that in general $(a^b)^c$ has more values than a^{bc} and that $a^b a^c$ has more values than a^{b+c}.

In a simply connected region in which f is analytic and never zero, we can define a branch of $\log f(z)$, and then use this to define $f(z)^{g(z)}$, when g is any analytic function, by $e^{g(z) \log f(z)}$.

Supplementary exercises

1. Find all values of

(*a*) $\log(-e)$
(*b*) $\log(ie)$
(*c*) $\log e^i$
(*d*) $\log \dfrac{1+i}{\sqrt{2}}$

2. Find all values of

(*a*) 2^i
(*b*) $(-i)^{-i}$
(*c*) $(\log i)^2$
(*d*) $(1+i)^{1/(1+i)}$

3. Find all values of

(*a*) $\sin^{-1} 2$
(*b*) $\cos^{-1}(8^{1/2}i)$
(*c*) $\sin^{-1}(i)$
(*d*) $\tan^{-1}(1+i)$
(*e*) $\sec^{-1}(\frac{1}{2})$

10D. Inverse trigonometric functions

We have seen that the values of $\sin z$, and hence those of $\cos z$, are not confined to the interval $[-1, 1]$. To find what values $\cos z$ can take, that is, to find the range of the cosine function, we have to solve the equation $w = \cos z$ for any given w. Let us do this first in a special case, for example, for $w = 3$ (chosen to emphasize that real numbers greater than 1 can be values of the cosine function). We have to solve

$$\frac{e^{iz} + e^{-iz}}{2} = 3,$$

that is,

$$e^{2iz} - 6e^{iz} + 1 = 0.$$

Solving this as a quadratic equation in e^{iz}, we get

$$e^{iz} = 3 \pm \sqrt{8},$$
$$z = -i \log(3 \pm \sqrt{8}),$$

where we are to take all values of the logarithm. Since $3 - \sqrt{8} = 1/(3 + \sqrt{8})$, we can write more simply

$$z = \pm i \log(3 + \sqrt{8}) = \pm i[\ln(3 + \sqrt{8}) + 2k\pi i], \qquad k = \text{integer},$$
$$= 2k\pi \pm i \ln(3 + \sqrt{8}).$$

EXERCISE 10.10 Imitate the preceding calculation with 3 replaced by an arbitrary complex number w, and show that the values of $\cos^{-1} w$ are given by the formula

$$\cos^{-1} w = \pm i \log(w + \sqrt{w^2 - 1})$$

with all values of the logarithm. Notice that $\cos z$ (and therefore $\sin z$) takes all complex values.

For $\sin^{-1} w$ we have

$$\begin{aligned}\sin^{-1} w &= (2k + \tfrac{1}{2})\pi - \cos^{-1} w \\ &= 2k\pi + \tfrac{1}{2}\pi \pm i \log(w + \sqrt{w^2 - 1}).\end{aligned}$$

This is often written in the equivalent form

$$\sin^{-1} w = -i \log(iw \pm \sqrt{1 - w^2});$$

the equivalence depends on the fact that $i\pi/2$ is a value of $\log i$.

EXERCISE 10.11 Find a formula for $\tan^{-1} w$ and notice that $\pm i$ are not in the range of the function $\tan z$.

Notes

1. Our definition of a complex number as just an ordered pair of real numbers goes back only to Hamilton in the middle of the nineteenth century, although (from a modern point of view) the definition of a complex number as a point in the plane is equivalent to defining it as the ordered pair of its coordinates.

2. Some authors interchange the meanings of the symbols log and ln.

11 *Additional Definite Integrals*

11A. Logarithms and powers

We can now evaluate some definite integrals that contain logarithms or general powers.[1] Our first example is particularly satisfying because the indefinite integral of the integrand is not an elementary function. Consider

$$I(\alpha) = \int_0^\infty \frac{x^{\alpha-1}}{1+x}\, dx, \qquad \text{where } 0 < \alpha < 1.$$

The convergence of the integral (as $\lim_{\varepsilon\to 0,\ R\to\infty} \int_\varepsilon^R$) will follow from the process of calculating $I(\alpha)$.

We use

$$\int_C \frac{z^{\alpha-1}}{1+z}\, dz,$$

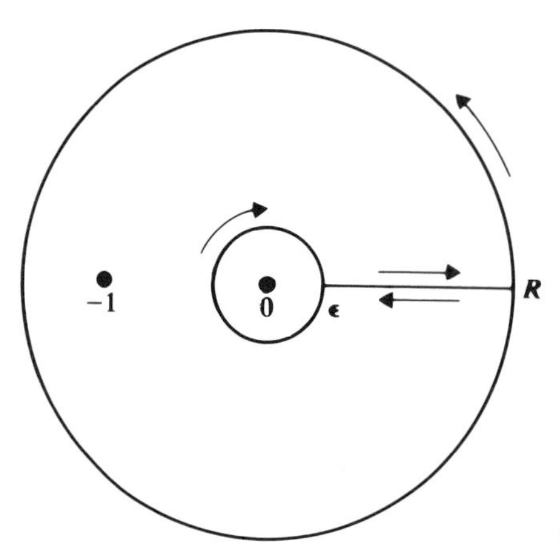

FIGURE 11.1

where C is the "keyhole" contour in Fig. 11.1; the ray from ε to R is described twice, once in each direction. Notice particularly that we do *not* use the principal value of the logarithm, but set $z^{\alpha-1} = e^{(\alpha-1)\log z} = e^{(\alpha-1)(\ln r + i\theta)}$, $0 \le \theta \le 2\pi$. We cannot apply the residue theorem directly because the contour is not in a simply connected region in which the integrand is analytic. However, we can consider two contours C_1 and C_2 as indicated in Fig. 11.2. The function $z^{\alpha-1} = \exp[(\alpha-1)(\log|z| + i\theta)]$ is analytic, except for a pole at -1, in a region in which $-\delta < \theta < \pi + \delta$: See Fig. 11.3. Then C_1 is inside D_1 and $\int_{C_1} = 0$ since there are no poles inside C_1. Similarly for C_2, where $z^{\alpha-1}$ is now defined with $\pi - \delta < \theta < 2\pi + \delta$. Then $\int_{C_2} = 0$. As $R \to \infty$ and $\varepsilon \to 0$, the integrals around both semicircles (centered at 0) will be shown below to approach 0. The integrals around the small semicircles centered at -1 combine to give $-2\pi i$ times the residue at -1. Along $(-R, 0)$, the values of $z^{\alpha-1}$ agree and the integrals in opposite directions cancel; along $(0, R)$, the values of $z^{\alpha-1}$ are different. The total effect is that we *can* integrate around the original contour just as if it were in a simply connected region of analyticity, by taking $z^{\alpha-1} = x^{\alpha-1}$ $(\theta = 0)$ along the upper side of the positive real axis and $z^{\alpha-1} = x^{\alpha-1}e^{2\pi i(\alpha-1)}$ on the lower side.[2]

Consequently, the sum of the integrals back and forth along the positive real axis equals $2\pi i$ times the residue at -1.

To see that the integrals around the large and small circles in the original figure actually do approach 0, we estimate the integral along the large circle by

$$\int_{-\pi}^{\pi} \frac{R^{\alpha-1}R\,d\theta}{R-1} \le 2\pi R^{\alpha-1}\frac{R}{R-1} \to 0 \qquad (R \to \infty),$$

and the integral around the small circle by

$$\int_{-\pi}^{\pi} \frac{\varepsilon^{\alpha-1}\varepsilon}{1-\varepsilon}\,d\theta \to 0 \qquad (\varepsilon \to 0).$$

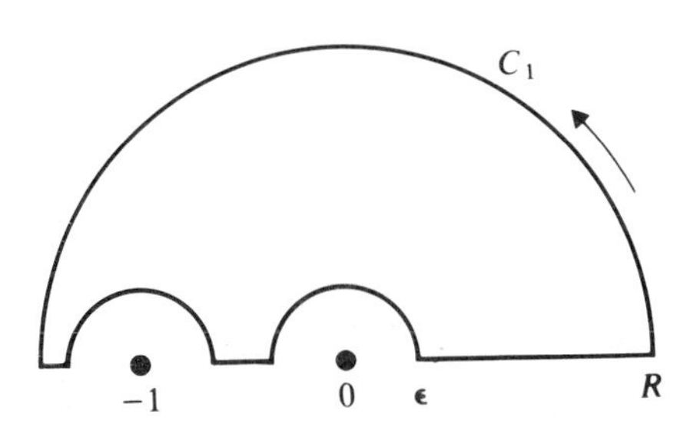

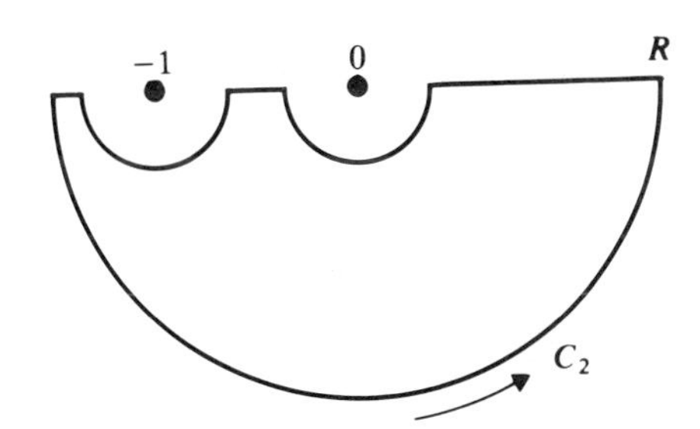

FIGURE 11.2

The residue at $z=-1$, with our choice of $z^{\alpha-1}$, is $e^{i\pi(\alpha-1)}$. Hence we have

$$\int_0^\infty \frac{x^{\alpha-1}\,dx}{1+x}+\int_\infty^0 \frac{x^{\alpha-1}e^{2\pi(\alpha-1)i}}{1+x}\,dx=\int_0^\infty \frac{x^{\alpha-1}}{1+x}[1-e^{2\pi i(\alpha-1)}]\,dx=2\pi i e^{i\pi(\alpha-1)},$$

or

$$\int_0^\infty \frac{x^{\alpha-1}}{1+x}\,dx[e^{-i\pi(\alpha-1)}-e^{i\pi(\alpha-1)}]=2\pi i,$$

that is,

$$\int_0^\infty \frac{x^{\alpha-1}}{1+x}\,dx=\frac{\pi}{\sin \pi\alpha}.$$

EXERCISE 11.1 Evaluate the following integrals by the methods of this section.

(*a*) $\displaystyle\int_0^\infty \frac{x^\lambda\,dx}{a^2+x^2},\ -1<\lambda<1,\ a>0$ (*b*) $\displaystyle\int_0^\infty \frac{\ln x\,dx}{x^\lambda(1+x)},\quad 0<\lambda<1$

(*c*) $\displaystyle\int_0^\infty \frac{\ln x\,dx}{(a+x)(b+x)},\quad a,b<0$: Consider $\displaystyle\int_C \frac{(\log z)^2\,dz}{(a+z)(b+z)}$

EXERCISE 11.2 Evaluate $\int_0^\infty (1+x^4)^{-1}\,dx$ by making the substitution $x^4=t$ and

FIGURE 11.3

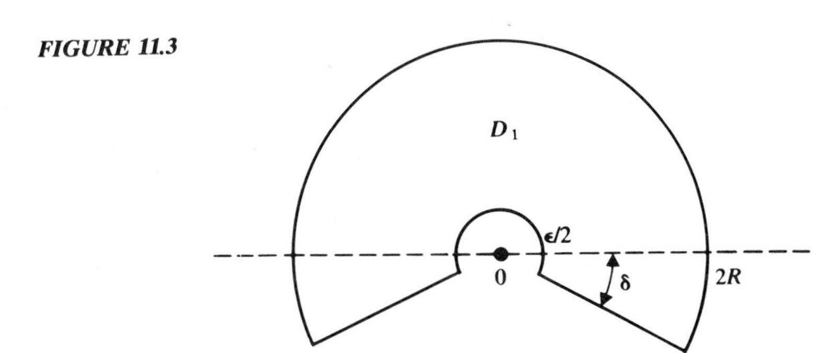

then using the method of this section. Notice that you have to calculate only one residue instead of two, in contrast to Exercise 9.3.

EXERCISE 11.3 Use the idea of Exercise 11.2 to evaluate the following integrals:

$$(a)\quad \int_0^\infty \frac{dx}{x^4+2x^2+2} \qquad\qquad (b)\quad \int_0^\infty \frac{x^3\,dx}{1+x^6}$$

$$(c)\quad \int_0^\infty \frac{dx}{(x^3+a^3)^3},\quad a>0 \qquad\qquad {}^*(d)\quad \int_0^\infty \frac{\log x\,dx}{1+x^4}$$

Supplementary exercises

1. $\displaystyle\int_0^\infty \frac{x^{1/2}\,dx}{(1+x)^2}$

2. $\displaystyle\int_0^\infty \frac{dx}{x^{1/2}(1+x^2)}$

3. $\displaystyle\int_0^\infty \frac{dx}{1+x^8}$

4. $\displaystyle\int_0^\infty \frac{dx}{x^{10}+10}$

5. $\displaystyle\int_0^\infty \frac{dx}{x^6+3x^3+2}$

***6.** PV $\displaystyle\int_0^\infty \frac{\ln x}{1-x^2}\,dx$; hence find $\displaystyle\int_0^1 \frac{\ln x}{1-x^2}\,dx$.

(*Hint*: Use a semicircle in the upper half plane.)

11B. Indefinite integrals

It is sometimes claimed that the method of contour integration has the drawback of applying only to definite integrals. This criticism is not entirely justified, because contour integration will evaluate at least the indefinite integrals of rational functions.[3] To illustrate the idea, I evaluate

$$\int_a^b \frac{dx}{x^2+2x+2}$$

this way (it is, of course, trivial by elementary methods). This is equivalent to finding the indefinite integral. It is enough to evaluate

$$\int_a^\infty \frac{dx}{x^2+2x+2}.$$

We consider

$$I=\int_C \frac{\log(z-a)}{z^2+2z+2}\,dz$$

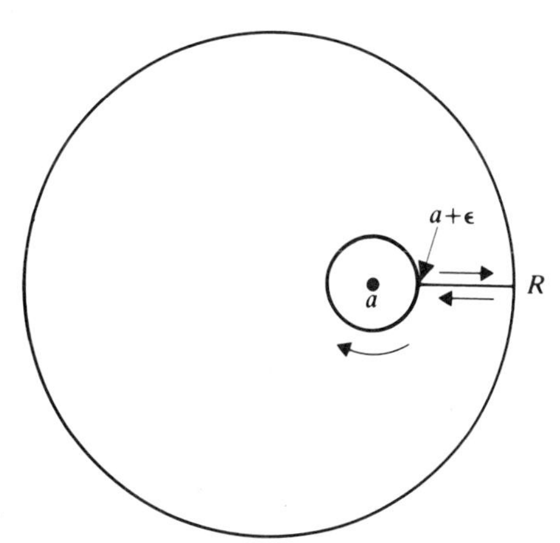

FIGURE 11.4

around a contour like that in Sec. **11A**, taking $0 \le \operatorname{Im} \log(z-a) \le 2\pi$; see Fig. 11.4. Arguing as in Sec. **11A**, we obtain

$$\lim_{R\to\infty,\ \varepsilon\to 0} I = \int_a^\infty \frac{\ln(x-a)\,dx}{x^2+2x+2} - \int_a^\infty \frac{\ln(x-a)+2\pi i}{x^2+2x+2}\,dx$$

$$= 2\pi i \text{ (sum of residues of the integrand at } -1 \pm i).$$

Hence

$$\int_a^\infty \frac{dx}{x^2+2x+2} = -\sum \text{(residues)}.$$

The residues are

$$\frac{\ln|-1-a+i| + i\arg(-1-a+i)}{2i} \quad \text{and} \quad \frac{\ln|-1-a-i| + i\arg(-1-a-i)}{-2i},$$

where the imaginary parts of the logarithms are the angles shown in Fig. 11.5 (p. 100), that is, $\pi - T$ and $\pi + T$, where T is the principal value of $\cot^{-1}(1+a)$. Thus the logarithmic terms in the sum of residues cancel (they are the logarithms of the moduli of conjugate complex numbers), and the sum of the residues is $-T$. Hence the original integral equals $T = \cot^{-1}(1+a)$, or $\frac{1}{2}\pi - \tan^{-1}(1+a)$.

EXERCISE 11.4 Evaluate

$$\int_a^\infty \frac{dx}{1+x^2}, \qquad a > 0,$$

by this method, and check that you get the correct result.

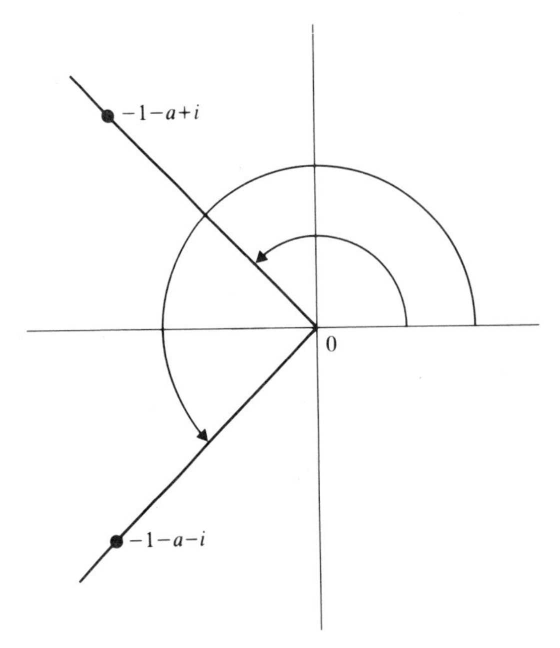

FIGURE 11.5

Supplementary exercise

***1.** By the method of this section, calculate

$$\int_0^\infty \frac{dx}{1+x^3}.$$

[The same method can be used to evaluate the integral over (a, ∞), but the computation is rather tedious.]

11C. Changing variables in improper integrals

This subsection is principally a warning against a trap that you might fall into by naïvely assuming that improper definite integrals will actually behave in the way that formal manipulations suggest.

Example 1 Consider

$$I_1 = \int_0^\infty e^{-t}\,dt.$$

Since the integral is so simple, you would be unlikely to try the formal change of variable $t = is$; but if you did, you would get

$$I_2 = i\int_0^\infty e^{-is}\,ds,$$

which evidently diverges. However, in a more complicated situation the error introduced by the naïve calculation might not be so obvious, and you might really be misled.

The way to be sure what is happening in cases like this is to treat both integrals as contour integrals. In the present case,

$$I_1 = \int_{C_1} e^{-z}\, dz, \qquad C_1 = \text{positive real axis;}$$

$$I_2 = \int_{C_2} e^{-z}\, dz, \qquad C_2 = \text{positive imaginary axis.}$$

You might be tempted to argue that $C_1 - C_2$ is a simple closed curve on the Riemann sphere, that the integrand is analytic inside this curve, and therefore that $I_1 - I_2 = 0$. However, the integrand is not even continuous on the closure of the region bounded by the curve; hence we should not expect Cauchy's theorem to apply to it. The only realistic approach is to treat both integrals as limits, close the contour in the finite plane, apply Cauchy's theorem, and see whether the extra integral approaches zero. It is not worth doing this in the present case, because I_2 is so obviously divergent.

Example 2 Start from the integral

$$J_1 = \int_{-\infty}^{\infty} e^{-x^2}\, dx,$$

which we know from Sec. **7H** is equal to $\pi^{1/2}$, and try setting $x = u + it$. Does

$$J_2 = \int_{-\infty}^{\infty} e^{-(u+it)^2}\, du$$

converge, and if so, to what value?

J_2 is the limit of $\int e^{-z^2}\, dz$ along the line segment $(-T + it,\ T + it)$. Since e^{-z^2} has no singular points inside the rectangle in Fig. 11.6, $\int e^{-z^2}\, dz = 0$ around the

FIGURE 11.6

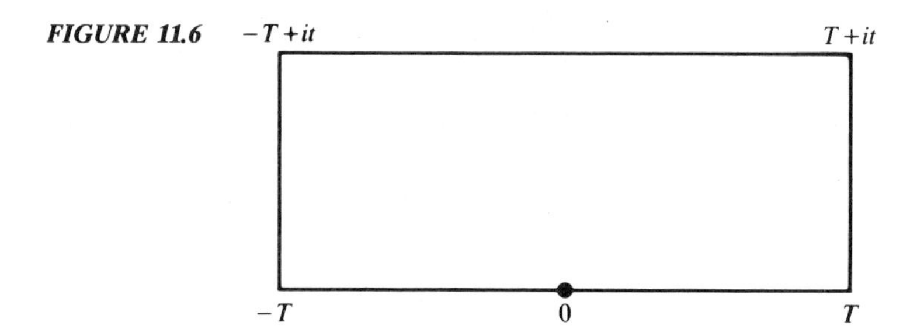

rectangle, and so

$$\int_{-T}^{T} e^{-u^2}\,du - \int_{-T}^{T} e^{-(u+it)^2}\,du + i\int_0^t e^{-(T+iy)^2}\,dy - i\int_0^t e^{-(-T+iy)^2}\,dy = 0.$$

Since $e^{-(T+iy)^2} = e^{-T^2}e^{-2iTy}e^{y^2}$, it is clear that

$$\int_0^t e^{-(T+iy)^2}\,dy \to 0$$

as $T\to\infty$; similarly for the integral at the other end of the rectangle; consequently,

$$\int_{-\infty}^{\infty} e^{-(u+it)^2}\,du = \int_{-\infty}^{\infty} e^{-u^2}\,du = \pi^{1/2},$$

just as the naïve argument suggests.

Example 3 Let

$$I = \int_{-\infty}^{\infty} e^{x-\exp(2x)}\,dx.$$

The integral converges because $e^{-\exp(2x)}$ overwhelms e^x for positive x and e^x overwhelms $e^{-\exp(2x)}$ when x is negative. Formally replace x by $u+i\pi$. We get

$$\int_{-\infty}^{\infty} e^{u+i\pi-\exp(2u+2i\pi)}\,du = -\int_{-\infty}^{\infty} e^{u-\exp(2u)}\,du = -I.$$

This would imply that $I=0$, which is impossible, because the integrand is positive.

Example 4 In Sec. **7H** we defined

$$\Gamma(z) = \int_0^{\infty} t^{z-1}e^{-t}\,dt, \qquad \operatorname{Re} z > 0.$$

If we take z real and naïvely replace t by is, we get

$$\Gamma(x) = e^{i\pi x/2}e^{-i\pi/2}\int_0^{\infty} s^{x-1}e^{-is}i\,ds,$$

$$(\cos\tfrac{1}{2}\pi x - i\sin\tfrac{1}{2}\pi x)\Gamma(x) = \int_0^{\infty} s^{x-1}\cos s\,ds - i\int_0^{\infty} s^{x-1}\sin s\,ds,$$

which would yield

$$\Gamma(x)\cos\left(\frac{\pi x}{2}\right) = \int_0^{\infty} s^{x-1}\cos s\,ds,$$

$$\Gamma(x)\sin\left(\frac{\pi x}{2}\right) = \int_0^{\infty} s^{x-1}\sin s\,ds.$$

To see whether these formulas are correct, consider

$$\int_C t^{x-1}e^{-t}\,dt, \qquad 0<x<1,$$

around the rectangle with vertices at $t=0$, R, iR, and $R+iR$. Then, as $R\to\infty$, the integrals along the real and imaginary axes respectively approach $\Gamma(x)$ and

$$e^{i\pi x/2}\int_0^\infty s^{x-1}e^{-is}\,ds,$$

and the other two integrals approach zero, as in Exercise 9.10. Hence our formulas are correct [and could be used as alternative definitions of $\Gamma(x)$]. We can extend these formulas to all z for which the integrals on the right converge by using the coincidence principle (Sec. **7G**).

EXERCISE 11.5 The formal change of variable $x=iy$ in

$$\int_0^\infty \frac{dx}{x^2-i-1}$$

yields

$$\int_0^\infty \frac{-i\,dy}{y^2+i+1}.$$

Is this a correct result?

After the preceding examples, you may be afraid to make any changes of variable at all in contour integrals. Actually, there is no problem as long as you are dealing with continuously differentiable bounded curves. Let C be defined by $z=C(t)$, $a\le t\le b$. Then

$$I=\int_C f(z)\,dz=\int_a^b f[C(s)]C'(s)\,ds.$$

What we are likely to want to do is to replace $\int_C f(z)\,dz$ by $\int_\gamma f[g(w)]g'(w)\,dw$, where g is a univalent analytic function (see p. 112), $g^{-1}(z)=w$, and $\gamma(s)=g^{-1}[C(s)]$ is the curve in the w plane corresponding to C. Then

$$\int_\gamma f[g(w)]g'(w)\,dw=\int_a^b f\{g[\gamma(s)]\}g'[\gamma(s)]\gamma'(s)\,ds=\int_C f(z)\,dz,$$

with $z=g(w)$.

This is just what we did in Sec. **8G** when we computed a residue at ∞.

Supplementary exercises

1. Consider $\int_C (z^4+16)^{-1}\,dz$, where C consists of the real axis from 0 to R, the quarter circle Q: $|z|=R$ from R to iR; and the y axis from iR to 0. The integral

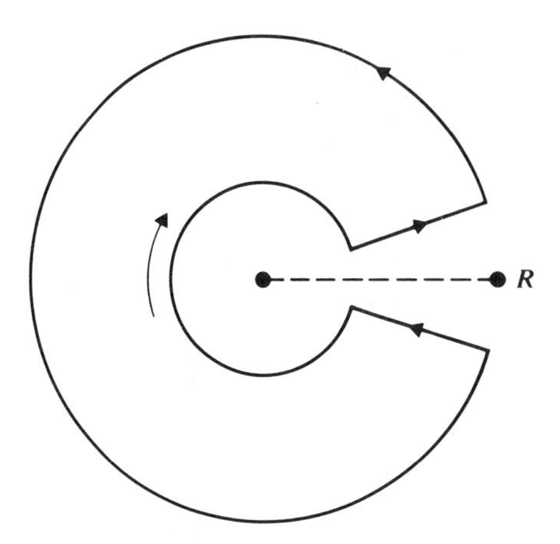

FIGURE 11.7

along Q approaches 0 as $R \to \infty$; the integral along $(0, R)$ approaches a limit L; the integral from iR to 0 is

$$\int_R^0 \frac{i\,dy}{y^4+16} = -i\int_0^R \frac{dy}{y^4+16} \to -iL.$$

This appears to show that $L = -iL$. Explain what went wrong.

2. Let $I = \int_0^\infty \exp(-x^4)\,dx$. Take $x = iy$ to get $i\int_0^\infty \exp(-y^4)\,dy = iI$. Explain.

Notes

1. For many more integrals that can be evaluated by contour integration, see Mitrinović and Kečkić. In particular, $\int_0^\infty e^{-x^2}\,dx$ can be evaluated (in several ways) by contour integration; see Mitrinović and Kečkić, pp. 158 ff; also Remmert, p. 297.

2. Another, perhaps more intuitive, way of treating the contour in Fig. 11.1 is to think of it as a limit of contours like the one shown in Fig. 11.7. However, the details of carying out the limiting process seem to me to be more tedious than the proof in the text.

3. For further developments of this (very old) idea, see R. P. Boas and L. Schoenfeld, "Indefinite Integration by Residues," *SIAM Rev.* *8* (1966): 173–183.

12 *Zeros of Analytic Functions*

12A. The argument principle

We showed in Sec. **7G** that the zeros of analytic functions are isolated (except for the identically zero function). We are now going to obtain more specific information about the zeros by finding a formula for the number of zeros of a function inside a closed curve. Information of this kind is useful in theoretical problems and also in some quite practical ones.[1]

It is just as easy to study meromorphic functions and find a formula for the number of zeros minus the number of poles. Suppose, then, that C is a simple closed curve and that f is meromorphic in a bounded region D that contains C and the interior of C, and that f has neither zeros nor poles on C. The closure of the interior of C is a compact set Δ. There are only a finite number of zeros inside C, because if there were infinitely many zeros, they would have a limit point in Δ, hence in D, and f would be the zero function by Sec. **7G**. Similarly, if there were an infinite number of poles, $1/f$ would be the zero function.

Now consider how f behaves in the neighborhood of a zero z_0. If $|z - z_0|$ is small enough, we have $f(z) = (z - z_0)^n \varphi(z)$, where φ is analytic at z_0 and $\varphi(z_0) \neq 0$ (Sec. **7G**). Consequently, $\log f(z) = n \log(z - z_0) + \log \varphi(z)$. If z describes a small circle γ around z_0, in the positive sense, then $\log f(z) = n \log|z - z_0| + ni \arg(z - z_0) + \log \varphi(z)$, and $\arg(z - z_0)$ increases by 2π. The other terms in $\log f(z)$ return to their starting values. Consequently, $\log f(z)$ increases by $2\pi i n$ as z goes around γ; or, what is the same thing, $\arg f(z)$ increases by $2\pi n$. Similarly, if f has a pole of order m at z_0, we see that $\arg f(z)$ decreases by $2\pi m$. It is then rather plausible that if z goes around C in the positive direction, $\log f(z)$ changes by $2\pi(N - M)i$, where N is the number of zeros of f inside C, and M is the number of poles; or, $\arg f(z)$ increases by $2\pi(N - M)$. This is an informal statement of the **argument principle**, which we shall prove rather indirectly because of the complications inherent in giving a precise definition of the change in $\arg f(z)$.

Whatever branch of $\log f(z)$ we choose, its derivative is $f'(z)/f(z)$, so it is natural to look at the integral

$$\int_C \frac{f'(z)}{f(z)}\,dz.$$

We can evaluate it by the residue theorem, since the integrand has only a finite number of poles inside C. What are its residues?

EXERCISE 12.1 Show that the residue of $f'(z)/f(z)$ at a zero of f is the order of the zero, and at a pole is the negative of the order of the pole.

Consequently,

$$\int_C \frac{f'(z)}{f(z)}\,dz = 2\pi i(N - M) = 2\pi i(\text{number of zeros inside } C \text{ minus number of poles inside } C).$$

Now choose a point z_1 on C and consider

$$F(z) = \int_{z_1}^{z} \frac{f'(w)}{f(w)}\,dw,$$

with integration along C in the positive direction. We are to think of F as a function on the trace of C. Then we can define $\arg f(z)$ as the imaginary part of $F(z)$. This will be a continuous function on C until z reaches z_1 again, when it will have a jump of amount $-2\pi(N-M)$.

If, then, we want to find the number of zeros of the analytic function f inside C, knowing that there are finitely many zeros, and none on C, we can either calculate

$$\frac{1}{2\pi i}\int_C \frac{f'(z)}{f(z)}\,dz,$$

or calculate the increase of $\arg f(z)$ as z goes around C in the positive direction.

The first process can, in principle, be done numerically; indeed, it does not even require great accuracy in computation, since we are calculating an integer, and if we can approximate it within $\frac{1}{2}$, we know its exact value. This is actually done in practice when a zero of a complicated function has to be found quite accurately.

12B. Finding the number of zeros of a polynomial in a region

When f is a relatively simple function, it may be easy to describe a branch of $\arg f(z)$ and follow it around C to determine its increase. A situation that arises in applications is that in which C is a line $x = \text{constant}$ and we want to know how many zeros a given polynomial has in one of the half planes bounded by C. We can replace C by a large semicircle in the half plane of interest, large enough so that all the zeros are inside it. If we plot the curve described by $f(z)$ as z describes the boundary of the semicircular region, the number of times that this curve goes around 0 is the change in $(2\pi)^{-1}\arg f(z)$ around C. Such a plot is called a **Nyquist diagram.**[2] It is not usually necessary to draw the graph accurately; a qualitative picture will serve.

In many applications the function in question is a polynomial, but of high enough degree so that it is not trivial to find the location of its zeros explicitly.[3] To illustrate the method, let us take the relatively simple example $P(z) = z^3 + 3z^2 + 4z + 2$, and ask how many zeros P has in the half plane $\operatorname{Re} z > 0$. Since P is real on the real axis, and obviously not zero on the positive real axis, it must have a zero on the negative real axis and either a pair of conjugate complex zeros or three negative real zeros. It is sensible to dispose of the second possibility first. If P had three negative real zeros, its derivative would have two, but $P'(z) = 3z^2 + 6z + 4$ does not have real zeros. Hence we have only to decide whether the nonreal zeros of P are in the right-hand or left-hand half plane. Since P takes conjugate values at conjugate points, we may consider only the region cut out of the first quadrant by a large quarter circle (center at 0). Let us start at 0 with $\arg P(z) = 0$ and follow a continuous branch of $\arg P(z)$, first along the real axis from 0 to a large positive R. Since $\log P(z)$ is continuous on $(0, R)$, we have to have $\arg P(R) = 0$ because the only positive values of $\arg P(R)$ are integral multiples of 2π. Now let z follow the quarter-circle $|z| = R$, $0 \le \theta \le \pi/2$, from R to iR. We have

$$P(z) = z^3\left(1 + \frac{3}{z} + \frac{4}{z^2} + \frac{2}{z^3}\right).$$

If $|z|$ is large, the expression in parentheses stays in a disk centered at 1 with radius less than 1; hence the argument of this expression starts near 0 and remains near 0. On the other hand, $\arg(z^3)$ increases by $3\pi/2$ when R is large and z goes from R to iR, so we have $\arg P(R)$ close to $3\pi/2$ when R is large and $z = iR$ ("z near $i\infty$").

Next we follow the change in $\arg P(z)$ as $z = iy$ traces the imaginary axis from $i\infty$ to 0. We have

$$P(iy) = -iy^3 - 3y^2 + 4iy + 2$$

and $\arg P(iy)$ is some value of $\Theta = \tan^{-1} T$, where

$$T = \frac{y(y^2 - 4)}{3y^2 - 2}$$

[the ratio of Im $P(iy)$ to Re $P(iy)$].

The variation of Θ as z goes down the positive imaginary axis is shown in the following table.

y	$\tan\Theta$	*quadrant of* Θ	Θ
Near $+\infty$	$+$	III	Near $3\pi/2$
$\infty > y > 2$	$+$	III	$3\pi/2 > \Theta > \pi$
2	0	III to II	π
$2 > y > (\frac{2}{3})^{1/2}$	$-$	II	$\pi > \Theta > \pi/2$
$(\frac{2}{3})^{1/2}$	∞	II to I	$\pi/2$
$(\frac{2}{3})^{1/2} > y > 0$	$+$	I	$\pi/2 > \Theta > 0$

In more detail, for large positive y, T is large and positive and Θ is near $3\pi/2$, so Θ must be an angle in the third quadrant, near $3\pi/2$. As y decreases, Θ must remain in the third quadrant until T crosses either 0 or ∞, where T changes sign. This happens first when $y = 2$; here T passes through 0. Since Θ has been in the third quadrant, it has to go into the second quadrant. There it remains until y reaches $(\frac{2}{3})^{1/2}$, where T becomes infinite and changes sign. Then Θ leaves the second quadrant and must go into the first quadrant. Finally, as $y \to 0$, $T \to 0$ also and $\Theta \to 0$. The path of $P(z)$ (in a P plane) is shown schematically in Fig. 12.1. We see that it does not go around the origin. Hence the net change in $\arg P(z)$ is zero. Thus there is no zero of $P(z)$ in the first quadrant; by symmetry, there is none in the fourth quadrant; hence there are conjugate complex zeros in the second and third quadrants.

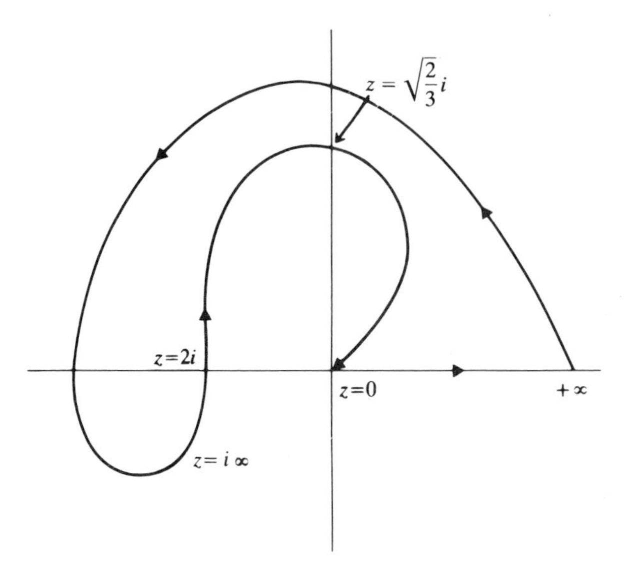

FIGURE 12.1

EXERCISE 12.2 Find which quadrants contain the zeros of

(a) $z^4 - 4z^3 + 11z^2 - 14z + 10$

(b) $z^4 + 2z^3 + 3z^2 + z + 2$

Supplementary exercises

Find which quadrants contain zeros of each polynomial.

1. $z^3 - 3z^2 + 4z - 1$ (There is a single real zero.)

2. $z^4 + z^3 + 6z^2 + 3z + 5$

3. $z^4 - 2z^3 + 2z^2 + 1$

4. $z^4 - 3z + 5$

5. $z^6 - z^3 - 9z + 64$ (no real zeros)

12C. Rouché's theorem

If f is analytic inside and on a simple closed curve C, and we change f by a sufficiently small amount on C, it is plausible that we would not change the *number* of zeros of f inside C (since that number is an integer and cannot change by a small amount); we would, however, change the location of the zeros. The following theorem makes this idea precise.

ROUCHÉ'S THEOREM *Let f and g be analytic inside and on a simple closed curve C and suppose that $|g(z)| < |f(z)|$ for z on C. Then $f(z)$ and $f(z) + g(z)$ have the same number of zeros inside C.*

Notice that the strict inequality in the hypothesis implies that f has no zeros on C. Also notice that nothing is said about the size of g inside C.

I give two proofs, one geometric and the other computational.

(*a*) The argument principle says that the number of zeros of $f+g$ inside C is $(2\pi)^{-1}$ times the change in $\arg[f(z)+g(z)]$ as z goes once around C in the positive direction. We can write

$$\arg[f(z)+g(z)]=\arg\left\{f(z)\left[1+\frac{g(z)}{f(z)}\right]\right\}=\arg f(z)+\arg\left[1+\frac{g(z)}{f(z)}\right].$$

As z goes around C, $|g(z)/f(z)|<1$ by hypothesis. This means that the point $g(z)/f(z)$ remains in the open disk of center 0 and radius 1. Consequently, $1+g(z)/f(z)$ remains in the open disk of center 1 and radius 1, and cannot go around 0 as z goes around C. Hence $\arg[1+g(z)/f(z)]$ returns to its original value. Therefore the change in $\arg[f(z)+g(z)]$ equals the change in $\arg f(z)$, which is 2π times the number of zeros of f inside C.

(*b*) For an alternative proof, we use the integral form of the argument principle. If N is the number of zeros of $f+g$ inside C, we have

$$2\pi iN=\int_C \frac{f'(z)+g'(z)}{f(z)+g(z)}\,dz.$$

Since we want to bring in $f'(z)/f(z)$, let's subtract this from the integrand:

$$\frac{f'(z)+g'(z)}{f(z)+g(z)}-\frac{f'(z)}{f(z)}=\frac{f(z)g'(z)-f'(z)g(z)}{f(z)[f(z)+g(z)]}.$$

Now notice that

$$\frac{f(z)g'(z)-f'(z)g(z)}{[f(z)]^2}$$

is the derivative of $\varphi(z)=g(z)/f(z)$. Hence we have

$$\begin{aligned}\frac{f'(z)+g'(z)}{f(z)+g(z)}&=\frac{f'(z)}{f(z)}+\frac{\varphi'(z)f(z)}{f(z)+g(z)}\\&=\frac{f'(z)}{f(z)}+\frac{\varphi'(z)}{1+\varphi(z)}.\end{aligned}$$

Consequently,

$$2\pi iN=\int_C \frac{f'(z)}{f(z)}\,dz+\int_C \frac{\varphi'(z)}{1+\varphi(z)}\,dz.$$

The first integral is $2\pi i$ times the number of zeros of f. We now need only to show that the second integral is 0. This is not an application of Cauchy's theorem, since we do not know that φ is analytic inside C (because of the zeros of f). However, *on* C,

$$\frac{1}{1+\varphi(z)}=\sum_{n=0}^{\infty}(-1)^n[\varphi(z)]^n,$$

with uniform convergence on C. We may therefore multiply by $\varphi'(z)$ and then integrate term by term. Since

$$\int \varphi'(z)[\varphi(z)]^n\,dz=\frac{1}{n+1}[\varphi(z)]^{n+1},$$

and φ is analytic on C, all the integrals are 0. Here we have applied the principle (Sec. **5D**) that the integral of a derivative around a closed curve is zero.

Here is an illustration of the use of Rouché's theorem for a specific function. Let us show that e^z+z^3 has no zeros for $|z|<0.7$ but has three zeros for $|z|<2$.

If $|z|<0.7$, take $f(z)=e^z$ in Rouché's theorem, $g(z)=z^3$. For $|z|=0.7$, we have $|z|^3<0.35<|e^z|$ because $|e^z|$ is smallest at $z=-0.7$ and $e^{-0.7}>0.49$. On the other hand, if $|z|=2$, take $f(z)=z^3$ and $g(z)=e^z$. Then $|f(z)|=8>e^2=\max_{|z|=2}|e^z|$.

EXERCISE 12.3 How many zeros does $2iz^2+\sin z$ have in the rectangle $|x|\le\pi/2$, $|y|\le 1$?

EXERCISE 12.4 Show that when f and g are meromorphic, the conclusion of Rouché's theorem becomes that the number of zeros of $f+g$ minus the number of poles of $f+g$ equals the number of zeros of f minus the number of poles of f. Find an example to show that $f+g$ may have more zeros than f, even when f has no poles.

EXERCISE 12.5 Prove the fundamental theorem of algebra by applying Rouché's theorem to a large circle with $f(z)=z^n$.

Here is another useful consequence of Rouché's theorem.

HURWITZ'S THEOREM *If the functions f_n are analytic in a region D, have no zeros in D, and converge in D to f, uniformly on each compact subset, then either $f(z)\equiv 0$ or f has no zeros in D.*

The sequence $\{e^z/n\}$ shows that it is possible to have $f(z)\equiv 0$ even though each f_n is zero-free. It also follows that the zeros of f in D are limits of sequences of zeros of $\{f_n\}$.

EXERCISE 12.6 Prove Hurwitz's theorem.

EXERCISE 12.7 Show by an example that a limit point of zeros of $f_n(z)$, on the boundary of D, need not be a zero of f.

Hurwitz's theorem lets us conclude that, for example, π is a limit of zeros of the partial sums of the power series of $\sin z$.

EXERCISE 12.8[4] Let f be analytic in a neighborhood of 0 except for an isolated essential singular point at 0. Suppose also that a sequence $\{z_n\}$ of zeros of f approaches 0. Then there are a subsequence T_{m_n} of the symmetric partial sums $T_m = \sum_{k=-m}^{m} a_k z^k$ of the Laurent series of f, and a sequence $\{w_n\}$, $w_n \to 0$, such that $f(w_n) \to 0$ and $T_{m_n}(w_n) = 0$.

Supplementary exercises

1. How many zeros does $3e^z - z$ have in $|z| < 1$?
2. How many zeros does $2z^2 - e^{z/2}$ have in $|z| < 1$?
3. Let $\{s_n\}$ be a sequence of analytic functions that converges uniformly in a region D containing 0. If $\exp[s_n(z)] \to f(z)$ and $f(0) = 0$, what can you say about f?
4. Justify the remark that immediately follows Exercise 12.7.
5. Since $[1 + (z/n)]^n$ has zeros, why doesn't its limit e^z have zeros?

12D. The open mapping theorem

When we study functions from the real line to the real line, we can visualize the correspondence between x and $f(x)$ by drawing a graph. For functions from the complex plane to the complex plane, however, we cannot even imagine a graph, since it would have to be drawn in four-dimensional space. Consequently, we have to use some other method of picturing the behavior of functions. The best that we can do is to consider various sets in the z plane and their images in a $w = f(z)$ plane, and ask questions like, What do the images of particular curves in the z plane look like in the w plane? What properties of a set remain valid for its image?[5] Questions of this kind will be discussed in the next chapter; however, there is one important property of the image of a set by an analytic function that we can easily establish by using Rouché's theorem.

OPEN MAPPING THEOREM *If f is analytic on an open set S, and not a constant, then the image of S under $w = f(z)$ is an open set.*

To prove the open mapping theorem, we start from a function f that is analytic and not constant in a region D, let $z_0 \in D$, and $w_0 = f(z_0)$. We are going to show that the image of D under $w = f(z)$ contains an open set that contains w_0. This will ensure that f carries every open subset of D to an open set. For, if Σ is the image of an open subset S of D, let z_0 be a point of the inverse image of $w_0 \in \Sigma$. The image of S contains an open set that contains w_0; that is, Σ contains an open set that contains w_0, which means that Σ is open.

We now consider the function g defined by $g(z) = f(z) - w_0$. Then g is a nonconstant analytic function with a zero at z_0. (Here z_0 is not necessarily a simple zero.) Since the zeros of g are isolated, D must contain a disk $|z - z_0| \le r$ on whose circumference γ we have $g(z) \ne 0$. Since g is continuous, $\inf_{z \in \gamma} |g(z)| = \delta > 0$. We shall show that the disk Δ: $|w - w_0| < \delta$ is in the image of D; that is, every w in Δ equals $f(z)$ for some $z \in D$. To do this, take any w in Δ and look at the function h defined by $h(z)=f(z)-w$. Since $h(z)=g(z)-(w-w_0)$, we can apply Rouché's theorem to h and g in Δ because $|w-w_0|<\delta<|g(z)|$ for z on γ (by the way δ was defined). Rouché's theorem then tells us that h has in D the same number of zeros that g has, namely, at least one. That is, $f(z)=w$ at least once for $z \in D$, which is to say that w is the image of some z under f. Since w was an arbitrary point of Δ, the open set Δ is the image of some open subset of D.

Notes

1. For example, the stability of a control system is decided by whether or not a polynomial related to the differential equation of the system has all its zeros in a particular half plane.

2. I am assured that nowadays every engineering student learns about Nyquist diagrams [A. Jaffe, "Ordering the Universe: The Role of Mathematics," *SIAM Rev. 26* (1984): 496]. The exact definition of a Nyquist diagram varies from one textbook to another.

3. Computer algorithms for finding zeros of polynomials are now so fast that I suspect that Nyquist diagrams may be obsolescent.

4. A. Abian, "On a Property of Finite Truncations of the Laurent Series of Analytic Functions," *Publ. Math. Debrecen 30* (1983): 129–131.

5. Imagine that you are not allowed to draw graphs of functions in calculus, but can only look at subsets of the real line and see what their images are like on a copy of the line. This is analogous to the situation that we face in studying functions from the complex plane to the complex plane, so it is no wonder that complex analysis is harder to understand than real analysis.

13 *Univalence and Inverses*

13A. Univalent functions

Since a point where $f(z)=w$ is a zero of $f(z)-w$, our discussion of zeros of functions is closely connected with the question of how many times an analytic function takes each of its values. We say that a function f (not necessarily analytic) is **univalent**[1] on a set S (usually a region) if f assumes, on S, each of its values just once; in other words, f is univalent if $f(z_1)=f(z_2)$, with z_1 and z_2 in S, implies that $z_1=z_2$. Notice that we do not consider multiplicities in defining "univalent"; for example, x^3 is univalent on the interval $[-1,1]$ even though 0 is a triple zero.

There is an essential difference between univalence for real-valued functions on a real interval and univalence of analytic functions in a region. In the real case a univalent differentiable function is monotonic and its derivative is either nonnegative or nonpositive, and can take the value 0. On the other hand, *if f is analytic and univalent in a region D, then f' has no zeros in D.* This would be a trivial statement if we counted a point where $f'(z)=0$ as being a point where f has a multiple value, but with our definition it requires proof. Let us show, then, that if $f'(z_0)=0$, then f is not univalent in any neighborhood of z_0.

Since f is represented in a neighborhood of z_0 by its Taylor series, we can find a disk centered at z_0 in which

$$f(z)-f(z_0)=(z-z_0)^k\varphi(z),$$

with $k\geq 2$ and φ analytic at z_0 with $\varphi(z_0)\neq 0$ (unless f is constant, in which case it is certainly not univalent). We can find a disk Δ, centered at z_0, so small that $f'\neq 0$ in Δ except at z_0 (the zeros of f' are isolated), and so that Δ also contains no zero of $h(z)=f(z)-f(z_0)$ except z_0; hence there is a positive δ such that $|h(z)|\geq 2\delta>0$ on the boundary of Δ. Now apply Rouché's theorem (Sec. **12C**) to h as the f and $-\delta$ as the g of that theorem. We conclude that $h(z)$ and $h(z)-\delta$ have the same number of zeros inside Δ, namely, k. Thus f takes the value $f(z_0)+\delta$ a total of k times inside Δ. However, f cannot take this value at z_0, because it already takes a different value at z_0; at other points of Δ, $h'(z)=f'(z)\neq 0$, so that h has only simple zeros. This means that f takes the value $f(z_0)+\delta$ in Δ at $k\geq 2$ different points, so that f is not univalent in Δ.

We have therefore shown that if f is univalent in a region D, the derivative f' is never 0 in D. Another—more geometric—proof of this is given in Sec. **22A**.

However, you must *not* jump to the conclusion that if f is analytic in D and $f'(z)\neq 0$, then f is univalent in D.

EXERCISE 13.1 Find a region D and a function f analytic in D such that $f'(z)\neq 0$ in D but f is not univalent in D.

Nevertheless, it *is* true that if f is analytic at z_0 and $f'(z_0)\neq 0$, then f is univalent in some disk centered at z_0. In other words, *if $f'(z)\neq 0$ in a region D, then f is locally univalent in D.*

This is perhaps most easily established by using power series. We have

$$f(z)=f(z_0)+a_1(z-z_0)+a_2(z-z_0)^2+\cdots,\qquad a_1=f'(z_0)\neq 0,$$

in a disk $|z-z_0|\leq s<r$ $(r>0)$. Let z_1 and z_2 be two different points of the disk Δ: $|z-z_0|\leq r$. Then

$$\begin{aligned} f(z_1)-f(z_2)&=a_1(z_1-z_2)+\sum_{k=2}^{\infty}a_k[(z_1-z_0)^k-(z_2-z_0)^k]\\ &=(z_1-z_2)\Big\{a_1+\sum_{k=2}^{\infty}a_k[(z_1-z_0)^{k-1}+\\ &\qquad (z_1-z_0)^{k-2}(z_2-z_0)+\cdots+(z_2-z_0)^{k-1}]\Big\},\end{aligned}$$

$$|f(z_1)-f(z_2)|\geq|z_1-z_2|\Big\{|a_1|-\sum_{k=2}^{\infty}k\,|a_k|s^{k-1}\Big\}.$$

The last series converges because its terms are the moduli of terms of the differentiated Taylor series of f about z_0; in fact, it converges if s is less than both $|z_1 - z_0|$ and $|z_2 - z_0|$. Hence the series defines a function of s that is $O(s)$ as $s \to 0$, and we have

$$|f(z_1) - f(z_2)| \geq |z_1 - z_2| \, [|a_1| - O(s)], \qquad s \to 0.$$

We therefore have $|f(z_1) - f(z_2)| > 0$ for small enough positive $|z_1|$, $|z_2|$. This means that f is univalent in a neighborhood of $z = z_0$.

EXERCISE 13.2 Show that a sufficient condition for an analytic function f to be univalent in the unit disk D: $|z - z_0| < 1$ is that $|f'(z) - f'(z_0)| < |f'(z_0)|$ for $0 < |z - z_0| < 1$.

13B. Inverse functions

A univalent (that is, one-to-one) analytic function f, without regard to any other properties that it may possess, automatically has an inverse function, customarily denoted[2] by f^{-1}.

The open mapping theorem (Sec. **12D**) shows that this inverse function is at least continuous, because f carries open sets to open sets, and so the inverse images of open sets under f^{-1} are open. What we would like, of course, is to show that f^{-1} *is analytic* (when f is univalent), and that the derivative of f^{-1} can be computed by the usual calculus formula, which says that the derivative of f^{-1} at w is $1/f'(z)$, where $w = f(z)$.

We show that f^{-1} is analytic simply by computing its derivative. If $w = f(z)$ and $w + \varepsilon = f(z + \delta)$, we have $z = f^{-1}(w)$ and $z + \delta = f^{-1}(w + \varepsilon)$. Therefore

$$\delta = f^{-1}(w + \varepsilon) - f^{-1}(w) \to 0$$

as $\varepsilon \to 0$, since f^{-1} is continuous. Then $f(z + \delta) - f(z) \neq 0$ and $f'(z) \neq 0$, since f is univalent. Consequently,

$$\frac{f^{-1}(w + \varepsilon) - f^{-1}(w)}{\varepsilon} = \frac{\delta}{f(z + \delta) - f(z)} \to \frac{1}{f'(z)}$$

as $\varepsilon \to 0$. That is,

$$\frac{d}{dw} f^{-1}(w) = \frac{1}{f'(z)},$$

as expected.

If you ever forget the formula, you can recapture it by the formal calculation

$$\frac{d}{dw} f^{-1}(w) = \frac{d}{dz} f^{-1}[f(z)] \frac{dz}{dw} = \left(\frac{dw}{dz}\right)^{-1} = \frac{1}{f'(z)},$$

which takes it for granted that f^{-1} is differentiable.

The power series of f^{-1} will be discussed in Sec. **15F**.

EXERCISE 13.3 Recall that an entire function is one that is analytic in the whole finite plane. Show that the only univalent entire functions are the affine functions, $f(z) = az + b$.

EXERCISE 13.4 Show that the uniform limit of a sequence of univalent functions is either univalent or constant.

Supplementary exercises

In these exercises, "univalent" means "analytic and univalent."

1. Show that $z(1-z)^{-2}$ is univalent in $|z|<1$.

2. If f and g are univalent, is $(f+g)/2$ necessarily univalent?

3. If f is univalent and $f(z) \neq 0$, is f^2 necessarily univalent?

4. If f is univalent in a neighborhood of 0, is $\int_0^z f(w)\,dw$ univalent?

5. Show that $z(1-z)^{-3}$ is univalent for $|z|<\frac{1}{2}$ but not in any larger disk.

13C. Univalence on a curve implies univalence inside

It is often much easier to verify that a function is univalent on a curve than that it is univalent in a region. Fortunately we have the following theorem:

If f is analytic on and inside a simple closed curve C, and f is univalent on C, then f is also univalent inside C.

This is a simple special case of **Darboux's theorem**, which we will meet in a more general form in Sec. **22D**.

For the proof, let $w = f(z)$; since f is univalent on C, the image of C is a simple closed curve Γ in the w plane, oriented in the positive sense (Sec. **12A**). Let w_0 be a point inside Γ; then $\arg(w - w_0) = \arg[f(z) - w_0]$ increases by 2π as z describes C, so that, by the argument principle, $f(z) - w_0$ has one zero inside C. Similarly, if w_0 is outside Γ, then $f(z) - w_0$ has no zeros inside C. That is, f takes each value inside Γ once inside C and takes no value outside Γ when z is inside C. There remains the possibility that f might, at some point inside C, take a value w_0 that is *on* Γ. But we know that the mapping $w = f(z)$ carries open sets to open sets, so in this case the image of a disk centered at z_0 would contain a disk centered at w_0, and would consequently contain both points inside Γ and points outside Γ; we have just seen that this is impossible.

We have therefore shown not only that f is univalent on the region bounded by C, but that it is a map *onto* the region bounded by Γ.

EXERCISE 13.5 Show that the converse of Darboux's theorem is false: Find a curve C and a function f such that f is univalent inside C but not on C.

Notes

1. Synonyms for "univalent" are "one-to-one," "simple," "injective," "biuniform," and "schlicht." "Schlicht" means "smooth" in German. In the early 1930s I once heard J. L. Coolidge, after a lecture on schlicht functions, ask plaintively whether there was an English equivalent. Osgood replied that you *could* call them "univalent functions," and everybody would know that you meant "schlicht."

2. Other notations for inverse functions have been proposed, but none has become widely accepted. Our notation is consistent with the convention that (for example) $\log^2 x$ means $\log(\log x)$, in contrast to $\sin^2 x = (\sin x)^2$. I have always supposed that f^{-1} stems from "thinking" that if $y = f(x)$, then we "divide" by f to get x.

14 *Laurent Series*

14A. Representation of analytic functions in an annulus

We have used power series to represent functions that are analytic in a disk $|z - z_0| < R$, but these series are naturally not available for functions that are analytic except for an isolated singular point at z_0. Since we found the Taylor series for an analytic function by starting from Cauchy's integral formula and expanding the kernel in a series, it is plausible that we should be able to do something similar for a function with an isolated singular point. In fact we can do more: We can find a series expansion for a function f that is analytic in any closed annulus between two concentric circles. To keep the formulas simple, I start with the case when $z_0 = 0$.

Assume, then, that f is analytic for $0 < r \le |z| \le R$. Although this annulus is not simply connected, we can still represent $f(z)$ for z in the annulus between the circles C_1: $|z| = r$ and C_2: $|z| = R$ by Cauchy's formula with integration in the positive sense around C_2 and in the negative sense around C_1:

$$f(z) = \frac{1}{2\pi i}\int_{C_2} \frac{f(w)\,dw}{w - z} - \frac{1}{2\pi i}\int_{C_1} \frac{f(w)\,dw}{w - z}.$$

This should seem very plausible if we think of introducing a cut to make the annulus simply connected and integrating as in Fig. 14.1. However, the proof goes through more simply if we use the same sort of device that we used in Secs. **7A** and **11A**. We dissect the annulus into two pieces. Then if z is between C_1 and C_2,

$$\frac{1}{2\pi i}\int \frac{f(w)\,dw}{w - z} = 0$$

around either the upper or lower half of Fig. 14.2 (the upper half, as the figure

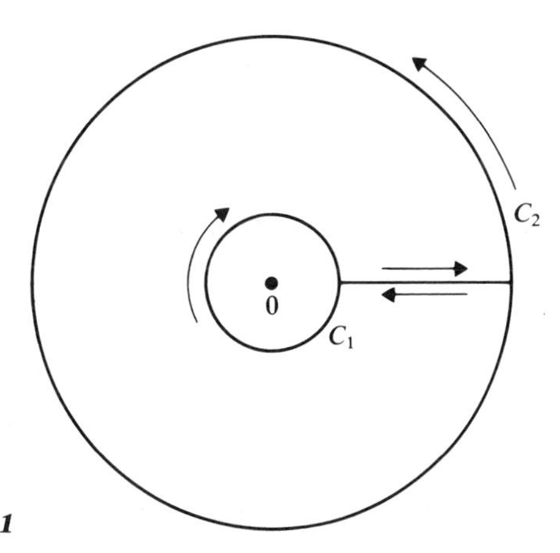

FIGURE 14.1

is drawn), and

$$\frac{1}{2\pi i}\int \frac{f(w)\,dw}{w-z} = f(z)$$

around the other half. The integrals along the horizontal line segments cancel. Thus

$$f(z) = \frac{1}{2\pi i}\int_{|w|=R} \frac{f(w)\,dw}{w-z} - \frac{1}{2\pi i}\int_{|w|=r} \frac{f(w)\,dw}{w-z} = f_2(z) + f_1(z).$$

Here the first integral represents a function f_2 that is analytic *inside* C_2 (compare Sec. **7C**), and the second integral represents a function f_1 that is analytic *outside* C_1. Then:

EXERCISE 14.1 Show that

$$f_2(z) = \sum_{n=0}^{\infty} c_n z^n$$

FIGURE 14.2

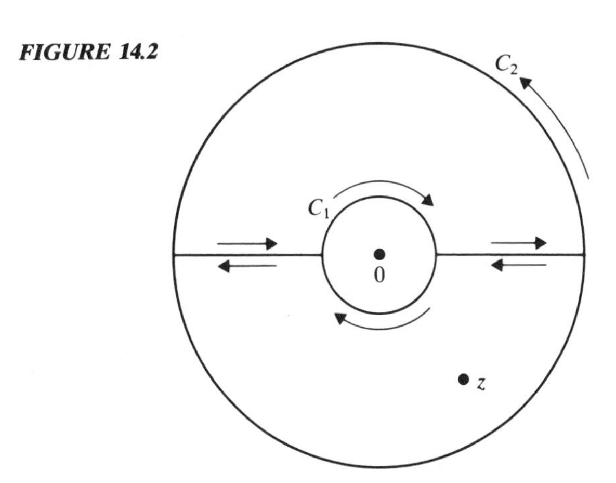

with

$$c_n = \frac{1}{2\pi i} \int_C \frac{f(w)\, dw}{w^{n+1}},$$

where C is any simple closed curve inside C_2 and surrounding C_1, and similarly,

$$f_1(z) = \sum_{n=1}^{\infty} c_{-n} z^{-n} = \sum_{n=-\infty}^{-1} c_n z^n$$

with

$$c_{-n} = \frac{1}{2\pi i} \int_C w^{n-1} f(w)\, dw \qquad (n \geq 1)$$

(integration in the positive direction in both cases).

We now have

$$f(z) = \sum_{n=-\infty}^{\infty} c_n z^n$$

for every z between C_1 and C_2. This series is the **Laurent series** for f in the annulus between C_1 and C_2; we cannot speak just of "the Laurent series for f," since f may have other Laurent series in other annuli. That a function has only one Laurent series in a given annulus will be shown in Sec. **14B**.

Notice that when $n \geq 0$, we do not necessarily have $c_n = f^{(n)}(0)/n!$; in fact, we generally don't, since f does not have to be analytic at $z = 0$. Also, for $n = -1$ we should not expect to have $c_{-1} = 0$, because f is not, in general, analytic inside C_1. If f happens to be analytic inside C_1 except for a pole at 0, then c_{-1} is the residue of f at 0. This is an easy way to find the residue if the Laurent series has been found in some other way.

If we consider an annulus centered at z_0 instead of at 0, the *Laurent series* is

$$\sum_{n=-\infty}^{\infty} c_n (z - z_0)^n$$

with

$$c_n = \frac{1}{2\pi i} \int_C f(w)(w - z_0)^{-n-1}\, dw, \qquad -\infty < n < \infty,$$

where C is any simple closed curve inside C_2 and surrounding C_1.

As a limiting case, when $R \to \infty$ we can take C_2 to be an arbitrarily large circle, and the Laurent series converges to $f(z)$ outside C_1.

EXERCISE 14.2 Show that if f has a pole of order n at z_0 and is analytic for $0 < |z - z_0| < R$, then the Laurent series for f in this punctured disk has no terms with index less than $-n$, and $c_{-n} \neq 0$.

EXERCISE 14.3 Let the Laurent series of f that converges outside a disk containing all the finite singular points of f be $\sum_{n=-\infty}^{\infty} c_n z^n$. Show that $-c_{-1}$ is the residue of f at ∞ (see Sec. **8G**).

14B. Methods for finding Laurent series

Using the integral formulas for the coefficients is not always (perhaps, not usually) the best way to find the coefficients for a function that is defined by an explicit formula. Before we go into methodology, we had better know that a given function has only one Laurent series in a specific annulus (it may have other Laurent series in other annuli, as we remarked above). The condensed statement that *Laurent series are unique* must be understood in this sense.

The uniqueness of a Laurent series comes about because a Laurent series is the sum of a power series in $(z - z_0)$ and a power series in $1/(z - z_0)$. Each of these series can be multiplied by any power $(z - z_0)^{-m}$ (m, an integer) and integrated around a circle in the region of convergence of the series (see Secs. **7D** and **3D**).

EXERCISE 14.4 Prove the uniqueness of Laurent series; that is, show that a function has only one Laurent series in a specified annulus.

It now follows that if we can get a convergent Laurent series for a function f in an annulus, by any method at all, for instance algebraically, this series must be *the* Laurent series of f in that annulus.

As an example, let us find the three Laurent series of

$$f(z) = \frac{1}{z(z-1)(z-2)}$$

in two different ways. There are three series because we can expand $f(z)$ in the punctured disk $0 < |z| < 1$, in the annulus $1 < |z| < 2$, or in the region outside the disk $|z| \leq 2$.

Method 1 In this method we expand f by using algebraic transformations and the binomial theorem.

First write f as a sum of partial fractions,

$$f(z) = \frac{\frac{1}{2}}{z} - \frac{1}{z-1} + \frac{\frac{1}{2}}{z-2} = \frac{1}{2z} + \frac{1}{1-z} - \frac{\frac{1}{4}}{1-(z/2)}.$$

In the punctured disk $0 < |z| < 1$, we expand the second and third fractions by the binomial theorem. We then obtain

$$f(z) = \frac{1}{2z} + \sum_{n=0}^{\infty} z^n - \frac{1}{4}\sum_{n=0}^{\infty} 2^{-n} z^n = \frac{1}{2z} + \sum_{n=0}^{\infty} (1 - 2^{-n-2}) z^n.$$

This is the Laurent series of f for $0 < |z| < 1$.

When $1<|z|<2$, we write

$$f(z)=\frac{1}{2z}-\frac{1/z}{1-(1/z)}-\frac{\frac{1}{4}}{1-\frac{1}{2}z}$$

and obtain

$$\begin{aligned}f(z)&=\frac{1}{2z}-\frac{1}{z}\sum_{n=0}^{\infty}z^{-n}-\frac{1}{4}\sum_{n=0}^{\infty}2^{-n}z^{n}\\&=-\sum_{n=-\infty}^{-2}z^{n}+\frac{1}{2}z^{-1}-\frac{1}{4}\sum_{n=0}^{\infty}2^{-n}z^{n}\\&=-\frac{1}{2}z^{-1}-\sum_{n=2}^{\infty}z^{-n}-\frac{1}{4}\sum_{n=0}^{\infty}2^{-n}z^{n}.\end{aligned}$$

Notice particularly that the coefficient of $1/z$ in this series is *not* the residue of f at 0.

Similarly, for $|z|>2$, we write

$$f(z)=\frac{1}{2z}-\frac{1/z}{1-(1/z)}+\frac{1}{2}\frac{1/z}{1-(2/z)}$$

and expand the fractions in powers of $1/z$.

EXERCISE 14.5 Obtain the Laurent series in this case.

Method 2 In this method we define the coefficients c_n of the Laurent series by the integral formula,

$$c_n=\frac{1}{2\pi i}\int_{C_3}\frac{f(w)\,dw}{w^{n+1}}$$

and evaluate them by calculating residues.

In the case at hand, f is analytic for $|z|<1$ except for the pole at 0, so we can take C_3 to be any curve inside the circle $|z|=1$ and surrounding the origin; the c_n will then be the coefficients of the Laurent series of f that converges for $0<|z|<1$.

We see at once that c_n is the residue of $f(w)/w^{n+1}$ at the origin, that is, the residue of

$$\frac{1}{w^{n+2}}\frac{1}{w-1}\frac{1}{w-2}.$$

Notice that when $n<-1$, we have $-n-2>-1$, so $-n-2$ is positive or zero. Consequently,

$$\frac{w^{-n-2}}{(w-1)(w-2)},\qquad n<-1,$$

is analytic at 0, its residue is 0, and therefore the Laurent series in $0<|z|<1$ has only one term of negative index. This is characteristic for a pole of order 1.

The sum of the terms of negative index in the Laurent series of f in a neighborhood of an isolated singular point z_0 is called the **principal part** of f "at" z_0. By Exercise 14.2:

A pole is characterized by the occurrence of only finitely many (but not no) nonzero terms in the principal part of f. At an essential singular point there are infinitely many nonzero terms in the principal part. Finally, at a removable singular point the function has no principal part.

Returning to our example, we consider the term with $n=-1$, for which the coefficient c_{-1} is the residue of

$$\frac{1}{w(w-1)(w-2)}$$

at 0. The technique for evaluating residues at simple poles (Sec. **8F**) gives the value $\frac{1}{2}=c_{-1}$. When $n\geq 0$, we want the residue of

$$\frac{1}{w^{n+2}}\frac{1}{(w-1)(w-2)},$$

which is

$$\lim_{w\to 0}\frac{1}{(n+1)!}\left(\frac{d}{dw}\right)^{n+1}\frac{1}{(w-1)(w-2)}.$$

This is most easily calculated as

$$\begin{aligned}\lim_{w\to 0}\frac{1}{(n+1)!}\left(\frac{d}{dw}\right)^{n+1}&\left(\frac{1}{w-2}-\frac{1}{w-1}\right)\\ &=\lim_{w\to 0}\frac{1}{(n+1)!}\left[\frac{(n+1)!(-1)^{n+1}}{(w-2)^{n+2}}-\frac{(n+1)!(-1)^{n+1}}{(w-1)^{n+2}}\right]\\ &=-2^{-n-2}+1,\end{aligned}$$

in agreement with what we found by Method 1; in fact, Method 1 seems to be rather faster.

However, Method 2 shows up much better for the second Laurent series, the one for $1<|z|<2$. Here we have

$$c_n=\frac{1}{2\pi i}\int_{C_4}\frac{f(w)\,dw}{w^{n+1}},$$

where C_4 is a circle $|z|=r_4$, $1<r_4<2$. We have

$$\int_{C_4}\frac{f(w)\,dw}{w^{n+1}}=\int_{C_3}\frac{f(w)\,dw}{w^{n+1}}+\text{residue of }[f(w)w^{-n-1}]\text{ at }w=1,$$

where the residue is the residue of

$$\frac{1}{(w-1)(w-2)w^{n+2}}$$

at $w=1$. Here the residue is simply

$$\lim_{w\to 1}\frac{1}{(w-2)w^{n+2}}=-1,$$

and so the Laurent series is obtained by adding $-z^n$, $-\infty<n<\infty$, to the term of index n of the Laurent series for $0<|z|<1$, resulting in the same series as before.

EXERCISE 14.6 Obtain the coefficients of the Laurent series for $|z|>2$ from those for $1<|z|<2$ by adding in the residues at $w=2$.

Either method will work for any function that is meromorphic in the extended plane. (We saw in Exercise 8.18 that this means that the function is rational.) However, for many elementary functions, such as $\log(1-\sin^{-1} z)$, $e^{\sin z}$, or $\tanh z$, it is usually easier to proceed by combining simpler series: see Sec. **15**.

EXERCISE 14.7 Find several terms of each Laurent series for

(*a*) $\dfrac{z+1}{z^3(z-2)}$

(*b*) $\dfrac{z^3}{(2z+1)(3z-2)}$

At the end of Sec. **8F** we showed that a meromorphic function that has real values on the real axis has conjugate residues at conjugate poles. We can now give a simpler proof of this useful fact.

We know that in a neighborhood of z_0,

$$f(z)=\sum_{n=-\infty}^{\infty} c_n(z-z_0)^n \qquad \text{with residue } c_{-1}.$$

In a neighborhood of $\overline{z_0}$,

$$f(z)=\sum_{n=-\infty}^{\infty} \gamma_n(z-\overline{z_0})^n, \qquad \text{with residue } \gamma_{-1}.$$

Then we have

$$f(\bar z)=\sum_{n=-\infty}^{\infty} \gamma_n(\bar z-\overline{z_0})^n.$$

But we also have

$$f(z)=\overline{f(\bar z)}=\sum_{n=-\infty}^{\infty} \overline{\gamma_n}(z-z_0)^n,$$

and so the residue of f at z_0 is $\bar\gamma_{-1}$.

EXERCISE 14.8 If $\log z = \ln r + i\theta$, $0 < \theta < 2\pi$, find the residues of $(\log z)/(1+z^2)$ at $z = \pm i$. Why are they not complex conjugates?

EXERCISE 14.9 The function defined by

$$\frac{2}{z}+\frac{1}{z^2}+\frac{1}{z^3}+\cdots=\frac{2}{z}+\sum_{n=2}^{\infty}\frac{1}{z^n}=\frac{1}{z}+\frac{1}{z-1}$$

has a pole at 0. Is the residue there equal to 2, 1, or something else? Explain.

EXERCISE 14.10 Let a and b be real numbers, $a < b$, and let $f(z) = (z-a)^{1/2}(z-b)^{1/2}$, where z is not on the interval $[a, b]$ and the square roots are taken to be positive for large real positive z. Find the residue of f at ∞, and hence evaluate[1]

$$\int_a^b \sqrt{(x-a)(b-x)}\,dx.$$

Supplementary exercises

Find the Laurent series (in powers of z) for:

1. $\dfrac{z}{(z+1)(3-z)}$ (3 series)

2. $\dfrac{1}{z^2(4z-1)}$ (2 series)

3. $\dfrac{z^2+1}{z(2z+3)}$ (2 series)

4. $\dfrac{z^2+1}{z(z-2)(z+5)}$ (3 series)

Note

1. Of course the integral can be looked up in a good table. The technique for evaluating it by "real" methods used to be taught in calculus courses. Similar integrals arise in physics, where it is apparently regarded as easier to use contour integration than to do the necessary algebra. Compare Goldstein, p. 302, 1st ed.; p. 474, 2nd ed.

15 *Combinations of Power Series and Laurent Series*

15A. Linear combinations

If we want the Maclaurin series for $2\sin z - 3\cos z$, it would be absurd to differentiate the function repeatedly in order to find the coefficients. Since we are familiar with the series

$$\sin z = z - \frac{z^3}{3!} + \frac{z^5}{5!} - \cdots$$

and

$$\cos z = 1 - \frac{z^2}{2!} + \frac{z^4}{4!} - \cdots$$

we can at once write down

$$2 \sin z - 3 \cos z = -3 + 2z + \frac{3}{2} z^2 - \frac{1}{3} z^3 + \cdots.$$

The algebra of series tells us that we can combine the two convergent series in this way to get a convergent series, and the uniqueness of the power series for a given function in a given region (Sec. **4F**) tells us that the equation is correct.

More generally, if the power series or Laurent series of two functions f and g have a common region of convergence, then *at least* in that region a linear combination $af + bg$ with constant coefficients a and b is represented by the power or Laurent series whose coefficients are the same linear combination of the coefficients of f and g. In symbols, $a \sum f_n z^n + b \sum g_n z^n = \sum (af_n + bg_n) z^n$.

Naturally, if f and g are series of powers of $(z - z_0)$ with the same z_0 for both series, we can proceed in exactly the same way.

Notice that f and g are not required to have the same region of convergence; we require only that there is a disk or annulus centered at z_0 in which they both converge. On the other hand, it is quite possible that the linear combination converges in a larger region than either of the series: for an extreme example, think of $f(z) - f(z)$.

15B. Multiplication of power series

Given $f(z) = \sum_{n=0}^{\infty} f_n z^n$, $|z| < r_1$, and $g(z) = \sum_{n=0}^{\infty} g_n z^n$, $|z| < r_2$, both f and g are analytic in the disk $|z| < \min(r_1, r_2)$ and consequently so is their product. This product then has a power series that converges at least in $|z| < \min(r_1, r_2)$; how do we find its coefficients c_n?

In Sec. **4B** we gave the formal computation for the formula

$$c_n = \sum_{k=0}^{n} f_k g_{n-k}.$$

We can justify the result by appealing to the formula

$$c_n = \frac{1}{n!} \left(\frac{d}{dz} \right)^n f(z) g(z) \Big|_{z=0}$$

and using Leibniz's rule for differentiating a product:

$$\frac{d}{dz}(fg) = \sum_{k=0}^{n} \binom{n}{k} f^{(k)}(z) g^{(n-k)}(z).$$

Then we have

$$\begin{aligned} f(z)g(z) &= \sum_{n=0}^{\infty} \frac{1}{n!} [f(z)g(z)]^{(n)}\big|_{z=0}\, z^n \\ &= \sum_{n=0}^{\infty} \frac{1}{n!} \sum_{k=0}^{n} \frac{n!}{k!(n-k)!} f^{(k)}(0) g^{(n-k)}(0) z^n \\ &= \sum_{n=0}^{\infty} z^n \sum_{k=0}^{n} f_k g_{n-k}, \end{aligned}$$

at least in $|z| < \min(r_1, r_2)$, and possibly in a larger disk.

Notice that the formula gives c_n as a finite sum, so that we can get any desired number of terms of the product series explicitly.

In specific cases, it is often more convenient simply to write down the two series and multiply them as if they were polynomials, retaining only terms up to a convenient power of z (and being careful not to overlook any terms). The technique was illustrated in Sec. **4B**.

We can multiply Laurent series in the same way if they have a common annulus of convergence and both series have only finitely many terms of negative index. In the general case we have the problem that the coefficients of the product are infinite series, so that the formal product may not be very helpful.

Supplementary exercises

Find the first few terms of the Maclaurin series, or Laurent series valid in a neighborhood of 0, for:

1. $\dfrac{\sin z}{z(1-z)}$

2. $z^{-4}(e^{z^2} - 1)$

3. $e^{-z} \sin z$

4. $[\log(1-z)]^2$

5. $2^{\sin z}$

6. Find the Laurent series for $e^z/(1-z)$ in a neighborhood of ∞. (Compare Exercise 4.8.)

7. If $f(z) = \sum_{n=0}^{\infty} a_n z^n$ for z in a neighborhood of 0, what are the Maclaurin coefficients of $f(z)/(1-z)$?

15C. Application to the Maclaurin series of $\tan z$

This is a rather nontrivial, but special, application of the algebra of power series. We start by finding the Maclaurin series of

$$A(t) = \frac{t}{e^t - 1} = \sum_{n=0}^{\infty} Q_n t^n.$$

Some observant person noticed that

$$A(t) + \frac{1}{2}t = \frac{t}{e^t - 1} + \frac{t}{2} = \frac{1}{2}t\frac{e^t + 1}{e^t - 1} = \frac{1}{2}t\frac{e^{t/2} + e^{-t/2}}{e^{t/2} - e^{-t/2}} = \frac{1}{2}t \coth\left(\frac{t}{2}\right).$$

It is then clear that $A(t) + \frac{1}{2}t$ is an even function, so all the coefficients of odd powers in $A(t)$, except the first, are zero. Now

$$e^t A(t) = A(t) + t,$$

that is,

$$e^t \sum_{k=0}^{\infty} Q_k t^k = \sum_{k=0}^{\infty} Q_k t^k + t.$$

If we write $e^t = \sum_{n=0}^{\infty} t^n/n!$, multiply the power series on the left-hand side of the preceding equation, and compare powers of t on the two sides, we obtain

$$Q_k = \sum_{j=0}^{k} \frac{Q_j}{(k-j)!}, \qquad k > 1,$$

or

$$k!Q_k = \sum_{j=0}^{k} \binom{k}{j} j!Q_j, \qquad k > 1.$$

If we define $B_k = k!Q_k$, we then have

$$B_k = \sum_{j=0}^{k} \binom{k}{j} B_j$$

and

$$A(t) = \sum_{n=0}^{\infty} \frac{B_n}{n!} t^n.$$

The B_k are the **Bernoulli numbers**; the first few are $B_0 = 1$, $B_1 = -\frac{1}{2}$, $B_2 = \frac{1}{6}$, $B_4 = -\frac{1}{30}$, $B_6 = \frac{1}{42}$, $B_8 = -\frac{1}{30}$; $B_{2k+1} = 0$ for $k > 0$. Note, however, that some books use different notations, so you should always check the first few Bernoulli numbers before using formulas from a book. The Bernoulli numbers can be computed recursively from the formula above; several hundred are already available explicitly.[1] The simple form of the first few Bernoulli numbers is deceptive: for example,

$$B_{20} = -\frac{174611}{330};$$

the numerator of B_{60} (written as a fraction in lowest terms) has 43 decimal digits.

Now we can write

$$\frac{t}{e^t+1} = \frac{t}{e^t-1} - \frac{2t}{e^{2t}-1}.$$

Hence

$$\frac{t}{e^t+1} = \sum_{k=0}^{\infty} \frac{B_k}{k!}(1-2^k)t^k,$$

and consequently,

$$\frac{2t}{e^{2t}+1} = \sum_{k=0}^{\infty} \frac{B_k}{k!}(1-2^k)2^k t^k.$$

Therefore

$$t\frac{e^{-t}-e^t}{e^t+e^{-t}} = \sum_{k=0}^{\infty} \frac{B_k}{k!}(1-2^k)2^k t^k - t.$$

Now take $t = iz$ and remember that $B_1 = -\frac{1}{2}$:

$$iz\frac{e^{-iz}-e^{iz}}{e^{iz}+e^{-iz}} = \sum_{k=2}^{\infty} \frac{B_k}{k!}(1-2^k)2^k i^k z^k.$$

The terms of odd index are all zero, so we get

$$\tan z = \sum_{k=1}^{\infty} \frac{B_{2k}}{(2k)!}(1-2^{2k})2^{2k}(-1)^k z^{2k-1}, \qquad |z| < \frac{\pi}{2}.$$

Similarly, we can express the Laurent coefficients of $\cot z$ and $\csc z$ (but not of $\sec z$) in terms of Bernoulli numbers. Explicitly,

$$\tan z = z + \frac{z^3}{3} + \frac{2z^5}{15} + \frac{17}{315}z^7 + \frac{62}{2835}z^9 + O(z^{11}), \qquad \left(|z| < \frac{\pi}{2}\right).$$

Supplementary exercises

Find a few terms of the Laurent series in a neighborhood of $z = 0$ for:

1. $\dfrac{1}{e^z-1}$

2. $\dfrac{e^z}{\sin z}$

3. $\dfrac{3z}{e^{3z} - 1 - 3z}$

4. $\dfrac{1}{\log(1 - z)}$

Use Laurent series to find the residues of

5. $\dfrac{z^{-2}(\cos z)}{\log(1 + z)}$ at $z = 0$

6. $z^{-6}(\tan z)e^{z^2}$ at $z = 0$

15D. Division

If f and g are defined by power series and we want the power series of f/g, we need not only to have a common disk of convergence for the series, but also to have $g(0) \neq 0$. However, if $g(0) = 0$ [and $g(z) \not\equiv 0$] we can write $g(z) = z^m h(z)$ with $h(0) \neq 0$, divide f by h to get a quotient q, and then write $f(z)/g(z) = z^{-m} f(z)/h(z)$. This gives us a Laurent series for f/g, convergent in some punctured disk $0 < |z| < r$.

In principle, if $f/g = q$, then $f = gq$, and the coefficients of the Maclaurin series of q can be calculated recursively from the multiplication formula. However, when only a small number of terms of the quotient series are needed, it is usually more convenient to divide one series by the other as if they were polynomials.

EXERCISE 15.1 Find a few terms of the power series of $\tan z$ again by dividing $\sin z$ by $\cos z$.

This division process is sometimes a convenient way of finding the residue of f/g at a zero of g. Suppose, for example, that we want the residue at 0 of $e^{z^2}/\sin(z^3)$. By the method of Sec. **8F**, we would have to differentiate $z^3 e^{z^2}/\sin(z^3)$ twice. However, if we divide the series for $z^{-3}\sin(z^3)$ into the series for e^{z^2}, we find a quotient $1 + z^2 + \cdots$; dividing this by z^3, we get a term $1/z$, and so the residue is 1.

EXERCISE 15.2 Carry out the division here.

15E. Composition

It can be very difficult to find more than a few terms of the power series of a composite function by differentiation, because the derivatives of a composite function can become very complicated very fast.[2] Suppose that $g(z) = \sum_{m=0}^{\infty} g_m(z - a)^m$ and $f(w) = \sum_{n=0}^{\infty} f_n(w - b)^n$; we ask for the power series of $F(z) = f[g(z)]$ in powers of $(z - a)$. We can consider $f[g(z)]$ only for points z such that $g(z)$ is in the domain of f. It may not be convenient to find the range of g explicitly, but if we assume that $g(a)$ is inside the disk of convergence of f, then some disk centered at $g(a)$ contains the points $g(z)$ for which z is sufficiently close to a. That is, $f[g(z)]$ is defined at least in a neighborhood of the point $z = a$. A natural way to proceed would be to write

$$F(z) = \sum_{n=0}^{\infty} f_n[g(z) - b]^n = \sum_{n=0}^{\infty} f_n\left[\sum_{m=0}^{\infty} g_m(z-a)^m - b\right]^n$$

and to try to rearrange this as a series of powers of $(z - a)$. If we expand the nth powers as series of powers of $(z - a)$ and collect terms, we obtain, at least formally, $F(z)$ as a series

$$F(z) = \sum_{n=0}^{\infty} c_n(z-a)^n.$$

We will show below that this series for F really converges to $F(z)$ in *some* disk centered at $z = a$. (If the point 0 is in this disk, we could further rearrange the series for F in powers of z, but this is not necessarily what we would need in a specific problem.)

The simplest situation is that in which $g_0 = b$, since then each of the coefficients in the series $F(z) = \sum c_n(z-a)^n$ will be a finite sum, so that we can find the coefficients c_n one after another for as long as our patience holds out. The process may appear tedious, but it is much less so (and somewhat less susceptible to mistakes) than calculating $F^{(n)}(a)/n!$.

Now we must justify the rearrangement of the double series

$$\sum_{n=0}^{\infty} f_n\left\{\sum_{m=0}^{\infty} g_m(z-a)^m - b\right\}^n$$

into a series of powers of $(z - a)$. Although the inner series (one series for each n) are absolutely convergent, it does not seem evident that the double series is itself absolutely convergent. We proceed indirectly.

For each n, the function $[g(z) - b]^n$ is analytic for $|z - a| < r$ (say). Hence $[g(z) - b]^n$ has a Taylor expansion

$$[g(z) - b]^n = \sum_{k=0}^{\infty} c_{kn}(z-a)^k, \qquad |z-a| < r.$$

This expansion could be found by multiplying out

$$\left\{\sum_{m=0}^{\infty} g_m(z-a)^m - b\right\}^n$$

because, as we know, a finite product of convergent power series can correctly be rearranged into a single power series. Let $|z - a| < s \le r$ be a disk in which the values of $|g(z)|$ are in the disk of convergence of f. Then in $|z - a| < s$ we have

$$F(z) = \sum_{n=0}^{\infty} f_n\left\{\sum_{k=0}^{\infty} c_{kn}(z-a)^k\right\}.$$

On the other hand, F has a Taylor expansion

$$F(z)=\sum_{k=0}^{\infty}\lambda_k(z-a)^k,$$

and what we want to show is that

$$\lambda_k=\sum_{n=0}^{\infty}f_n c_{kn}.$$

If we take $z=a$, we have

$$\lambda_0=F(a)=\sum_{n=0}^{\infty}f_n c_{0n},$$

which is what we want for $k=0$.

Now if $f(w)=\sum f_n(w-b)^n$ for $|w|<R$, we have $\sum f_n(w-b)^n$ converging uniformly for $|w|<R_1<R$. Hence, since we assumed that $|g(z)-b|<R$, we have

$$\sum f_n\{g(z)-b\}^n=\sum_n f_n\left\{\sum_{k=0}^{\infty}c_{kn}(z-a)^k\right\},$$

with uniform convergence for $|z-a|<s$. This is not a power series, but by the uniform convergence theorem (Sec. **7D**), it can be differentiated term by term. If we differentiate it, we get

$$F'(z)=\sum_{n=0}^{\infty}f_n\left[\sum_{k=0}^{\infty}c_{kn}k(z-a)^{k-1}\right].$$

Taking $z=a$, we see that

$$\lambda_1=F'(a)=\sum_{n=0}^{\infty}f_n c_{1n}.$$

Continuing in this way, we get

$$\lambda_k=\sum_{n=0}^{\infty}f_n c_{kn},$$

as expected.

Substitution of one Laurent series into another is not always possible. If $f(w)=\sum_{n=-\infty}^{\infty}c_n w^n$ and $g(z)=\sum_{n=-\infty}^{\infty}\gamma_n z^n$, the first for $r_1<|w|<r_2$ and the second for $s_1<|z|<s_2$, we would have to know that the values w of $g(z)$ in some subannulus of $s_1<|z|<s_2$ are in $r_1<|w|<r_2$, and there seems to be no simple way of guaranteeing this, as there was in the case of power series.

EXERCISE 15.3 Find several terms of the power series, about $z=0$, of the following functions:

(*a*) $e^{\sin z}$ (*b*) $e^{\cos z}$ (*c*) $\log(1+ze^z)$

15F. The power series of an inverse function

Let
$$w=f(z)=\sum_{k=0}^{\infty} a_k z^k$$
in a neighborhood of $z=0$, with $a_1 \neq 0$. The inverse function $f^{-1}(w)$ is defined in a neighborhood of $w=a_0=f(0)$ (see Sec. **13B**) and has a power series
$$f^{-1}(w)=\sum_{n=0}^{\infty} b_n(w-a_0)^n.$$
Then $b_0=f^{-1}(a_0)=f^{-1}[f(0)]=0$. We have
$$\begin{aligned} z &= f^{-1}[f(z)] \\ &= \sum_{n=1}^{\infty} b_n\left\{\sum_{k=1}^{\infty} a_k z^k\right\}^n. \end{aligned}$$
In principle, we can expand the powers and then compare the coefficients of powers of z on the two sides of the equation to find as many coefficients b_k as we like. Since the series have their constant terms equal to 0, we have only a finite number of terms to consider for each b_n, so we can compute the b_n recursively. Thus
$$a_1 b_1 = 1, \qquad b_1=\frac{1}{a_1},$$
$$0=b_1a_2+b_2a_1^2=\frac{a_2}{a_1}+b_2a_1^2,$$
$$b_2=-\frac{a_2}{a_1^3}. \qquad \text{Then } b_3=-\frac{a_3}{a_1^4}+2\frac{a_2^2}{a_1^5},$$
and so on. However, the algebra gets complicated very fast as n increases. Formulas for the b_n are available.[3]

Supplementary exercises

Find several terms of the Maclaurin series of

1. $\sin(e^z-1)$

2. $\log(1+\sin z)$

3. $\exp[z/(1-z)]$

4. $\sin(\sin z)$

Notes

1. D. E. Knuth and T. J. Buckholtz, "Computation of Tangent, Euler and Bernoulli Numbers," *Math. of Comp. 21* (1967): 663–688.

2. Francisco Faà di Bruno's formula for the nth derivative of a composite function can be looked up in books on combinatorics, for example, Comtet, pp. 138 ff. There is a recent rather simple discussion by S. Roman, "The Formula of Faà di Bruno," *Amer. Math. Monthly 87* (1980): 805–809.

3. The formulas are given in many books; Comtet gives the coefficients explicitly for $n \leq 8$ (pp. 148 ff.). The process is, or used to be, known as "reversion of series." For recent elegant discussions in terms of purely formal power series, see P. Henrici, "An Algebraic Proof of the Lagrange-Bürmann Formula," *J. Math. Analysis Appl. 8* (1964): 218–224; "Die Lagrange-Bürmannsche Formel bei formalen Potenzreihen," *Jahresber. Deutsch. Math. Verein. 86* (1984): 115–134.

16 *The Maximum Principle*

16A. The simplest versions

An informal statement of the **maximum principle** (or maximum modulus theorem) is that "an analytic function takes its maximum on the boundary." This may serve as a mnemonic, but otherwise it won't do, both because a complex-valued function can't literally have a maximum and because it doesn't say what set the boundary is bounding. There are a number of ways of making the vague statement precise.

1. Let f be analytic in a bounded region D, continuous in $\bar{D}$ (the closure of D), and let $|f(z)| \leq M$ for $z \in \partial D$ (the boundary of D). Then $|f(z)| \leq M$ in D; moreover, if $|f(z)| = M$ for some $z \in D$, then f is constant in D.

2. The maximum of $|f(z)|$ is attained on the boundary of whatever region we are considering, and only there except in the case of a constant f.

3. Let f be analytic in a region D (not necessarily bounded). Then $|f(z)|$ cannot have a proper local maximum in D, and can have a local maximum (of any kind) in D only if f is constant in D.

Recall that z_0 is a *proper* local maximum of $|f|$ if $|f(z_0)| > |f(z)|$ for all z in some punctured neighborhood of z_0.

It is clear that (1) follows from (2) because $|f|$, as a continuous function on a compact domain, has a maximum in $\bar{D}$, and (2) says that this maximum cannot occur inside D unless f is constant. However, the emphasis is different in (1) and (2): in (1) we are more obviously concerned with inferring something about f in D from information about the values of f on the boundary, as we might expect from Cauchy's integral formula.

Similarly, (3) implies (2) (and is much stronger because it excludes local as well as global maxima from D). In fact, the hypotheses of (2) imply that $|f|$ has a maximum, and (3) says that the maximum cannot occur at a point of D. Statement (3) also has the advantage of applying to unbounded regions, whereas (1) and (2), as they stand, do not (see below).

We can prove (2) and (3) simultaneously, by contradiction, starting from Cauchy's formula; (2) is somewhat easier, and illustrates the method in its simplest form. Suppose that $|f(z)|$ does have a maximum value M at $z_0 \in D$ and that $|f(z)| \le |f(z_0)| = M$ in a disk $|z - z_0| < R$. Let $r < R$ and represent $f(z_0)$ by Cauchy's formula:

$$f(z_0) = \frac{1}{2\pi i} \int_{|w-z_0|=r} \frac{f(w)}{w - z_0}\, dw = \frac{1}{2\pi} \int_0^{2\pi} f(z_0 + re^{i\theta})\, d\theta,$$

$$M = |f(z_0)| \le \frac{1}{2\pi} \int_0^{2\pi} |f(z_0 + re^{i\theta})|\, d\theta.$$

Since $M = (2\pi)^{-1} \int_0^{2\pi} M\, d\theta$, we have

$$\frac{1}{2\pi} \int_0^{2\pi} [|f(z_0 + re^{i\theta})| - M]\, d\theta \ge 0.$$

Since $|f(z_0 + re^{i\theta})| - M \le 0$, the last inequality requires that $|f(z_0 + re^{i\theta})| = M$ for almost all θ and hence (since $|f|$ is continuous) for all θ.

This means, first of all, that $|f|$ cannot have a proper maximum at z_0, so that statement (2) follows; and second, that if $|f|$ has an improper maximum, then $|f(z)|$ is constant, and equal to M, on every sufficiently small circle $|z - z_0| = r$ and hence on a disk $|z - z_0| < R$. But by Exercise 2.8, this means that $f(z)$ is constant in a neighborhood of z_0, and hence in D, by Sec. **7G.**

EXERCISE 16.1[1] Prove a weak form of the maximum principle by applying Cauchy's formula to $[f(z)]^m$, taking mth roots of moduli, and letting $m \to \infty$.

EXERCISE 16.2 Show that $\operatorname{Re} f$ and $\operatorname{Im} f$ both obey the maximum principle.

There is no minimum principle in general because $|f(z)|$ attains the smallest possible value whenever $f(z) = 0$. However:

EXERCISE 16.3 Show that if a nonconstant analytic function f has no zeros in a region D, then $|f|$ cannot have a local minimum inside D.

EXERCISE 16.4 Show that the real part (or imaginary part) of a nonconstant function, analytic in a region D, cannot have a minimum inside D.

The following application of the maximum principle is extremely useful.

SCHWARZ'S LEMMA *If f is analytic in a closed disk Δ of radius 1, centered at z_0; if $f(z_0) = 0$; and if $|f(z)| \le M$ on the boundary of Δ, then $|f(z)| \le |z - z_0| M$ for z inside Δ. There is equality at some interior point only if $f(z) \equiv e^{i\lambda}(z - z_0)$ (λ real).*

EXERCISE 16.5 Prove Schwarz's lemma. [Consider $f(z)/(z - z_0)$.]

EXERCISE 16.6 Under the hypotheses of Schwarz's lemma, what can you say about $f'(z_0)$? (See also Exercise 25.6.)

EXERCISE 16.7[2] Let f be analytic in $|z| \le 1$ and let $|f(z)|$ be maximized for $|z| \le 1$ at z_0 with $|z_0| = 1$. Show that $f'(z_0) \ne 0$ unless f is constant.

Supplementary exercises

1. Find an example to show that form (1) of the maximum principle (at the beginning of Sec. **16A**) fails for unbounded regions.
2. The largest value of $\sin x$ on $0 \le x \le 2\pi$ is at the interior point $x = \pi/2$. Why does this not contradict the maximum principle?
3. Reformulate Schwarz's lemma for a disk of radius R.
4. Use Supplementary Exercise 3 to prove Liouville's theorem (Sec. **7C**).
5. Use Supplementary Exercise 3 to prove the Casorati-Weierstrass theorem (Sec. **8E**) for transcendental entire functions (i.e., those that are not polynomials).
6. Let f be analytic in the closed unit disk, $f(0) = 0$, and $|f(z)| \le |e^z|$ for $|z| = 1$. How large can $|f(\log 2)|$ be?

16B. Refinements and counterexamples

We can generalize statement (1) above by replacing the hypothesis that $|f(z)| \le M$ on ∂D by the hypothesis that $\limsup |f(z)| \le M$ as z approaches each point of ∂D (so that f does not even have to be defined on the boundary). In other words, we are to show that *if, for each $z_0 \in \partial D$, and each $\varepsilon > 0$, there is a neighborhood U of z_0 such that $|f(z)| \le M + \varepsilon$ in $U \cap D$, then $|f(z)| \le M$ inside D, and $|f(z)| = M$ for some z only if f is constant in D.*

These hypotheses do not even imply, in any obvious way, that $|f|$ is bounded in D. Let us, then, begin by showing that $|f|$ is at least bounded in D. Suppose that it isn't. Then there are points z_n of D such that $|f(z_n)| > n$, so that $f(z_n) \to \infty$ as $n \to \infty$. The sequence $\{z_n\}$, being bounded, contains a convergent subsequence $\{z_{n_k}\}$ with a limit z_∞. The point z_∞ cannot be in D, since $f(z_\infty)$ would have to be infinite. Consequently, $z_\infty \in \partial D$, and for each positive ε we have $|f(z_{n_k})| \le M + \varepsilon$ if k is sufficiently large. But this contradicts $|f(z_{n_k})| \to \infty$. Therefore $|f|$ is indeed bounded in D.

Now let N be the least upper bound of $|f(z)|$ for z in D, so that there are points z_n (not necessarily the same points as before) in D such that $|f(z_n)| \to N$. Again select a convergent subsequence and call its limit z_∞. If $z_\infty \in D$, $|f(z_\infty)| = N$ and $|f(z)|$ is maximized at z_∞, which can happen only if $f(z)$ is a constant c; but then $|c| \le M$. On the other hand, if $z_\infty \in \partial D$, we have $|f(z_{n_k})| \le M + \varepsilon$ for sufficiently large k, and therefore $N \le M$.

Now let us consider the question of whether we can relax the hypothesis that $\limsup |f(z)| \le M$ at *every* point of ∂D. We might well suppose that a single point cannot make much difference, but in fact even one exceptional point can destroy the theorem. An example is provided by $f(z)=\exp[(1+z)/(1-z)]$ in the disk $|z|<1$. If $z=re^{i\theta}$, $r<1$, we find that when $\theta=0$, we have $|f(z)|\to\infty$ as $r\to 1$. However, if $\theta\neq 0$, we have $f(z)\to\exp[-i\cot(\theta/2)]$ as $r\to 1$. That is, $|f(z)|\to 1$ as $r\to 1$ for all values of θ except 0; but $|f|$ is unbounded.

The inadmissibility of an exceptional point in the maximum principle is even clearer when the exceptional point is at ∞, that is, when we consider functions in an unbounded region. It is still true that a nonconstant analytic function in an unbounded region D cannot have a local maximum in D unless the function is constant in D. However, the failure of version (1) of the maximum principle is shown by, for example, $f(z)=e^z$ in the right-hand half plane $x\geq 0$: on the boundary (the imaginary axis) $|f(z)|=1$; but $|f(z)|=e^x$ is unbounded.

In Sec. **33**, we will see that the maximum principle continues to hold when there is an exceptional point if we impose an extra condition on the function in a neighborhood of the exceptional point.

16C. Carathéodory's inequality

There is a quite different generalization of the maximum principle that depends on the idea that knowing an upper bound for the real part of an analytic function ought to tell us something about the size of the absolute value of the function. Whatever it tells us cannot be altogether straightforward: We get no information if f is a pure imaginary constant. Also, in the unit disk the function $f(z)=i\log(z+1)$ is unbounded but its real part is bounded; this shows that a bound for $\operatorname{Re} f(z)$ in an open disk cannot imply a bound for $|f(z)|$ in the whole disk. However, we can say something useful if we do not try to say too much.

CARATHÉODORY'S INEQUALITY *Let f be analytic for $|z|\leq R$ and let $A(r)$ be the maximum of* $\operatorname{Re} f(z)$ *on the circle $|z|=r<R$ (hence—by Sec.* ***16A****—in $|z|\leq r$). Then*

$$|f(re^{i\theta})|\leq |f(0)|+\frac{2r}{R-r}[A(R)-\operatorname{Re} f(0)], \qquad 0<r<R.$$

If f is a constant, there is nothing to prove. If f is not a constant, we start with the case when $f(0)=0$. Consider the function g, defined by

$$g(z)=\frac{f(z)}{2A(R)-f(z)}.$$

Since the real part of the denominator cannot be 0, the denominator itself cannot be 0. Consequently, g is analytic in $|z|\leq R$, and $g(0)=0$. Moreover, in the present case $A(R)>0$, since $f(0)=0$ and f is not a constant. Consequently, we have

$$|g(z)|^2 = \frac{|f(z)|^2}{[2A(R) - \operatorname{Re} f(z)]^2 + [\operatorname{Im} f(z)]^2}$$
$$= \frac{|f(z)|^2}{\{A(R) + [A(R) - \operatorname{Re} f(z)]\}^2 + [\operatorname{Im} f(z)]^2} \le 1$$

because $A(R) - \operatorname{Re} f(z) > 0$ by the maximum principle and $A(R) > 0$. Now we can apply Schwarz's lemma to $g(z)$:

$$|g(z)| \le \frac{|z|}{R} (<1) \qquad \text{for } |z| < R.$$

But $2A(R)g(z) - f(z)g(z) = f(z)$ by the definition of g. Consequently,

$$f(z)[1 + g(z)] = 2A(R)g(z).$$

Remembering that $A(R) > 0$, we have

$$|f(z)| = \frac{2A(R)\,|g(z)|}{|1 + g(z)|} \le 2A(R)\frac{|z|}{R}\frac{1}{1 - |g(z)|}$$
$$\le 2A(R)\frac{|z|}{R}\frac{1}{1 - (|z|/R)}$$
$$= 2A(R)\frac{|z|}{R - |z|}.$$

This establishes the inequality when $f(0) = 0$. If $f(0) \ne 0$, we consider $f_1(z) = f(z) - f(0)$ and apply what we have done so far to f_1.

As an application of Carathéodory's theorem, let us prove the analog of Liouville's theorem for the real part of an entire function.[3] *Let f be entire and let $\operatorname{Re} f$ be bounded above; then f is constant.*

To prove this, take any (large) R and let $r = R/2$. Then Carathéodory's inequality says that, for $|z| \le r$, we have

$$|f(z)| \le |f(0)| + 2\left[\max_{|z|=R} \operatorname{Re} f(Re^{i\theta}) - \operatorname{Re} f(0)\right].$$

Since the right-hand side is bounded above, independently of R, it follows that $|f|$ is bounded, and so f is constant by Liouville's theorem.

EXERCISE 16.8 Prove the preceding theorem by applying Liouville's theorem to the entire function e^f.

EXERCISE 16.9 Show that if f is entire and $\operatorname{Re} f(re^{i\theta}) = o(r^\lambda)$ (uniformly in θ), then f is a polynomial of degree less than λ.

Notes

1. This proof originated with E. Landau.
2. This seems to have first been noticed by K. Löwner; cf. Pólya and Szegő, problems III 290, 291, pp. 162, 373.
3. This was first proved by J. Hadamard.

CHAPTER 3

Analytic Continuation

17 The Idea of Analytic Continuation

17A. Definitions

The series

$$\frac{1+i}{2}\sum_{n=0}^{\infty}\frac{(-1)^n}{(1-i)^n}(z+i)^n$$

converges uniformly on each compact subset of the disk $|z+i|<\sqrt{2}$, and hence represents an analytic function in this disk. The integral

$$\int_0^{\infty} e^{-(z+1)t}\,dt$$

converges uniformly in compact subsets of the half plane $\operatorname{Re} z > -1$, and represents an analytic function there. In fact, both functions have the same values $1/(1+z)$ in their respective domains, and it is natural to think of their being the same function, even though they are defined differently in different regions.

This line of thought leads to the following definition. If f_1 is analytic in a region D_1 and f_2 is analytic in D_2, if D_1 and D_2 overlap in a nonempty set (necessarily open), and if $f_1(z)=f_2(z)$ for z in the intersection $D_1 \cap D_2$, we say that f_2 is a **direct analytic continuation** (or extension) of f_1 from D_1 to $D_1 \cup D_2$, or equally that f_1 is a direct analytic continuation of f_2. What makes this a viable definition is that if we make the extension in two ways, the result is the same; that is, if f_1 is analytic in D_1, and f_2 and f_3 are direct analytic continuations of f_1 to D_2, then $f_2(z)=f_3(z)$ at all points of D_2. This is a consequence of the coincidence principle (Sec. **7G**), since $f_2(z)=f_3(z)$ throughout the open set $D_1 \cap D_2$.

If we continue f_1 analytically from D_1 to D_2, we obtain a function that is analytic in $D_1 \cup D_2$. This function might have a direct analytic continuation to a region D_3, and so on. An analytic continuation of f_1 (without qualification) means the result of a finite number of direct analytic continuations. It can perfectly well happen that analytic continuations of the same function lead eventually to different values in some region; in particular, we could return to the original region with different values. The result of performing all possible analytic continuations of a function is known as a **complete analytic function**; since it might have different values at the same point, it is not a function in the strict sense. We shall return to this topic in connection with functions like $\sqrt{z}$ and $\log z$ in Sec. **29**.

17B. Continuation by power series

We showed in Sec. **7F** that if f is analytic in a disk Δ: $|z - z_0| < R$, and we expand f about the point z_1 in Δ as a Taylor series $\sum_{n=0}^{\infty} c_n(z - z_1)^n$, then the new series will converge in the largest open disk, centered at z_1, that is a subset of Δ. Nothing says that the new series might not converge in a still larger disk; if it does, it will provide an analytic continuation of f outside Δ.

For example, consider

$$f(z) = (1 - z)^{-1} = \sum_{n=0}^{\infty} z^n, \qquad |z| < 1.$$

If we expand f about the point $z = -\frac{1}{2}$, we obtain

$$\sum_{k=0}^{\infty} \left(z + \frac{1}{2}\right)^k \left(\frac{2}{3}\right)^{k+1},$$

with radius of convergence $\frac{3}{2}$. We have thus extended f to a domain that actually contains the original domain as a subset. Since f is really analytic in the whole plane except the point $z = 1$, all continuations obtained by repeated expansions in power series will represent the same function.

On the other hand, if we were to start from

$$(1 - z)^{1/2} = \sum_{n=0}^{\infty} \binom{\frac{1}{2}}{n} (-z)^n$$

in $|z| < 1$, and continue through a chain of disks so as to return to points in the original disk, we could end up with a different branch of the square root, as indicated in Fig. 17.1 (p. 140).

Continuation through a chain of disks is a basic method of analytic continuation when the continuation is possible at all; in Sec. **17F** we shall see examples of power series for which no continuation beyond the original disk is possible.

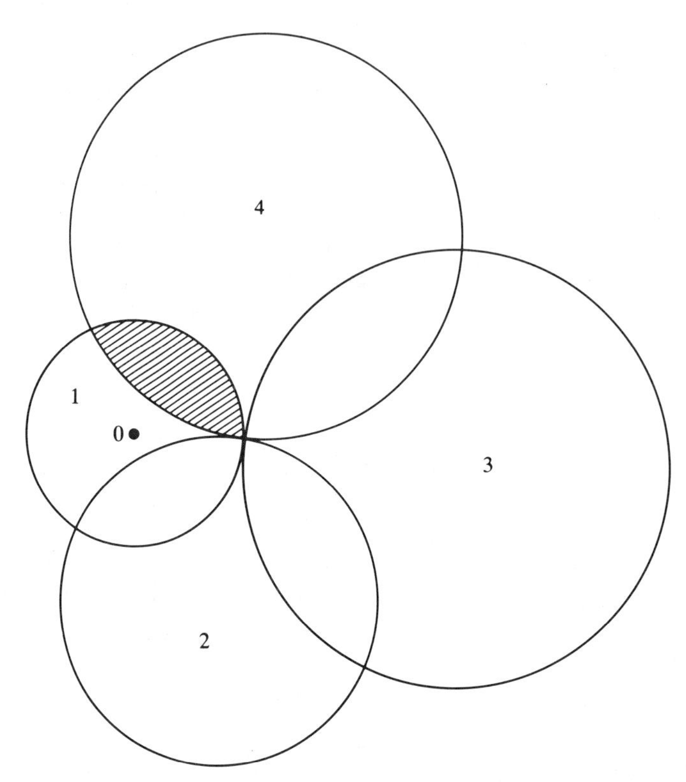

FIGURE 17.1

17C. Continuation by reflection

The following theorem is the **Schwarz reflection principle**. It deals with a special, but very useful, situation, and has the advantage of producing an analytic continuation explicitly.

Let f be analytic in a region D in the upper half plane; let the boundary ∂D of D intersect the real axis in a single segment L; let f be continuous on $D \cup L$; and let f take only real values on L. If D^* is the reflection of D with respect to the real axis (see Fig. 17.2), then f can be continued analytically across L into D^*, by taking $f(z) = \overline{f(\bar{z})}$ (z in D^*).

The theorem can be generalized to situations where D is reflected across a different line, or across an arc of a circle; and the continuity condition can be weakened.[1]

The Schwarz principle follows at once from a more general result (which again is a special case of something more general):

With D and L as in the statement of the Schwarz principle, suppose that f_1 is analytic in D and continuous in $D \cup L$; suppose also that f_2 is analytic in

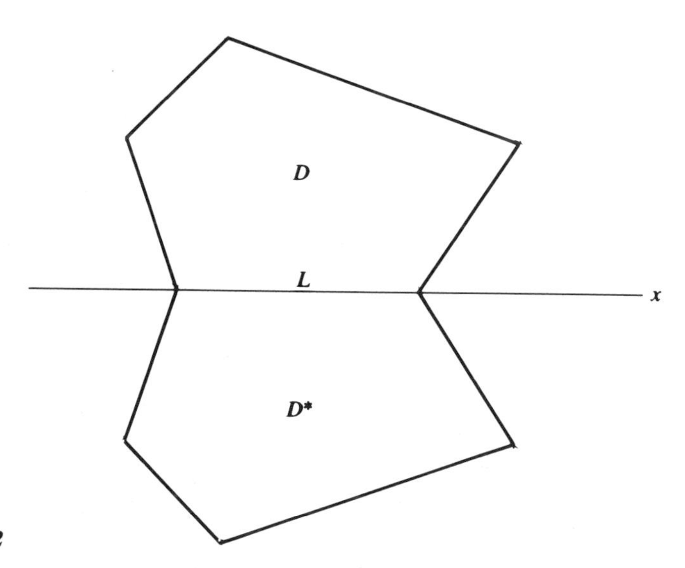

FIGURE 17.2

D^* and continuous in $D^* \cup L$. Finally, suppose that $f_1(z) = f_2(z)$ for all $z \in L$. Then f_2 is an analytic continuation of f_1 across L.

The Schwarz principle is the case when $f_2(z) = \overline{f_1(\bar{z})}$.

The preceding proposition would be easy to prove if we had the stronger form of Cauchy's theorem that does not require that f is analytic on C, but only that f is analytic inside C and continuous in the closed region bounded by C. This not easy to prove in general,[2] but the special case that we need here is rather simple. Since all that we need to know is that f can be continued across the real axis, it is enough to consider a disk Δ, centered on L, with its upper half in D, and show that f can be continued across the diameter L_1 of Δ. We then have the situation shown in Fig. 17.3 (p. 142), where A is a line segment parallel to L and so close to L that $\int_A f(z)\,dz$ and $\int_{L_1} f(z)\,dz$ are close to each other, which we can arrange since f is uniformly continuous in $D \cup L$. If C is the upper half of $\partial\Delta$ (the boundary of Δ) and C_1 is the part of $\partial\Delta$ above A, we have $\int_{A \cup C_1} f(z)\,dz = 0$ by Cauchy's theorem, and we can make

$$\left| \int_A f(z)\,dz - \int_{L_1} f(z)\,dz \right|$$

arbitrarily small. Hence $\int_{L_1 \cup \partial\Delta} f(z)\,dz = 0$.

Now let C^* be the lower half of $\partial\Delta$ and form

$$F(z) = \frac{1}{2\pi i} \int_{C \cup C^*} \frac{f(w)}{w - z}\,dw.$$

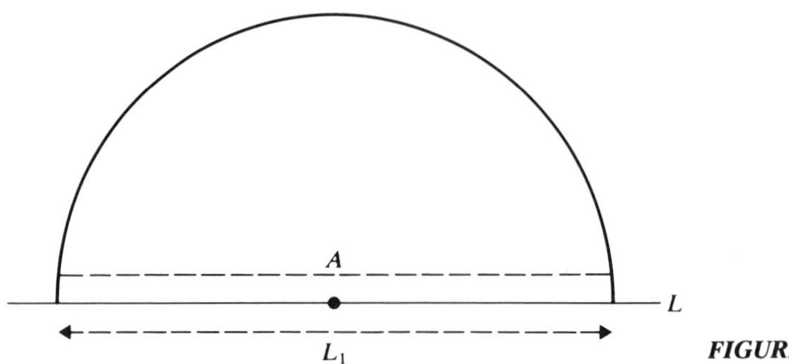

FIGURE 17.3

Then F is analytic inside Δ. Suppose that $z \in \Delta \cap D$. We can write

$$\begin{aligned} F(z) &= \frac{1}{2\pi i}\left[\int_C \frac{f(w)\,dw}{w-z} + \int_{L_1} \frac{f(w)\,dw}{w-z}\right] \\ &\quad + \left[\int_{C^*} \frac{f(w)\,dw}{w-z} - \int_{L_1} \frac{f(w)\,dw}{w-z}\right] \\ &= f_1(z) + f_2(z). \end{aligned}$$

We have $f_1(z) = f(z)$ and $f_2(z) = 0$ (because $z \in \Delta \cap D$). Similarly, $f_1(z) = 0$ when $z \in \Delta \cap D^*$, the reflection of $\Delta \cap D$ across L. Hence F provides an analytic continuation of f from $\Delta \cap D$ to $\Delta \cap D^*$. The same argument applies to any other disk constructed like Δ, and hence f has an analytic continuation to D^*.

A stronger version of Schwarz's principle will be given in Sec. **20F**.

*17D. Overconvergence

When a function is defined by a power series with a finite radius of convergence, it is sometimes possible to extend the function beyond the disk of convergence by grouping the terms of the series, that is, by considering only a subsequence of the partial sums. A power series for which this can be done is said to be **overconvergent**. We are going to construct an example of an overconvergent power series; actually it is more convenient to construct the grouped series first and then remove the parentheses to obtain the power series.

The key to the construction is a simple lemma that we shall need again later (Sec. **17F**); I present it as an exercise.

EXERCISE 17.1 Let $\{n_k\}$ be a sequence of positive integers such that $n_{k+1}/n_k \geq \lambda > 1$; let p be a positive integer, $p > 1/(\lambda - 1)$. Then the polynomial $P_k(z) = [z^p(1+z)]^{n_k}$, when expanded in powers of z, contains no power of z that appears in any other P_j $(j \neq k)$.

We define a function f by $f(z)=\sum_{k=1}^{\infty} a_k P_k(z)$, where $1/a_k$ is the coefficient with largest modulus in the binomial expansion of $(1-z)^{n_k}$. By Exercise 17.1, (with $-z$ instead of z), $P_k(z)$ contains no powers that appear in any other $P_j(z)$. Consequently, if we replace each $P_k(z)$ by its expansion in powers of z, we get a power series $S=\sum c_k z^k$ whose partial sums of order $(p+1)n_k$ are just the partial sums of the series $\sum P_k(z)$. Furthermore, S has an infinite number of terms z^m with coefficient 1, and none with coefficient of modulus greater than 1. Hence the radius of convergence of S is 1. We now show that $\sum P_k(z)$ converges in a set that extends outside the circle $|z|=1$; this will mean that a sequence of partial sums of S (namely, those of index n_k) converges in a set that contains points z with $|z|>1$; in other words, S is overconvergent. This will follow if there are points w outside the unit disk for which $|w^p(1-w)|<\sigma<1$, since then $\sum a_k P_k(w)$ will be dominated by $\sum |a_k|\, \sigma^{n_k}$ with $|a_k|\le 1$. The existence of such points is clear, since if $|1-w|<\varepsilon$ $(0<\varepsilon<1)$, we have $|w^p(1-w)|<|w|^p\, \varepsilon<\sigma$ if $|w|<(\sigma/\varepsilon)^{1/p}$, which allows $|w|=1+\delta$ with $\delta>0$ if $1>\sigma>\varepsilon$.

We conclude, therefore, that *our series S is overconvergent in a neighborhood of* 1 *provided that* $n_{k+1}/n_k>\lambda>1$ *and* $p>1/(\lambda-1)$. The series S has long gaps (runs of zero coefficients): The ratio of the lowest exponent following a gap to the highest exponent preceding that gap is $(n_{k+1}/n_k)p/(p+1)>1$.

Notice that the sequence of partial sums that converges outside the unit disk consists of the partial sums that end at a gap.

It can be shown that the existence of infinitely many gaps of the kind we have been considering is sufficient for overconvergence in a neighborhood of every point of the circle of convergence at which the sum of the series is analytic. (We shall see in Exercise 17.4 that it is not possible to have the sum of the series analytic at *every* point of the circle of convergence.) The existence of long gaps is also necessary for overconvergence, in the sense that an overconvergent series whose radius of convergence is 1 is always the sum of a series that converges in a larger disk and a series with long gaps.[3]

EXERCISE 17.2 Show that when $p=1$ and $\lambda>2$, the overconvergent series just constructed has a sequence of partial sums that converges at least in the union of the unit disk and the disk $|z-1|<1$ (see Fig. 17.4, p. 144).

17E. An obstacle to analytic continuation

We obviously cannot continue $1/(1-z)$ through the point 1, because of the pole at 1. What is less obvious is that no power series with positive coefficients can be directly continued through the real positive point on its circle of convergence. More precisely:

If $f(z)=\sum_{k=0}^{\infty} a_k z^k$ with $a_k\ge 0$ and radius of convergence R, then we cannot make a direct analytic continuation of f from a point on $(0, R)$ to any points of modulus greater than R.

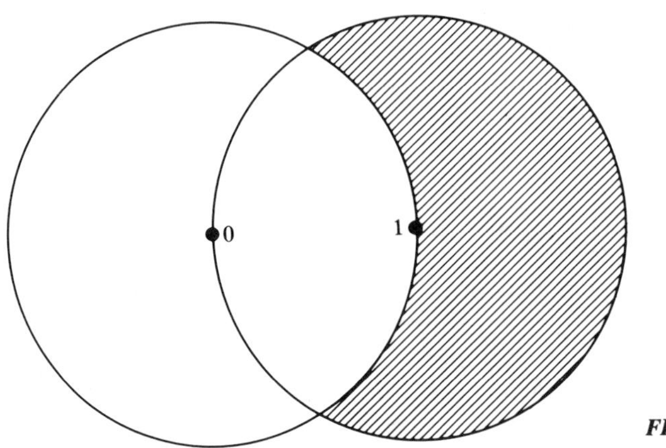

FIGURE 17.4

This is the **Vivanti-Pringsheim theorem**.

Thus when $a_k \geq 0$, the point R is necessarily a singular point of f, even if $\sum a_k R^k$ converges. (This point is not necessarily a singular point of the kind that we have been discussing up to now.) For a series $\sum_{k=0}^{\infty} a_k(z - z_0)^k$, the singular point will of course be at $z_0 + R$.

There are also functions that are analytic for $|z| < R$ but cannot be continued analytically outside this disk. Such a function is said to have a **natural boundary** on the circle $|z| = R$. We shall see an example after we have proved the Vivanti-Pringsheim theorem.

EXERCISE 17.3 Show that if $f(z) = \sum_{k=0}^{\infty} a_k z^k$ is analytic in $|z| < R$ and also at the point $z = R$, then it is always possible to make a direct analytic extension of f to some points outside $|z| \leq R$.

EXERCISE 17.4 Continuing Exercise 17.3, show that if f is analytic in $|z| \leq R$, this cannot be the disk of convergence of f.

The statement in this exercise is often expressed by saying that f *must have at least one singular point on its circle of convergence.*

EXERCISE 17.5 Show that if $\sum a_n z^n$ has radius of convergence 1, if $a_n \geq 0$, and if $\sum a_n$ diverges, then $f(x) \to \infty$ as $x \to 1$ along the radius $(0, 1)$.

The proof of the Vivanti-Pringsheim theorem is particularly easy when $\sum a_n$ diverges, as Exercise 17.5 shows. However, the simple example $f(z) = \sum_{n=1}^{\infty} z^n/[n(n+1)]$ shows that there can be a singular point at $z = 1$ even when $\sum |a_n|$ converges.

To prove the Vivanti-Pringsheim theorem in its general form, we suppose that it is false and deduce a contradiction. Suppose, then, that $f(z) = \sum_{n=1}^{\infty} a_n z^n$ has $a_n \geq 0$ and radius of convergence 1, and a direct analytic continuation past the point 1. Then the extended function would be analytic in some disk D centered at 1, and a disk centered at $\frac{1}{2}$, with radius r just greater than $\frac{1}{2}$, would be contained in the union of $\{|z| < 1\}$ and D (see Fig. 17.5). Then we would

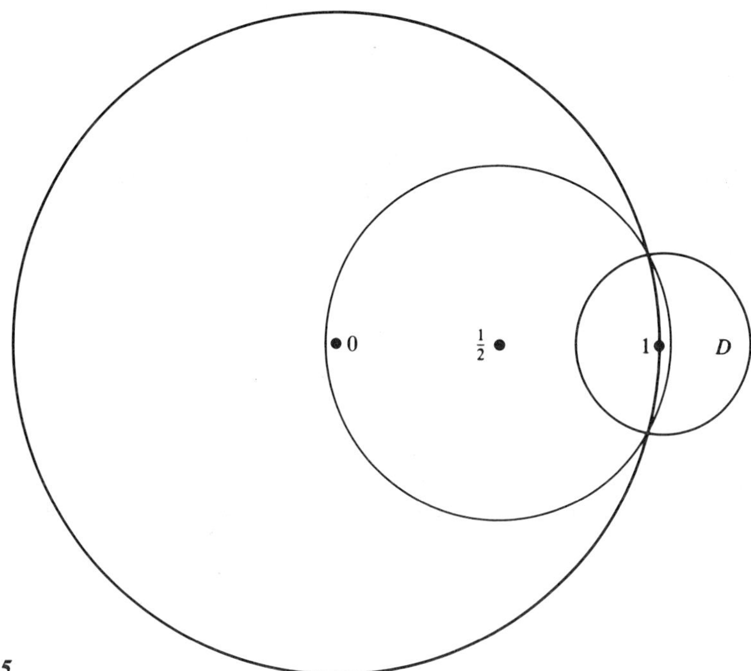

FIGURE 17.5

have

$$f^{(n)}\left(\frac{1}{2}\right)=\sum_{k=n}^{\infty}\frac{k!}{(k-n)!}a_k\left(\frac{1}{2}\right)^{k-n},$$

and the Taylor series of f about $\frac{1}{2}$ for a real $x>1$ would be

$$\begin{aligned} f(x)&=\sum_{n=0}^{\infty}\frac{f^{(n)}(\frac{1}{2})}{n!}\left(x-\frac{1}{2}\right)^n\\ &=\sum_{n=0}^{\infty}\frac{1}{n!}\left(x-\frac{1}{2}\right)^n\sum_{k=n}^{\infty}\frac{k!}{(k-n)!}a_k\left(\frac{1}{2}\right)^{k-n}. \end{aligned}$$

Here everything is positive and therefore we can sum in the opposite order to obtain

$$f(x)=\sum_{k=0}^{\infty}a_k\sum_{n=0}^{k}\frac{1}{n!}\left(x-\frac{1}{2}\right)^n\frac{k!}{(k-n)!}\left(\frac{1}{2}\right)^{k-n}.$$

The inner series is the binomial series for

$$\left[\left(x-\frac{1}{2}\right)+\frac{1}{2}\right]^k=x^k,$$

so we would have

$$f(x)=\sum_{k=0}^{\infty} a_k x^k, \qquad x>1.$$

But this is the original power series for f, now with $x>1$, although we assumed to begin with that f was not analytic in any disk of radius greater than 1, centered at 0 (remember that convergence for some $x>1$ implies convergence for all z with $|z|<x$.) This contradiction shows that f cannot be continued directly across $x=1$. [The example $f(z)=1/(1-z)$ shows that it may well happen that f can nevertheless be continued indirectly to points on the ray $x>1$.]

The argument is unaffected if we assume that all the coefficients, except for a finite number, are positive.

EXERCISE 17.6 Show that if $\sum a_n z^n$ and $\sum (\operatorname{Re} a_n) z^n$ have the same radius of convergence R, then the Vivanti-Pringsheim theorem continues to hold if we assume only that $\operatorname{Re} a_n \geq 0$. (In other words, instead of assuming that $a_n \geq 0$, it is enough to assume that $\operatorname{Re} a_n \geq 0$ unless $\sum (\operatorname{Re} a_n) z^n$ has a larger radius of convergence than $\sum a_n z^n$.)

*17F. Nowhere-continuable power series

We have just seen that the sum of a power series with positive coefficients cannot be continued directly past the real positive point on the circle of convergence. Now we shall be concerned with conditions that make it impossible to continue the sum of a power series beyond the disk of convergence in any direction at all, so that the circle of convergence is a natural boundary. Let us first verify that such functions do exist.

EXERCISE 17.7 Show that $f(z)=\sum_{n=1}^{\infty} z^{n!}$ cannot be continued outside the unit disk. (The Vivanti-Pringsheim theorem shows that f cannot be continued past the point 1. Now consider $f(ze^{i\pi p/q})$, where p and q are integers.)

A more general result is **Hadamard's gap theorem**, which says that $f(z)=\sum a_n z^n$, *with a finite radius of convergence, cannot be continued beyond its circle of convergence if* $a_n=0$ *except for* $n=n_k$, *where* $n_{k+1}/n_k \geq \lambda > 1$. We shall give a proof below.

First, however, let us notice that if we had proved the theorem that the existence of long gaps is sufficient for overconvergence in a neighborhood of every point on the circle of convergence at which the sum of the series is analytic, then Hadamard's theorem would follow at once.

EXERCISE 17.8 Prove this.

Hadamard's theorem is, however, not the last word on noncontinuable power series. A much deeper result is **Fabry's gap theorem**, which says that the much weaker gap condition $n_k/k \to \infty$ suffices to make $\sum a_k z^{n_k}$ noncontinuable.[4]

To prove Hadamard's theorem, we may suppose that the radius of convergence is 1. It is enough to show that f cannot be analytic at $z = 1$, since $f(ze^{i\alpha})$ has the same gaps in its power series as f does. Suppose then that f is analytic at $z = 1$; we try to obtain a contradiction. Consider the function $g(z) = f\{[\frac{1}{2}z^p(1+z)]^{n_k}\}$ with $p > \lambda/(\lambda - 1)$. Exercise 17.1 shows that each $[\frac{1}{2}z^p(1+z)]^{n_k}$ expands into a sum of powers of z that do not occur in the expansions of the corresponding sum for any other value of n_k. Hence the series by which we defined g is just the Maclaurin series of g with its terms grouped; that is, its partial sums are a subset of the partial sums of the Maclaurin series of g. Consequently, g is analytic wherever the Maclaurin series of g converges.

Now, since $\sum a_k z^{n_k}$ has radius of convergence 1, the set S where g is analytic contains at least the set where $|\frac{1}{2}z^p(1+z)| < 1$. This set contains at least the intersection of the sets where $|z| < 1$ and where $|z+1| < 2$. The second of these sets is a disk of radius 2 centered at -1, and contains all the points of the disk $|z| \leq 1$, except for $z = 1$. But since we assumed that f is analytic at 1, the point 1 also belongs to S, because $\frac{1}{2}z^p(z+1) = 1$ at $z = 1$. Since S contains 1, it contains a disk of positive radius ε centered at 1, and hence a disk, centered at 0, of radius $1 + \delta > 1$ (see Fig. 17.6). This means that the Maclaurin series of g has radius of convergence greater than 1 and hence that $\sum a_k[\frac{1}{2}r^p(r+1)]^{n_k}$

FIGURE 17.6

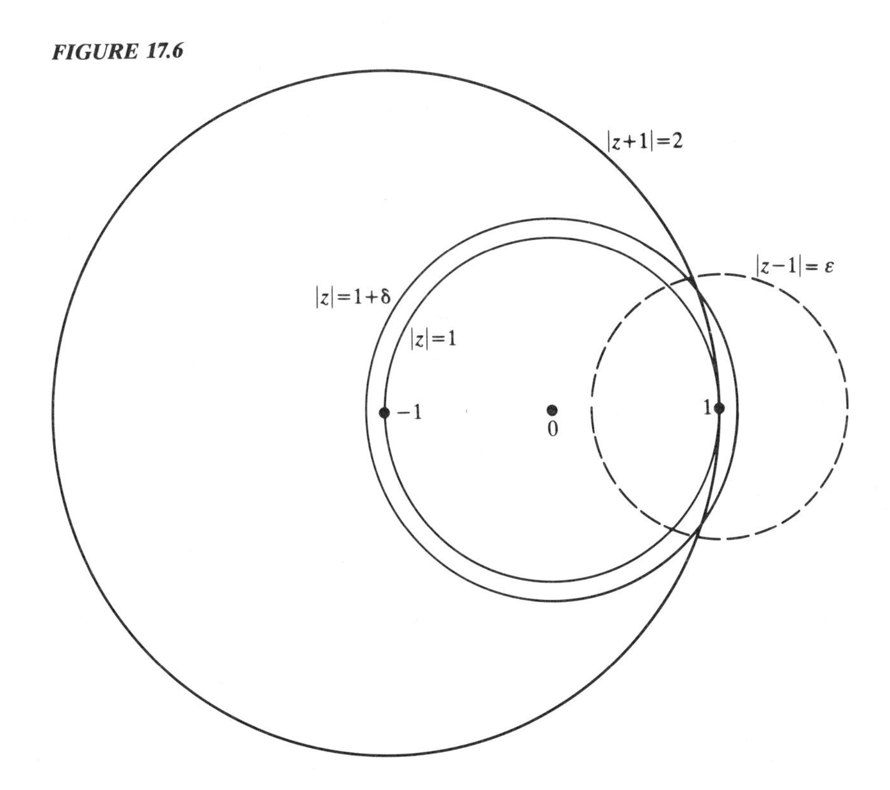

converges for some $r>1$. However, $\frac{1}{2}r^p(r+1)>1$ when $r>1$, and so $\sum a_k w^{n_k}$ converges for some $w>1$, contradicting the assumption that this series had radius of convergence 1. Hence Hadamard's theorem is established.

Supplementary exercises

1. Verify that $1/(1-z)$ can be continued outside the unit disk by expanding it in a Taylor series about $z=-h$, $0<h<1$.
2. Does the expansion of $f(z)=1/(1-z)$ in powers of $z+1$ provide an analytic continuation of f?
3. Let s_n be the partial sums of $\sum_{k=0}^{\infty} a_k$, where $\sum_{k=0}^{\infty} a_k z^k$ has radius of convergence 1. Suppose that it happens that $s_n=0$ except when n is a power of 2. Show that, in this case, $\sum_{k=0}^{\infty} a_k z^k$ cannot be continued outside the unit disk.
4. Let $f(z)=(e^{z-1}-1)^{-1/2}=\sum_{n=0}^{\infty} a_n z^n$, $|z|<1$. Does $\sum a_n$ converge?
5. Find the fallacy in the following argument. Let f be analytic in a closed disk bounded by a circle C. According to Cauchy's theorem,

$$f(z)=\frac{1}{2\pi i}\int_C \frac{f(w)\,dw}{w-z}.$$

We can write

$$f(z)=\frac{1}{2\pi i}\int_{C_1} \frac{f(w)\,dw}{w-z}+\frac{1}{2\pi i}\int_{C_2} \frac{f(w)\,dw}{w-z},$$

where C_1 and C_2 are the upper and lower halves of the circle. By the arguments of Sec. **7E**, f_1 is analytic in the whole plane except on C_1, and f_2 is analytic except on C_2. But $f_1(z)+f_2(z)=0$ outside C by Sec. **7B**, and $f_1(z)+f_2(z)=f(z)$ inside C. Hence $f(z)\equiv 0$.

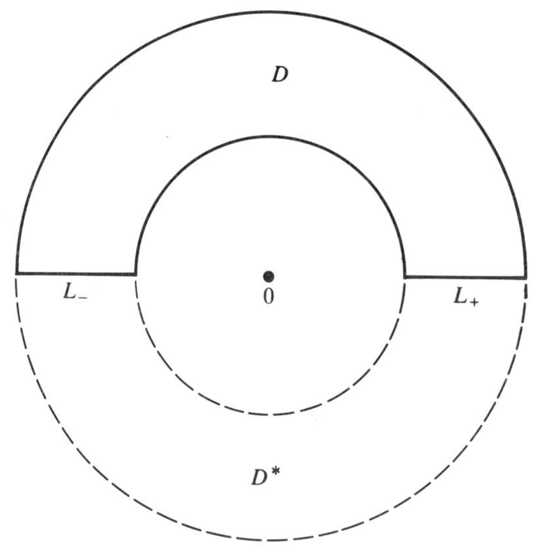

FIGURE 17N, for Note 1

Notes

1. For stronger versions of the reflection principle, see Hille, vol. 1, pp. 184–187.
   Notice that the continuation by Schwarz's principle does not necessarily lead to a function that is analytic in the region formed by D, D^* and the intervals of the real axis between them. For example, if D is as in Fig. 17N and $f(z)=z^{1/2}$ (principal square root), then f is analytic in D and real on L_+ but its continuation across L_+ does not provide a function that is analytic in the union of D, D^*, L_+, and L_-, even though f can be continued analytically across L_-.
2. See Sec. **6C** and note 3, Sec. **6**.
3. Overconvergence was discovered by M. B. Porter in 1906; the converse theorem was proved by A. Ostrowski considerably later. For details, see Dienes, p. 370.
4. For more details on gap theorems, see Dienes, chaps. 10, 11, and 14; Pólya; and Levinson.

18 *Power Series on the Circle of Convergence*

18A. Abel's theorem

It is natural to suppose that since $\sum_{n=1}^{\infty}(-1)^{n-1}/n$ converges, the sum of the series should be the value of $\sum_{n=1}^{\infty}(-1)^{n-1}z^n/n=\log(1+z)$ at $z=1$, that is, $\ln 2$. However, nothing up to now has said that we can use a Taylor series *on* its circle of convergence, even when it happens to converge at a point of this circle. Nevertheless, intuition is correct in this case. More generally, we have the following theorem:

ABEL'S THEOREM *If $\sum_{n=0}^{\infty}a_nz^n$ converges to $f(z)$ for $|z|<R$, and if $\sum_{n=0}^{\infty}a_nR^n$ converges to sum A, then $\lim_{x\to R-}f(x)=A$. The notation $x\to R-$ means that the limit is taken along the radius, from inside the disk $|z|<R$.*

It follows from this that if $\sum_{n=0}^{\infty}a_nR^ne^{in\theta}=A$, then $f(z)\to A$ along the radius $z=re^{i\theta}$ as $r\to R-$. Of course there is a corresponding result if the series is centered at z_0 instead of at 0.

Notice that the converse of Abel's theorem cannot be true. For example, if $f(z)=(1+z)^{-1}=\sum_{n=0}^{\infty}(-1)^nz^n$ for $|z|<1$, then $\lim_{x\to 1-}f(x)$ exists but $\sum(-1)^n$ diverges.

Abel's theorem is true for somewhat more general paths than radii, but not for arbitrary paths that end at $z=R$; illustrations of these phenomena are rather complicated, and will not be considered here.[1]

In proving Abel's theorem, we may suppose that $R=1$ [otherwise consider $f(Rz)$]. It is also convenient to assume that $\sum_{n=0}^{\infty}a_n=0$; this is legitimate because we can consider $f(z)-\sum_{n=0}^{\infty}a_n$ instead of $f(z)$. Now write $s_n=a_0+a_1+\cdots+a_n$, with s_{-1} defined to be 0, so that we have $s_n-s_{n-1}=a_n$. For

$|x|<1$, we have

$$f(x) = s_0 + (s_1 - s_0)x + (s_2 - s_1)x^2 + \cdots$$
$$= \lim_{N\to\infty} [s_0 + (s_1 - s_0)x + \cdots + (s_N - s_{N-1})x^N].$$

The sum in brackets can be rearranged into

$$s_0(1-x) + s_1(x - x^2) + \cdots + s_{N-1}(x^{N-1} - x^N) + s_N x^N.$$

Since $s_N x^N \to 0$, we now have

$$f(x) = (1-x)\sum_{k=0}^{\infty} s_k x^k.$$

It may now look plausible that $f(x)\to 0$ as $x\to 1-$, but it is not immediately obvious that the sum of the series doesn't overweigh the factor $(1-x)$. In such circumstances we can use the "divide-and-rule" principle: We try to break the sum into two parts and show that each part, after being multiplied by $1-x$, approaches zero for a different reason. To do this, we choose an arbitrary (small) positive number ε and write

$$f(x) = (1-x)\sum_{k=1}^{M} s_k x^k + (1-x)\sum_{k=M+1}^{\infty} s_k x^k = S_1 + S_2,$$

where M is chosen so that $|s_k| < \varepsilon$ when $k > M$ (possible because $s_k \to 0$). Then

$$|S_2| \le (1-x)\varepsilon \sum_{M+1}^{\infty} x^k \le (1-x)\varepsilon \sum_{k=0}^{\infty} x^k = \varepsilon.$$

Now, with M fixed, $S_1 \to 0$ as $x\to 1$, because S_1 is a sum of finitely many bounded terms, all multiplied by $1-x$. Consequently, $|S_1| < \varepsilon$ when x is close enough to 1, so that we have $|f(x)| \le 2\varepsilon$ for x sufficiently close to 1, which is to say that $f(x)\to 0$.

The process that we used in transforming the series for $f(x)$ is known as **partial summation**. It is a discrete analog of integration by parts, and can be used in many problems.

EXERCISE 18.1 Prove **Abel's convergence theorem:**[2] *If a_n decreases to 0 and $\sum b_n$ has bounded partial sums B_n, then $\sum a_n b_n$ converges.* [If $b_n = (-1)^n$, we have the familiar alternating series test for convergence.]

The same arguments apply when the b_n are not necessarily real, as long as the partial sums B_n are bounded. For example, consider the series $\sum_1^\infty n^{-1} i^n$, and take $a_n = 1/n$, $b_n = i^n$. Then B_n is either 1, $1+i$, i, or 0, so $B_n = O(1)$ and hence the given series converges.

EXERCISE 18.2 Prove the theorem mentioned in Sec. **4C**, that if series A and B converge, C is their formal product, and C converges, then the sum of C is the product of the sums of A and B.

EXERCISE 18.3 Using Exercise 4.17, show that when a_n and b_n decrease to 0, the series $\sum_{n=1}^{\infty} a_n \cos nx$ and $\sum_{n=1}^{\infty} b_n \sin nx$ converge for $0 < x \leq \pi$ and $0 \leq x \leq \pi$, respectively.

EXERCISE 18.4 If we weaken the hypothesis in Abel's theorem, we get a weaker conclusion. Show that if $f(z) = \sum_{n=0}^{\infty} a_n z^n$ converges for $|z| < R$ and $\sum a_n R^n$ has bounded partial sums, then $f(x)$ is bounded on $(0, R)$.

EXERCISE 18.5 If $f(z) = \sum_{n=0}^{\infty} a_n z^n$ for $|z| < R$ and $\sum a_n R^n = +\infty$, then $f(x) \to \infty$ as $x \to R-$. (This is a sharper form of Exercise 17.5.)

Supplementary exercises

1. Use Abel's theorem to show that

$$\sum_{n=1}^{\infty} \frac{(-1)^n}{n} = -\log 2.$$

2. Let $\binom{n}{k}$ denote, as usual, the coefficient of x^k in the binomial expansion of $(1+x)^n$. Appeal to Abel's theorem to show (without any "test for convergence") that

$$\sum_{n=0}^{\infty} (-1)^k \binom{-\frac{1}{2}}{k}$$

diverges.

3. Show that

$$\frac{1}{\sqrt{1}} + \frac{1}{\sqrt{2}} - \frac{1}{\sqrt{3}} - \frac{1}{\sqrt{4}} + \frac{1}{\sqrt{5}} + \frac{1}{\sqrt{6}} - \cdots$$

converges.

4. If a_n decreases to zero, for which real values of λ does $\sum a_n \cos(nx + \lambda)$ converge?

*18B. Summation of divergent series

Abel's theorem can be interpreted as saying that if a series is convergent, its sum can be obtained by inserting factors x^n and then letting $x \to 1$, instead of by taking the limit of the partial sums. Aside from its occasional use for summing convergent series such as $\sum (-1)^n/n$, Abel's theorem is also interesting because it suggests a method for attaching sums to some *divergent* series $\sum a_n$. For example, $\sum (-1)^n$ diverges, but $\sum_{n=0}^{\infty} x^n(-1)^n = 1/(1+x) \to \frac{1}{2}$ as $x \to 1$ along the radius $0 < x < 1$ of the unit disk. We say that $\sum_{n=0}^{\infty} (-1)^n$ is **summable** to $\frac{1}{2}$ by Abel's method, or is **Abel summable**. This method will sum many other divergent series.

EXERCISE 18.6 Sum $\sum_{n=1}^{\infty} n(-1)^n$ by Abel's method.

However, Abel's method will not sum a power series outside its disk of convergence.

EXERCISE 18.7 Prove the preceding statement.

Any systematic method of attaching a sum to a divergent series is called a **summation method**.[3] (Usually we want a summation method also to attach the usual sum to a convergent series.) Perhaps the simplest method that will sum some divergent series is the Y method,[4] which takes the limit of running averages of consecutive partial sums, that is, $\lim_{n\to\infty}(s_n + s_{n+1})/2$.

EXERCISE 18.8 Sum $\sum_{n=0}^{\infty}(-1)^n$ by the Y method.

A more frequently used method is the $(C, 1)$ method (C stands for Cesàro), in which we average the first n partial sums; that is, we form $n^{-1}(s_1 + s_2 + \cdots + s_n)$ and hope that this approaches a limit as $n \to \infty$. It does approach a limit whenever $\sum a_n$ converges. This proposition follows at once from the following exercise, which is sometimes called **Kronecker's lemma**.

EXERCISE 18.9 If $\{s_n\}$ is a sequence of numbers and $s_n \to L$ as $n \to \infty$, then $n^{-1}(s_1 + s_2 + \cdots + s_n) \to L$.

Expressed geometrically, Exercise 18.9 says that the centroid of the first n elements of a convergent sequence has the same limit as the sequence.

It can be shown along the lines of the proof of Abel's theorem that every $(C, 1)$ summable series is Abel summable to the same sum. Consequently, the Abel method is more powerful than the $(C, 1)$ method, but it may be harder to apply. In some sense the $(C, 1)$ method is more like convergence than the Abel method is, so it may seem more satisfying as a generalization of convergence. If the notation $(C, 1)$ makes you suspect that there are also (C, k) methods, you are right; but we shall not discuss them here.

EXERCISE 18.10 Show that the $(C, 1)$ method sums $\sum_{n=1}^{\infty}(-1)^{n+1}$ but will not sum $\sum_{n=1}^{\infty}(-1)^{n+1}n$.

Although the Abel method will not sum a power series outside its (closed) disk of convergence (and not always on its circle of convergence), there are more complicated methods that will sum a power series if its sum can be continued analytically outside the disk. One method that will do this was suggested by Borel. To sum $\sum a_n$ by this method, form

$$\int_0^{\infty} e^{-u} \sum_{n=0}^{\infty} \frac{a_n u^n}{n!}\, du.$$

If the series in the integrand converges for all positive u and the integral is finite, then the value of the integral is the **Borel sum** of the series. The motivation of this definition is the observation that if we formally change the order of summation and integration, we will get precisely $\sum_{n=0}^{\infty} a_n$. Let us apply Borel's method to $\sum_{n=0}^{\infty}(-1)^n z^n$. First we have to form

$$\sum_{n=0}^{\infty} \frac{(-1)^n z^n u^n}{n!},$$

which we fortunately can sum explicitly to e^{-uz}. Then Borel's integral is

$$\int_0^\infty e^{-u(1+z)}\,du,$$

which converges and equals $1/(1+z)$ when $\operatorname{Re} z > -1$. Hence the Borel method sums $\sum_{n=0}^\infty (-1)^n z^n$ to $1/(1+z)$ when $\operatorname{Re} z > -1$, that is, in a half plane that contains the open disk of convergence.

A more direct generalization of Abel's method is to take the sum of $\sum a_n$ to be $\lim_{\delta\to 0} \sum_{n=0}^\infty a_n e^{-\delta\phi(n)}$, where $\delta \to 0$ through positive values. When $\phi(n) = n$, this is Abel's method. When $\phi(n) = n \ln n$, it is **Lindelöf's method**, which is much more powerful; it can be shown (not easily) that it sums every power series in its **Mittag-Leffler star**, which is obtained by drawing rays from the center through every singular point of the sum of the series and of its analytic continuations, and removing from the plane the parts of the rays beyond the singular points. The Mittag-Leffler star of $1/(1+z)$ is the whole plane except for the ray from -1 to $-\infty$.

One would not expect to be able to sum a series that diverges to $+\infty$. Nevertheless, it has recently been shown[5] that there is a method which sums $\sum_{n=0}^\infty (-1)^n z^n$ all over the plane except at $z = -1$, and even sums the series to $+\infty$ at $z = -1$.

There are many other summation methods that are useful for special purposes. Their existence may make us wonder why so much emphasis is customarily placed on ordinary convergence of series. One possible explanation is that the algebra of convergent series (adding or deleting terms, rearranging, etc.) can fail to apply to summable series.

EXERCISE 18.11 Inserting zero terms into a convergent series does not change its sum. Show, however, that the series

$$1-1+0+1-1+0+1-1+0+\dots$$

is $(C, 1)$ summable, but that its sum is not $\frac{1}{2}$.

Another reason for sometimes preferring ordinary convergence is that when a summation method is applied to a convergent series, it may initially give a poorer approximation to the sum than the partial sums of the series do.

18C. Tauberian theorems

We have noticed that the converse of Abel's theorem is not true; that is, if $f(z) = \sum_{n=0}^\infty a_n z^n$ and $f(x) \to L$ as $x \to 1$, then it does not necessarily follow that $\sum_{n=0}^\infty a_n = L$. There are, however, valid *conditional* converses, in which we impose additional hypotheses on $\{a_n\}$ that, together with $\lim_{x\to 1-} f(x) = L$, will imply that $\sum_{n=0}^\infty a_n = L$. Theorems of this kind are useful whenever we are more interested in finding the sum of a series of numbers than in showing that a function approaches a limit.

As a first example of such a conditional converse, we prove that *if* $f(x)=\sum_{n=0}^{\infty}a_nx^n$ for $|x|<1$, *if* $\lim_{x\to 1-}f(x)=L$, *and if* $a_n\geq 0$, *then* $\sum_{n=0}^{\infty}a_n=L$. This is an interesting counterpart to the Vivanti-Pringsheim theorem.

Since $a_n\geq 0$, we have

$$\sum_{k=0}^{n}a_kx^k\leq\sum_{k=0}^{\infty}a_kx^k\leq L,\qquad 0\leq x<1.$$

Hence $s_n=\sum_{k=0}^{n}a_k\leq L$, and therefore $\limsup_{n\to\infty}s_n\leq L$. On the other hand, $s_n\geq\sum_{k=0}^{n}a_kx^k=f(x)-\sum_{k=n+1}^{\infty}a_kx^k=f(x)-R_n(x)$. Choose x so that $f(x)>L-\varepsilon$; then take n so large that $R_n(x)<\varepsilon$ (possible because $\sum_{k=0}^{\infty}a_kx^k$ converges for this x). It follows that $s_n\geq L-2\varepsilon$, and since ε is arbitrary, $\liminf_{n\to\infty}s_n\geq L$. Hence $s_n\to L$.

If we take $L=\infty$, this theorem becomes Exercise 17.5.

A more difficult theorem imposes the condition $a_n=o(1/n)$ instead of $a_n\geq 0$.

TAUBER'S THEOREM *If* $f(x)=\sum_{n=0}^{\infty}a_nx^n$ *converges for* $|x|<1$ *and* $f(x)\to L$ *as* $x\to 1-$, *and* $na_n\to 0$, *then* $\sum_{n=0}^{\infty}a_n=L$.

Theorems of this kind, that is, conditional converses to limit theorems, have come to be called **Tauberian**. The theorems without additional hypotheses are called **Abelian**; Abel's theorem is a typical Abelian theorem; Tauber's theorem was the first example of a Tauberian theorem, or at least the first one to attract attention. The extra hypothesis in Tauber's theorem is more restrictive than is essential; it is actually enough to have na_n bounded. This is **Littlewood's theorem**;[6] since it is a considerably more difficult result, we shall not prove it here.

To prove Tauber's theorem, we again assume, as we may, that $L=0$, so that we have to show that $s_n\to 0$. Since $na_n\to 0$, Exercise 18.9, with s_n (there) $=n\,|a_n|$ (here), shows that if ε is an arbitrary (small) positive number, we can choose N so large that when $n>N$ we have $|na_n|<\varepsilon$ and also $N^{-1}\sum_{k=1}^{N}k\,|a_k|<\varepsilon$. Now write

$$\left|f(x)-\sum_{n=0}^{N}a_n\right|=\left|\sum_{n=0}^{\infty}a_nx^n-\sum_{n=0}^{N}a_n\right|=\left|\sum_{n=N+1}^{\infty}a_nx^n-\sum_{n=0}^{N}a_n(1-x^n)\right|$$

$$\leq\left|\sum_{n=N+1}^{\infty}a_nx^n\right|+\left|\sum_{n=0}^{N}a_n(1-x^n)\right|=S_1+S_2,$$

and choose x (near 1) so that $(N+1)^{-1}<1-x\leq N^{-1}$. We then have

$$S_1=\left|\sum_{n=N+1}^{\infty}na_n\frac{x^n}{n}\right|\leq\frac{\varepsilon}{N+1}\sum_{n=N+1}^{\infty}x^n\leq\frac{\varepsilon}{N+1}\cdot\frac{1}{1-x}<\frac{\varepsilon}{N+1}(N+1)=\varepsilon,$$

and

$$S_2 \le (1-x) \sum_{n=0}^{N} |a_n| (1 + x + \cdots + x^{n-1})$$

$$\le (1-x) \sum_{n=1}^{N} n\, |a_n| \le N(1-x) \frac{1}{N} \sum_{n=1}^{N} n\, |a_n| \le \varepsilon N(1-x) \le \varepsilon.$$

Thus

$$\left| f(x) - \sum_{n=0}^{N} a_n \right| \le 2\varepsilon,$$

and $f(x) \to 0$ by hypothesis as $x \to 1-$; therefore $\sum_{n=0}^{N} a_n \to 0$ as $N \to \infty$.

EXERCISE 18.12 With a weaker hypothesis than in Tauber's theorem, we can get a weaker conclusion: Show that if $a_n = O(1/n)$ and $f(x)$ is bounded as $x \to 1-$, then $\sum a_n$ has bounded partial sums.

EXERCISE 18.13 Show that if $a_n = O(1/n)$ and $f(x) \to +\infty$ as $x \to 1-$, then $\sum_{n=0}^{\infty} a_n = +\infty$.

Another interesting Tauberian theorem, which will not be proved here, is **Fatou's theorem**,[7] that $\sum a_n$ *converges if* $a_n \to 0$ *and* $f(z) = \sum_{n=0}^{\infty} a_n z^n$ *can be continued analytically past* $z = 1$.

Notes

1. For more information about the subject of this section, see the books by Hardy, Pitt, and Postnikov.
2. This is one of the most useful tests for convergence; unfortunately, it is not mentioned in most calculus books.
3. See particularly Hardy's book.
4. The Y method was so named in Szász's book.
5. L. Tomm, "A Regular Summability Method Which Sums the Geometric Series to its Proper Value in the Whole Complex Plane," *Canadian Math. Bull. 26* (1983): 171–180.
6. Proofs are given in books on Tauberian theorems; for a proof by complex analysis, see W. B. Jurkat, "Uber die Umkehrung des Abelschen Stetigkeitssätzes mit funktionentheoretischen Methoden," *Math. Z. 67* (1957): 211–222.
7. For two different proofs, see Titchmarsh, p. 218: Dienes, p. 467.

CHAPTER 4

Harmonic Functions; Conformal Mapping

19 *Harmonic Functions*

19A. Definitions

To say that $u = u(x, y)$ is **harmonic** in a region D means that u satisfies Laplace's equation,

$$\frac{\partial^2 u}{\partial x^2} + \frac{\partial^2 u}{\partial y^2} = 0,$$

which it will be more convenient to write in subscript notation as $u_{11} + u_{22} = 0$; it is customary to consider only real-valued harmonic functions.

We shall eventually be able to show (Sec. **20E**) that if u is continuous in a region and has just enough differentiability so that it can and does satisfy Laplace's equation, then u has continuous partial derivatives of all orders. For the present, however, we adopt the working definition that a harmonic function has continuous first and second partial derivatives and satisfies Laplace's equation in a region.

The connection between harmonic functions and analytic functions is that, since analytic functions have continuous second derivatives (as we have known since Sec. **7C**), the real and imaginary parts u and v of an analytic function f satisfy Laplace's equation. This follows from differentiating the Cauchy–Riemann equations:

$$\frac{\partial u}{\partial x} = \frac{\partial v}{\partial y}, \qquad \frac{\partial^2 u}{\partial x^2} = \frac{\partial^2 v}{\partial x\, \partial y},$$

$$\frac{\partial u}{\partial y} = -\frac{\partial v}{\partial x}, \qquad \frac{\partial^2 u}{\partial y^2} = -\frac{\partial^2 v}{\partial y\, \partial x}.$$

Since the two mixed second derivatives are equal, u is harmonic. Similarly (or because v is the real part of $-if$), v is harmonic.

Since one always cherishes the hope that sufficient conditions will turn out to be necessary, we might hope that, conversely, every harmonic function is the real part of some analytic function. This is, however, not quite true. For example, $u(x, y) = \ln(x^2 + y^2)^{1/2}$ (with the positive square root) is harmonic in the annulus $1 < |z| < 2$, but there is no (single-valued) analytic function in this annulus whose real part is u. Indeed, u is the real part of $\log z$, which is not analytic in the annulus.

However, a function that is harmonic in a simply connected region D is indeed the real part of a function f that is analytic in D.

We can (and shall) prove this by actually constructing such an f. (Naturally, f is determined only up to an additive imaginary constant.) The imaginary part of f is called a **harmonic conjugate** of u ["a," not "the," because if c is any real number, $f(z) + ic$ has the same real part as f]. The word "conjugate" here is supposed to convey the meaning of "associated," as does the same word in "complex conjugates," although the kind of association is quite different in the two phrases.

Because harmonic functions are connected with analytic functions in the way just described, theorems that we have proved about real parts of analytic functions (for example, Exercises 16.2 and 16.4, or the analog of Liouville's theorem in Sec. **16C**) are really theorems about harmonic functions.

The term "harmonic" is not restricted to two dimensions, although we shall not be concerned with harmonic functions in space: A solution of Laplace's equation in any number of dimensions is called harmonic. Unfortunately, there is no theory of analytic functions in more than two dimensions that corresponds to harmonic functions in the same way as in two dimensions; consequently, the study of Laplace's equation, and of the physical problems that it models, is more difficult in higher dimensions. In one dimension, on the other hand, the harmonic functions are just the affine functions $ax + b$, about which there is little to say.

19B. Finding a harmonic conjugate

Suppose now that we have a harmonic function u in a simply connected region D. Since we are looking for a harmonic conjugate v such that $u + iv = f$ is analytic in D, we want to satisfy the Cauchy–Riemann equations $u_1 = v_2$ and $u_2 = -v_1$. The first equation suggests that we ought to be able to find v by integrating u_1 with respect to y, since v_2 $(= u_1)$ would be obtained by differentiating v with respect to y. This approach will work locally, but not necessarily all over D, because if we start from a given point (x_0, y_0) of D, we cannot necessarily reach all points of D by proceeding just in the y direction. We could construct v in a neighborhood of (x_0, y_0) in this way, and hence construct f locally and hope to proceed by analytic continuation; and, as we shall see, this approach works well for some specific functions. It is also true that we ought to be able to find v by integrating $-u_2$ with respect to x. Having progressed this far, it might occur to us to combine the two approaches and integrate $u_1 - u_2$ along an arc in D from (x_0, y_0) to (x, y), that is, to consider

$$v(x, y) = \int_{(x_0, y_0)}^{(x,y)} [u_1(s, t)\, dt - u_2(s, t)\, ds].$$

If this formula is to define a function v, then the integral must be independent of the path along which we integrate. Now Green's theorem shows that the integral is independent of the path (in a simply connected region) precisely if u satisfies Laplace's equation.

EXERCISE 19.1 Calculate the partial derivatives of v and so show that u and v are the real and imaginary parts of an analytic function in D.

19C. Formulas

Sometimes we are given a formula for a harmonic function, and we want to find a harmonic conjugate, or the associated analytic function, explicitly. If we have u and want v, it is often easier to do the integration locally and in stages. If we integrate u_1 "partially" with respect to y, we will get a function $f(x, y)$, which, after we add some function $\phi(x)$, will be equal to $v(x, y)$. Now differentiate with respect to x to get $v_1(x, y) = -u_2(x, y) + \phi'(x)$. We can now find $\phi(x)$ up to an additive constant, and hence obtain a formula for v. An example will clarify the procedure. Let $u(x, y) = x^2 - y^2$. Then $u_1 = 2x$, $\int u_1\, dy = 2xy + \phi(x) = v(x, y)$, $v_1(x, y) = 2y + \phi'(x) = -u_2(x, y) = 2y$. Hence $\phi'(x) = 0$ and ϕ is a constant, so $v(x, y) = 2xy + c$.

EXERCISE 19.2 Find conjugate harmonic functions of

(*a*) $x^3 - 3xy^2$, (*b*) $e^{-y} \cos x$, (c) $\log(x^2 + y^2)$, (*d*) $y/[(1-x)^2 + y^2]$.

Often we are less interested in finding v than in finding f explicitly in terms of z. In our example of $u = x^2 - y^2$, it is easy to guess that $f(z) = z^2 + \text{constant}$, but in more complicated cases it is not always easy to express $u + iv$ as $f(z)$ even when we have explicit formulas for u and v. There are several short-cut methods for doing this, and I list them below for reference. Caution: If you start from a u that is not harmonic and use the general procedure, you will be stopped automatically at some stage; but *short-cuts can produce spurious results if you do not start from a harmonic u.*

There are three special methods for finding $f(z)$. At first sight, they all look fishy; but they are really all right, as we shall show after describing and illustrating them. You may prefer to skip ahead and read the proofs first before trying out the methods.[1]

RULE A *If we know u, and it is harmonic in some disk that contains an interval I of the real axis, form*

$$f'(z) = u_1(z, 0) - iu_2(z, 0)$$

and integrate f' to get f in a neighborhood of I. Then extend f by analytic continuation.

Note that the indicated derivatives have to be computed *before* replacing (x, y) by $(z, 0)$.

RULE B *If u is harmonic in a neighborhood of z_0, then in this neighborhood we have, up to an additive imaginary constant,*

$$f(z) = 2u\left(\frac{z + \bar{z}_0}{2}, \frac{z - \bar{z}_0}{2i}\right) - u(x_0, y_0).$$

We may take $z_0 = 0$ if u is harmonic in a neighborhood of 0.

RULE C *If we know both u and v, and know that they are harmonic conjugates in a neighborhood of* 0, *then*

$$f(z) = u(z, 0) + iv(z, 0).$$

These rules look outrageous at first sight because we don't yet know that it makes any sense to replace x and y by complex numbers. However, let us begin by giving some illustrations, which will indicate that the rules are useful enough to justify some effort in proving that they really work.

Example 1 $u(x, y) = x^2 - y^2$. Here $u_1 = 2x$, $u_2 = -2y$, and Rule A gives

$$f'(z) = 2z, \qquad f(z) = z^2 + \text{constant}.$$

By Rule B, with $z_0 = 0$,

$$f(z) = 2u\left(\frac{z}{2}, \frac{z}{2i}\right) = 2\left(\frac{z^2}{4} + \frac{z^2}{4}\right) = z^2.$$

For Rule C we are supposed to know that $v(x, y) = 2xy$; then

$$f(z) = z^2.$$

Example 2 $u(x, y) = e^{-y} \sin x$.

By Rule A,

$$f'(z) = \cos z + i \sin z = e^{iz},$$
$$f(z) = -ie^{iz} + ic, \qquad c \text{ real}.$$

By Rule B, with $z_0 = 0$,

$$f(z) = 2\exp\left(\frac{-z}{2i}\right)\sin\left(\frac{z}{2}\right)$$
$$= \frac{1}{i}\exp\left(\frac{iz}{2}\right)(e^{iz/2} - e^{-iz/2})$$
$$= -ie^{iz} + i.$$

By Rule C, $f(z) = \sin z + i(-\cos z) = -ie^{iz}$, if $v = -e^{-y}\cos x$.

Example 3

$$u = \frac{x}{x^2 + y^2}, \qquad u_1 = \frac{y^2 - x^2}{(x^2 + y^2)^2}, \qquad u_2 = \frac{-2xy}{(x^2 + y^2)^2}.$$

By Rule A,

$$f'(z) = -\frac{z^2}{z^4} = -\frac{1}{z^2},$$
$$f(z) = \frac{1}{z}.$$

By Rule B,

$$f(z) = 2\frac{(z + \bar{z}_0)/2}{[(z + \bar{z}_0)/2]^2 + [(z - \bar{z}_0)/(2i)]^2} - \frac{x_0}{x_0^2 + y_0^2}$$
$$= \frac{z + \bar{z}_0}{z\bar{z}_0} - \frac{x_0}{x_0^2 + y_0^2} = \frac{1}{z} + i\frac{y_0}{x_0^2 + y_0^2}.$$

Rule C is not applicable as it stands, but see the following exercise.

EXERCISE 19.3 Adapt Rule C to the case where u and v are harmonic conjugates in a neighborhood of $z_0 \neq 0$.

EXERCISE 19.4 Find analytic functions whose real parts are the functions in Exercise 19.2 and also $(x^2 + y^2)^{1/4}\cos[\frac{1}{2}\tan^{-1}(y/x)]$.

Supplementary exercises

Find analytic functions whose real parts are

1. $\tan^{-1}\left(\frac{y}{x}\right)$

2. $x^3y - xy^3$

3. $\frac{x^2 - y^2}{(x^2 + y^2)^2}$

4. $[\exp(x^2 - y^2)]\cos(2xy)$

5. $\dfrac{1 - x^2 - y^2}{(1-x)^2 + y^2}$

6. $\tan^{-1}\dfrac{2y}{1 - x^2 - y^2}$

19D. Representation of harmonic functions by Taylor series

We can establish Rules A, B, and C by an interesting combination of ideas from both real and complex analysis. At the same time we shall obtain the important result that *harmonic functions are represented by their double Taylor series*; that is, if u is harmonic in a disk centered at (x_0, y_0), then in this disk we have (in the subscript notation for partial derivatives)

$$u(x, y) = u(x_0, y_0) + (x - x_0)u_1 + (y - y_0)u_2$$
$$+\frac{1}{2}[(x - x_0)^2 u_{11} + 2(x - x_0)(y - y_0)u_{12} + (y - y_0)^2 u_{22}]$$
$$+ \ldots,$$

where all the derivatives are evaluated at (x_0, y_0).

To begin with, if u is harmonic in a neighborhood of $z_0 = (x_0, y_0)$, there is a harmonic conjugate v in this neighborhood, so that $f = u + iv$ is analytic at z_0 and therefore is represented by its Taylor series about z_0. To keep the formulas simple, we take $z_0 = 0$ [otherwise, we consider $g(z) = f(z + z_0)$]. Then

$$u(x, y) + iv(x, y) = f(z) = \sum_{n=0}^{\infty} c_n z^n,$$

where the series converges absolutely in a disk $|z| < 2R$ (the factor 2 is there to keep later formulas simple). That is,

$$\sum |c_n|\,|x + iy|^n = \sum |c_n|\,(x^2 + y^2)^{n/2}$$

converges when $x^2 + y^2 < (2R)^2$. Thus $\sum |c_n|\,|t|^n$ converges when $|t| < 2R$, and therefore $\sum |c_n|\,(|x| + |y|)^n$ converges when $|x| + |y| < 2R$. Consequently, we may expand the powers by the binomial theorem and rearrange the resulting series, since all the terms are nonnegative. Hence the series $\sum c_n(x + iy)^n$ may be rearranged in any way, after expanding the powers by the binomial theorem, *even if x and y are replaced by complex numbers of modulus less than R*. The same is true for $\sum \bar{c}_n(x - iy)^n$, since the absolute values of the terms are the same for both series.

We may therefore write

$$f(x + iy) = u + iv = \sum \operatorname{Re}[c_n(x + iy)^n] + i \sum \operatorname{Im}[c_n(x + iy)^n],$$

expand the powers, and rearrange to get an expansion

$$u(x, y)=\sum_{n=0}^{\infty}\sum_{m=0}^{\infty} a_{mn}x^m y^n + i\sum_{n=0}^{\infty}\sum_{m=0}^{\infty} b_{mn}x^m y^n,$$

where a_{mn} and b_{mn} are real, even with x and y complex, as long as $|x|$ and $|y|$ are less than R. We can do the same thing for $g(x, y)=\sum_{n=0}^{\infty}\bar{c}_n(x-iy)^n$.

We see then that u and v are represented by double power series when $|x|<R$ and $|y|<R$, and therefore have partial derivatives of all orders. Furthermore, the double series are the Taylor series of u and v because in two dimensions, as well as in one dimension, a function can have only one power series expansion about a given point. In fact, the coefficients in

$$f(x, y)=\sum\sum_{m,n=0}^{\infty} c_{mn}(x-x_0)^m(y-y_0)^n$$

are determined by the formula

$$f(x, y)=\sum_{n=0}^{\infty}\frac{1}{n!}\left[(x-x_0)\frac{\partial}{\partial x}+(y-y_0)\frac{\partial}{\partial y}\right]^n f(x, y)_{x_0, y_0}.$$

We shall take this for granted; a proof would require our knowing that a double power series can be differentiated term by term.

We have now shown that *every function* $u(x, y)$ *that is harmonic for* $|x|<R$ *and* $|y|<R$ *is represented by an absolutely convergent power series.* We have thus proved a "real" theorem by operating with complex analytic functions.

What we do next is an analog of our obtaining a *disk* of convergence (Sec. **3B**) for power series that were originally known only to have a real *interval* of convergence, simply by replacing x by z; or, more precisely, by considering the series in a complex region instead of on the real line. We now start from a double real power series, in powers of x and y, absolutely convergent for $|x|<R$ and $|y|<R$ (these are ordinary real absolute values). If we now replace x and y by complex numbers z and w of modulus less than R, the series continues to converge absolutely, since absolute convergence (of course) depends only on the absolute values of the terms. Now u and v have been extended to functions represented, in a complex neighborhood of z_0, by complex double power series. It then makes perfectly good sense to consider u and v as functions with complex arguments.

19E. Justification of the formulas for f in terms of u and v

We can now justify Rules A, B, and C of Sec. **19C**. Let us start with Rule C, where we suppose that u and v are harmonic conjugates, so that $u(x, y)+iv(x, y)=f(z)$. The function g, defined by $g(z)=u(z, 0)+iv(z, 0)$, has the power series obtained from $u(z, w)+iv(z, w)$ by setting $w=0$, and is therefore analytic in a neighborhood of $z=0$. For real x near 0, we have

$g(x)=f(x)$. Therefore $g(z)-f(z)$ is an analytic function with nonisolated zeros, and is therefore identically zero (Sec. **7G**); in other words, f and g are analytic continuations of each other. This is what Rule C says.

We can justify Rule A in a similar way. If $f(z)=u(x, y)+iv(x, y)$, then $f'(z)=u_1(x, y)+iv_1(x, y)=u_1(x, y)-iu_2(x, y)$ by the Cauchy–Riemann equations. When z is on the real axis, $z=x$, and

$$f'(x)=u_1(x, 0)-iu_2(x, 0).$$

Since both sides of the equation are analytic functions, it follows that $f'(z)=u_1(z, 0)-iu_2(z, 0)$.

Rule B is more difficult. We assume first that u is harmonic at 0, so that we may take $z_0=0$. Let $u(x, y)=\operatorname{Re} f(z)$, with $f(z)=\sum c_n z^n$. Define $g(z)=\sum \bar{c}_n z^n$. Then we have (with real x and y)

$$f(x+iy)+g(x-iy)=\sum (c_n z^n+\bar{c}_n \bar{z}^n)=2\operatorname{Re} f(x+iy)=2u(x, y).$$

We know from Sec. **19D** that both sides of the equation

$$2u(x, y)=f(x+iy)+g(x-iy)$$

are analytic functions of x for fixed y and of y for fixed x. Hence this equation persists for complex x and y. Taking $x=z/2$, $y=z/(2i)$, we then have

$$\begin{aligned} 2u\left(\frac{z}{2}, \frac{z}{2i}\right) &= f\left(\frac{z}{2}+\frac{z}{2}\right)+g\left(\frac{z}{2}-\frac{z}{2}\right) \\ &= f(z)+g(0). \end{aligned}$$

Taking $z=0$, we have in particular

$$u(0, 0)=\frac{1}{2}f(0)+\frac{1}{2}g(0).$$

Consequently,

$$\begin{aligned} 2u\left(\frac{z}{2}, \frac{z}{2i}\right)-u(0, 0) &= f(z)+g(0)-\frac{1}{2}f(0)-\frac{1}{2}g(0) \\ &= f(z)+\frac{1}{2}[g(0)-f(0)]. \end{aligned}$$

But $f(0)=c_0$ and $g(0)=\bar{c}_0$, so $g(0)-f(0)=g(0)-\overline{g(0)}$, a pure imaginary number. This is what Rule B says.

EXERCISE 19.5 Establish Rule B in the general case.

I emphasize again that before applying Rule A, B, or C, you must verify that the function u (or functions u and v) are actually harmonic, since there are no checks built into the rules. See the following exercise.

EXERCISE 19.6 Apply Rule B to $u(x, y) = 4x^2y$ and verify that the "answer" does not have u as its real part.

Note

1. For further discussion of these rules, see, for example, E. V. Laitone, "Relation of the Conjugate Harmonic Functions to $f(z)$," *Amer. Math. Monthly 84* (1977): 281–283 (together with the review of that paper in *Math. Rev. 55,* #10646). The proof of Rule B is given in more compact form, using analytic functions of two complex variables explicitly, in Cartan, pp. 126–127.

20 *Harmonic Functions in a Disk*

20A. Dirichlet problems

If you are at all familiar with ordinary differential equations, you will remember that one of the basic problems is to find, not just *a* solution of a differential equation, but one that satisfies some prescribed initial or boundary conditions. To do this, we often find the general form of the solutions of the differential equation and then specialize it to satisfy the additional conditions. For partial differential equations this approach does not work too well, because the set of general solutions is usually too large. For example, suppose that we want a solution of Laplace's equation in a disk, with prescribed boundary values on the circumference of the disk. This is the **Dirichlet problem** for the disk. Since the solutions of Laplace's equation are just the harmonic functions, it is not obvious how we are to pick out the one that has the prescribed boundary values; in fact, the next few sections will be devoted to ways of solving such problems.

Dirichlet problems arise in many physical contexts, as we shall see in more detail in Sec. **23**. To illustrate with a specific example, let's think of a bottle of beer lying on its side in a snowbank on a sunny day. The lower half of the bottle will be cold, the upper half will be warm, and we want to find the temperature distribution in the beer. To get a tractable mathematical model, we will think of the bottle as an infinite solid circular cylinder, with all transverse cross sections alike, so that any one cross section will be representative. See Fig. 20.1. This idealization reduces the problem to a two-dimensional one, so that our model requires us (as is shown in physics) to find a harmonic function T inside a disk Δ, which we may take to be the unit disk, when the upper half of the boundary is held at temperature T_1 and the lower half at temperature T_2; reasonable values are $T_1 = 100$ and $T_2 = 32$ (in degrees Fahrenheit); T is not necessarily harmonic on the circumference of Δ. The boundary conditions are to be thought of as limits of T as z approaches the boundary of Δ from the inside. What we shall be able to do is to get a

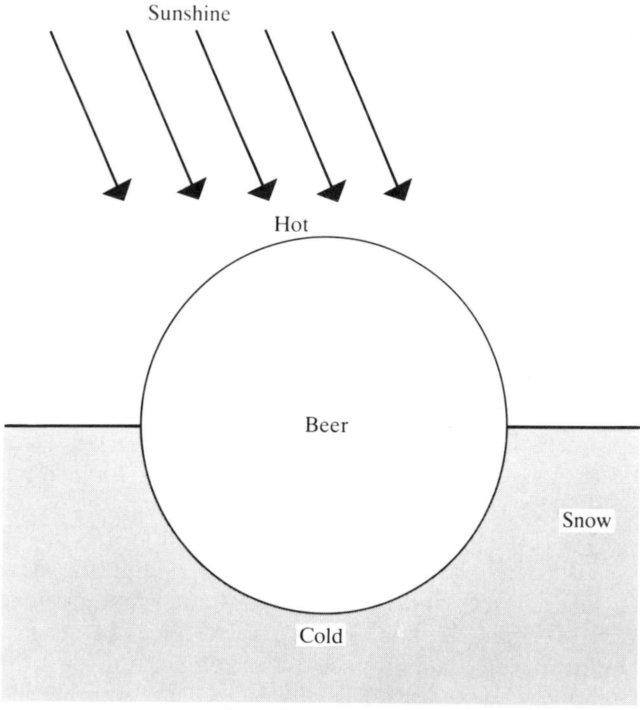

FIGURE 20.1 Cross Section of Bottle

formula for T at points *inside* the disk, and show that the values of T at points z inside Δ actually approach the given boundary values as z approaches the boundary. Fortunately, it can be shown that this always happens for Dirichlet problems with "reasonable" boundaries and "reasonable" sorts of regions, at least for most boundary points. Specifically, we will find (Exercise 21.1) that this particular Dirichlet problem is solved by

$$T = T(r, \theta) = \frac{68}{\pi} \tan^{-1} \frac{2r \sin \theta}{1 - r^2} + 66, \qquad 0 \le \theta \le 2\pi.$$

For $0 < \theta < \pi$, we have $T \to 100$ as $r \to 1-$; for $\pi < \theta < 2\pi$, $T \to 32$ as $r \to 1-$. At $\theta = 0$ or π, it happens (in this case) that $T = 66$ independently of r, so that we see that the limit of T is the average of its boundary values on the two sides of the points $z = \pm 1$. Notice that we cannot check the boundary values merely by substituting $r = 1$ in the formula for T.

It is important that there is only one solution of a Dirichlet problem for a region bounded by a simple closed curve. The solution is unique because the difference of two solutions would be a harmonic function with boundary values

0 and hence would vanish identically by the maximum and minimum principles for harmonic functions (Exercises 16.2 and 16.4).

In unbounded regions this is not necessarily true.

EXERCISE 20.1 Compute the imaginary part $v(x, y)$ of $\exp(1/z)$ (defined to be 0 at 0) in the upper half plane. Then $v(x, y)$ is harmonic for $y > 0$, and $v(x, 0) = 0$. Thus v appears to solve the Dirichlet problem for the upper half plane with boundary values 0 on the real axis. However, 0 is also harmonic in the upper half plane with boundary values 0. Show that $\lim_{y \to 0} v(0, y) \neq 0$.

In this exercise, the two "solutions" are not identical because a single exceptional boundary point can destroy the maximum principle (compare Sec. **16B**).

Uniqueness can fail in another way. For example, in the strip $D: 0 \leq y \leq 1$ in the (x, y) plane, the function $u(x, y) = y$ is a solution of the Dirichlet problem for D with boundary values 0 on $y = 0$ and 1 on $y = 1$. But so is $y + e^{\pi x} \sin \pi y = \operatorname{Im}(z + e^{\pi z})$. In physical applications, the second solution would be rejected as being too large for large x. To see why, consider the physical interpretation that the solution of this problem represents the steady-state distribution of temperatures in a thin metal plate covering D, with insulated faces and the edges $y = 0$, $y = 1$ maintained at respective temperatures 0 and 1. It should seem implausible, even to a pure mathematician, that under these circumstances the temperature in the plate would become extremely large for $y = \frac{1}{2}$ as $x \to \infty$. This example illustrates the point that *a Dirichlet problem for an unbounded region has a unique solution only if some restriction is imposed on how fast the solution can grow as* $z \to \infty$. We shall show in Sec. **33** that for our specific problem it would be enough to require that the solution has to be *bounded* in D.

We shall study mainly two special cases of the Dirichlet problem: the problem with general boundary values for a disk, and the problem when the boundary values are piecewise constant (see pp. 184 *ff*).

20B. Integral representation for harmonic functions in a disk

We are going to find *a representation, analogous to Cauchy's integral formula, for a function that is harmonic in a disk, in terms of its values on the circumference.* This must not be confused with the Dirichlet problem, which is to find a function with given boundary values when we do not know a priori that there is any such function. However, finding the representation formula is a first step toward solving the Dirichlet problem.

We begin with the unit disk $|z| \leq 1$ and write Cauchy's integral formula for a function f that is analytic in the closed disk:

$$f(z) = \frac{1}{2\pi i} \int \frac{f(w)\, dw}{w - z}, \qquad C: |w| = 1.$$

Just taking real parts on both sides will not lead to a representation for $\operatorname{Re} f(z)$ in terms of $\operatorname{Re} f(w)$, because the kernel in Cauchy's formula is not real. However,[1] since the point $1/\bar{z}$ symmetric to z with respect to the unit circle (defined after Exercise 2.6) is outside the disk, we also have

$$0 = \frac{1}{2\pi i} \int_C \frac{\bar{z} f(w)}{1 - w\bar{z}}\, dw.$$

Adding the two integrals, we obtain

$$f(z) = \frac{1}{2\pi i} \int_C \left(\frac{1}{w - z} + \frac{\bar{z}}{1 - w\bar{z}} \right) f(w)\, dw.$$

Since $|w| = 1$, $\bar{w} = 1/w$, $w = e^{i\phi}$, and we have

$$\begin{aligned} f(z) &= \frac{1}{2\pi} \int_{-\pi}^{\pi} \frac{1 - |z|^2}{(w - z)(\bar{w} - \bar{z})} f(w)\, d\phi \\ &= \frac{1}{2\pi} \int_{-\pi}^{\pi} \operatorname{Re}\left(\frac{w + z}{w - z} \right) f(w)\, d\phi. \end{aligned}$$

If we now write $z = re^{i\theta}$ in the next to last integral, we find that

$$f(z) = \frac{1}{2\pi} \int_{-\pi}^{\pi} f(e^{i\phi}) \frac{(1 - r^2)\, d\phi}{1 + r^2 - 2r\cos(\theta - \phi)}.$$

This is **Poisson's integral formula** for $f(z)$, established so far only when f is analytic in the closed disk. We shall see later that it holds when f is merely analytic in the open disk and continuous in the closed disk.

Since the "kernel" $P(r, \theta - \phi)$ given by

$$\frac{1 - r^2}{1 + r^2 - 2r\cos(\theta - \phi)} = 2\pi P(r, \theta - \phi)$$

is real, we can now take real parts on both sides of Poisson's formula and use it to represent the real part of f in terms of its boundary values $u(1, \theta)$.

Since every function that is harmonic in the unit disk is the real part of an analytic function, we have therefore found a formula that produces the values of a harmonic function in the unit disk, given its values on the boundary.

EXERCISE 20.2 Write out Poisson's formula for a disk $|z| < R$.

20C. The Dirichlet problem for the unit disk

Now let us suppose that we have a set of real boundary values $\psi(\theta)$. Will Poisson's formula deliver a function V that is harmonic in the open disk and approaches the given boundary values as z approaches a boundary point?

(Remember that Cauchy's integral formula does not have this property, as we saw in Sec. **7C**). In the first place we must suppose that ψ has period 2π and that the integral in Poisson's formula exists.

We shall need several rather obvious facts.

1. Constant functions are harmonic, and therefore represented by Poisson's formula.

2. $P(r, \theta) > 0$.

3. $\int_{-\pi}^{\pi} P(r, \theta)\, d\theta = 1$. This follows from fact 1.

4. The operator L that takes ψ to U is linear and "positive"; that is, it takes nonnegative functions to nonnegative functions, and therefore preserves inequalities: if $\psi_1 \geq \psi_2$, then $L[\psi_1] \geq L[\psi_2]$.

Now let $U(z) = U(r, \theta) = \int_{-\pi}^{\pi} \psi(\phi)P(r, \theta - \phi)\, d\phi$. In the first place, U is harmonic in the disk $|z| < 1$. We can see this most easily by writing

$$U(z) = \frac{1}{2\pi}\int_{-\pi}^{\pi} \psi(\phi)\,\mathrm{Re}\,\frac{w+z}{w-z}\, d\phi$$

$$= \mathrm{Re}\,\frac{1}{2\pi}\int_{-\pi}^{\pi} \frac{w+z}{w-z}\,\psi(\phi)\, d\phi.$$

The integral represents an analytic function by the reasoning of Sec. **7E**, and its real part U is therefore harmonic. More generally,

$$\int_a^b P(r, \theta - \phi)\psi(\phi)\, d\phi,$$

where (a, b) is strictly interior to $(-\pi, \pi)$, represents a harmonic function everywhere off the arc $a < \theta < b$ of the unit circumference.

We now want to show that $U(z) \to \psi(\theta)$ when the limit is taken as $z \to e^{i\theta}$ from inside the unit disk, provided that ψ is continuous at θ. If we do this, we shall have a solution for the Dirichlet problem for a piecewise continuous boundary function ψ. We cannot expect $U(z)$ to approach $\psi(\theta)$ at points of discontinuity of ψ, although, if ψ is bounded, then U is bounded and we shall be able to show that there is no other solution of the same Dirichlet problem. The fact that

$$\lim_{r \to 1-} U(re^{i\theta}) = \frac{1}{2}[\psi(\theta+) + \psi(\theta-)],$$

when the limits on the right exist, is of interest in connection with Fourier series, but not very relevant to Dirichlet problems, and we shall not stop to establish it.

To establish the convergence of $U(z)$ at a point θ_0 of continuity of ψ, we begin by writing

$$U(z) - \psi(\theta_0) = \int_{-\pi}^{\pi} P(r, \theta - \phi)[\psi(\phi) - \psi(\theta_0)]\, d\phi$$

$$= \int_{\theta_0-\varepsilon}^{\theta_0+\varepsilon} P(r, \theta - \phi)[\psi(\phi) - \psi(\theta_0)]\, d\phi + \int_{\theta_0+\varepsilon}^{\pi} + \int_{-\pi}^{\theta_0-\varepsilon}$$

$$= I_1 + I_2 + I_3.$$

Now the integrals I_2 and I_3 represent, as we saw above, harmonic functions on the arc $(\theta_0 - \varepsilon, \theta_0 + \varepsilon)$ of $|z| = 1$, and their limits as $|z| \to 1$ are zero on that arc because, as we see from the explicit formula for $P(r, \theta - \phi)$, the Poisson kernel approaches 0 as $r \to 1$, uniformly on every arc where $\cos(\theta - \phi)$ is bounded away from 1.

It remains to show that $I_1 \to 0$ as $z \to e^{i\theta_0}$. Given any positive δ, we could have chosen ε to begin with so that $|\psi(\phi) - \psi(\theta_0)| < \delta$ when $|\phi - \theta_0| < \varepsilon$. Then by facts 2 and 3 above, we have

$$\left|\int_{\theta_0-\varepsilon}^{\theta_0+\varepsilon} P(r, \theta - \phi)[\psi(\phi) - \psi(\theta_0)]\, d\phi\right|$$

$$\le \delta \int_{\theta_0-\varepsilon}^{\theta_0+\varepsilon} P(r, \theta - \phi)\, d\phi \le \delta \int_{-\pi}^{\pi} P(r, \theta - \phi)\, d\phi = \delta.$$

Since δ is arbitrary, it follows that $I_1 \to 0$.

We have thus constructed a solution of the Dirichlet problem with piecewise continuous boundary values. In fact, it is the only bounded solution, since we shall show in Sec. **33A** that there is only one bounded harmonic function with boundary values prescribed except at a finite number of points. (An unbounded solution would be physically unsatisfactory.)

We can now go back to the end of Sec. **20B** and see that the Poisson representation holds for functions that are merely harmonic in $|z| < 1$ and continuous in $|z| \le 1$. Let us suppose that u is such a function; construct the function U by Poisson's formula with the boundary values of u. Then $u - U$ is harmonic in the open disk, continuous in the closed disk, and with boundary values 0; then $u - U \equiv 0$ by the maximum and minimum principles for harmonic functions.

The Dirichlet problem for disks is now in rather satisfactory shape, because the Poisson integral is easily adapted to a disk of any size (Exercise 20.2) and in any location. For other regions we do not yet have an analog of Poisson's formula, although you may suspect that there ought to be one. That there is indeed an analog is a rather deep result, which will have to be postponed until we have discussed conformal mapping in considerable detail. However, using the Poisson integral is not always the most efficient method for solving a specific Dirichlet problem. As we shall see in Sec. **24**, many applications of harmonic functions involve Dirichlet problems, with piecewise constant boundary values, which can sometimes be solved quite explicitly.

20D. Removable singular points for harmonic functions

If an analytic function is bounded in the neighborhood of an isolated singular point, the singularity is removable (Sec. **8C**). The analogous theorem for harmonic functions says that:

> If a function u is harmonic and bounded in a neighborhood of a point z_0, with z_0 deleted, then u can be (re)defined at z_0 so as to become harmonic at z_0 (that is, in a neighborhood of z_0).

There is no loss of generality from assuming that $z_0 = 0$.

The Poisson integral for $|z| < r$ with the values of u on $|z| = r$ provides a function h that is harmonic for $|z| < r$, with $h = u$ on $|z| = r$. If we can show that $u - h \leq 0$ for $0 < |z| \leq r$ and also that $h - u \leq 0$ for $0 < |z| \leq r$, it follows that $u = h$ if we define $u(0) = h(0)$.

Let ψ stand for either $u - h$ or $h - u$, and consider

$$\psi_\varepsilon(z) = \psi(z) + \varepsilon \log(|z|/r), \qquad 0 < |z| \leq r,\ \varepsilon > 0.$$

Then ψ_ε is harmonic for $0 < |z| < r$, $\psi_\varepsilon(z) = 0$ for $|z| = r$, and $\limsup_{z \to 0} \psi_\varepsilon(z) < 0$ because ψ is bounded. Hence $\psi_\varepsilon(z) \leq 0$ in some punctured disk $0 < |z| \leq \delta$. Since ψ_ε is harmonic in the annulus $\delta \leq |z| \leq r$ and nonpositive on the boundary of the annulus, we have $\psi_\varepsilon(z) \leq 0$ in the annulus. Since δ is arbitrary, $\psi_\varepsilon(z) \leq 0$ for $0 < |z| \leq r$. Letting $\varepsilon \to 0$, we have $\psi(z) \leq 0$ for $0 < |z| \leq r$. Therefore $u - h \leq 0$ and $h - u \leq 0$ for $0 < |z| \leq r$; that is, u and h coincide for $0 < |z| \leq r$. Taking $u(0) = h(0)$, we have $u \equiv h$ for $|z| \leq r$.

20E. Intrinsic characterization of harmonic functions

If $u(z)$ is harmonic at z_0, then u is the real part of an analytic function f in a neighborhood of z_0. We saw in Sec. **16A** that Cauchy's formula implies that for small r we have

$$f(z_0) = \frac{1}{2\pi} \int_0^{2\pi} f(z_0 + re^{i\theta})\, d\theta;$$

hence

$$u(z_0) = \frac{1}{2\pi} \int_0^{2\pi} u(z_0 + re^{i\theta})\, d\theta.$$

This says that:

> The value of a harmonic function at a point is the average of its values on each sufficiently small circle centered at the point.

This is the **averaging property** of harmonic functions. As in Sec. **16A** it follows that a function with the averaging property cannot have a proper maximum or minimum, and not any maximum or minimum unless it is a constant.

We can now show that *every continuous function f that has the averaging property is harmonic,* without any hypothesis about the partial derivatives of f. To establish this, construct, by using the Poisson integral, the harmonic function g that has the same values as f on a circle $|z - z_0| = r$. The difference $g - f$ also has the averaging property for the same r and hence, since it is zero on the circle, is zero inside (by the maximum and minimum principles for such functions). Consequently, f is harmonic in a neighborhood of z_0, and hence everywhere, since z_0 could be anywhere. Therefore f in fact has derivatives of all orders and satisfies Laplace's equation.

This means that we could define a function to be harmonic merely if it is continuous and has the averaging property. However, we would like to be able to say that *f is harmonic if it is merely continuous and has enough differentiability so that $f_{11} + f_{22}$ is defined and equal to* 0. To prove this, we construct the harmonic function g that has the same values as f on a circle C centered at z_0, and consider $h = g - f + p(x - x_0)^2$, where p is a small positive number. If h had a maximum, its second partial derivatives would be negative or zero at the maximum point, so $h_{11} + h_{22} \le 0$ at that point. But $h_{11} + h_{22} = g_{11} + g_{22} - f_{11} - f_{22} + 2p = 2p > 0$. Therefore h has no maximum (in the region we are considering). We conclude that, in a disk about z_0, of radius r, on whose circumference $g - f = 0$, we have $g(z) - f(z) + p(x - x_0)^2 \le pr^2$. If we now let $p \to 0$, we obtain $g - f \le 0$. In the same way, we can show that $g - f \ge 0$. Hence $f \equiv g$ and f is harmonic.

The proposition that a function with the averaging property is harmonic can be weakened in various ways. For example, if D is any region for which the Dirichlet problem can be solved, a real function that is continuous in $\bar{D}$ is harmonic if it has merely the weaker averaging property that every point of D is the center of some circle in $\bar{D}$ for which f has the averaging property.[2]

20F. Schwarz reflection principle for harmonic functions

Suppose, as in Sec. **17C**, that D is a simply connected region in the upper half plane, and that ∂D intersects the real axis in a single line segment L. Suppose also that β is a function harmonic in D, and that $\beta(z) \to 0$ as $z \to L$. If D^* is the reflection of D with respect to the real axis, then if we take $v(z) = \beta(z)$ for $z \in D$, $v(z) = -\beta(\bar{z})$ for $z \in D^*$, and $v(z) = 0$ for $z \in L$, we will have v harmonic in $S = D \cup D^* \cup L$.

Since v is harmonic in D and in D^*, we only have to show that v is harmonic on L, that is, in a neighborhood of each point of L. Consider a small disk Δ centered at a point P of L; then $\int_{\partial\Delta} v$ is the sum of the integrals around the upper and lower halves of Δ. The integrals back and forth along the diameter of Δ cancel. The values of v on the upper and lower arcs of Δ are negatives of each other, so $\int_{\partial\Delta} v = 0 = v(P)$. Therefore v has the averaging property in disks centered on L; but v also has that property on all small disks centered elsewhere in D; therefore, by what was just proved, v is harmonic in S.

It follows that in the Schwarz reflection principle, as stated in Sec. **17C**, we may suppose only that $\operatorname{Im} f$, rather than f, is continuous on $D \cup L$, so that we have a stronger version of the Schwarz principle. (This will be needed in Sec. **28E**.) In fact, let $\beta = \operatorname{Im} f$ and extend β to S as above. Then v has a conjugate harmonic function $-u$ in the symmetric region, determined up to an additive constant. Pick a point z_0 in D and choose the constant so that $u = \operatorname{Re} f(z_0)$; then $u = \operatorname{Re} f(z)$ throughout S.

Notes

1. This step admittedly seems unmotivated; the reason for it will appear in Sec. **26E**.
2. For a more precise statement, and discussion, see R. B. Burckel, "A Strong Converse to Gauss's Mean-Value Theorem," *Amer. Math. Monthly* 87 (1980): 819–820.

*21 *Harmonic Functions and Fourier Series*

21A. Series expansion of the Poisson kernel

This section is a digression. If you already know something about Fourier series, it will show you how Fourier series are connected with harmonic functions; if you do not know much about Fourier series, you can now learn what they are and how they can be used to calculate harmonic functions. Poisson's integral is an attractive formula, but it is usually easier to sum a series numerically than to evaluate an integral numerically; and incidentally we shall learn still another method (besides the methods of Sec. **19**) of finding the harmonic conjugate of a harmonic function.

We start by expanding the Poisson kernel in powers of r. This is not easy to do in any straightforward way that I am aware of. The simplest procedure is to start from the result and work back, but the following method lets us discover the formula without too much ingenuity.

We start from the original form of the Poisson kernel in Sec. **20B** (p. 167), with $w = e^{i\phi}$, $\phi = 0$, namely,

$$2\pi P(r, \theta) = \frac{1 - |z|^2}{(1 - z)(1 - \bar{z})} = \frac{1 - z\bar{z}}{(1 - z)(1 - \bar{z})}.$$

Then

$$
\begin{aligned}
2\pi P(r, \theta) - 1 &= \frac{1 - z\bar{z} - (1 - z - \bar{z} + z\bar{z})}{(1-z)(1-\bar{z})} \\
&= \frac{z(1-\bar{z}) + \bar{z}(1-z)}{(1-z)(1-\bar{z})} \\
&= \frac{z}{1-z} + \frac{\bar{z}}{1-\bar{z}} \\
&= \sum_{n=1}^{\infty} z^n + \sum_{n=1}^{\infty} \bar{z}^n,
\end{aligned}
$$

so that we have, with $z = re^{i\theta}$,

$$
\begin{aligned}
2\pi P(r, \theta) &= 1 + \sum_{n=1}^{\infty} r^n e^{in\theta} + \sum_{n=1}^{\infty} r^n e^{-in\theta} \\
&= \sum_{n=-\infty}^{\infty} r^{|n|} e^{in\theta}.
\end{aligned}
$$

Then the Poisson integral of an integrable function ψ can be written in the form

$$\frac{1}{2\pi} \int_{-\pi}^{\pi} \sum_{n=-\infty}^{\infty} r^{|n|} e^{in(\theta-\phi)} \psi(\phi)\, d\phi.$$

When $0 < r < 1$, the series can be integrated term by term, since the series $\sum r^{|n|} e^{in(\theta-\phi)}$ is uniformly convergent on $(-\pi, \pi)$ because of the factors $r^{|n|}$. We find that the Poisson integral of ψ can be presented as

$$\sum_{n=-\infty}^{\infty} r^{|n|} e^{in\theta} \cdot \frac{1}{2\pi} \int_{-\pi}^{\pi} \psi(\phi) e^{-in\phi}\, d\phi.$$

Here the integrals

$$c_n = \frac{1}{2\pi} \int_{-\pi}^{\pi} \psi(\phi) e^{-in\phi}\, d\phi$$

are independent of θ and r, so that the harmonic function with boundary values $\psi(\theta)$ (at every point of continuity of ψ) can be written as

$$\sum_{n=-\infty}^{\infty} c_n r^{|n|} e^{in\theta},$$

where the coefficients c_n (**Fourier coefficients**) are given by the integral formula just above. If we (formally) set $r = 1$, we obtain the series

$$\sum_{n=-\infty}^{\infty} c_n e^{in\theta}.$$

This series might not converge, but it is certainly Abel summable (Sec. **18B**) to $\psi(\theta)$, at least at the points of continuity of ψ. This series is the **Fourier series** of ψ. It is customary to write

$$\psi(\theta) \sim \sum_{n=-\infty}^{\infty} c_n e^{in\theta},$$

where the symbol "$\sim$" (read, "corresponds to"; it is often pronounced "twiddle") indicates that the series, whether convergent or not, has its coefficients defined by the integral formula

$$c_n = \frac{1}{2\pi} \int_{-\pi}^{\pi} \psi(\theta) e^{-in\theta}\, d\theta.$$

You should keep in mind that a series $\sum c_n e^{in\theta}$ is a Fourier series only if its coefficients are defined in this particular way; otherwise it is just a **trigonometric series**.

A Fourier series is often written in the "sine-cosine" form

$$\psi(\theta) \sim \frac{1}{2} a_0 + \sum_{n=1}^{\infty} (a_n \cos n\theta + b_n \sin n\theta).$$

To put our Fourier series into this form, define

$$\psi_c(\theta) = \frac{1}{2}[\psi(\theta) + \psi(-\theta)], \qquad \psi_s(\theta) = \frac{1}{2}[\psi(\theta) - \psi(-\theta)];$$

note that ψ_c is even and ψ_s is odd, and that $\psi = \psi_c + \psi_s$. Then (formally)

$$\begin{aligned} \psi_c(\theta) &\sim \frac{1}{2} \sum_{n=-\infty}^{\infty} c_n (e^{in\theta} + e^{-in\theta}) = c_0 + \sum_{n=1}^{\infty} (c_n + c_{-n}) \cos n\theta \\ &= \frac{1}{2} a_0 + \sum_{n=1}^{\infty} a_n \cos n\theta, \\ \psi_s(\theta) &\sim \frac{1}{2} \sum_{n=-\infty}^{\infty} c_n (e^{in\theta} - e^{-in\theta}) = \sum_{n=1}^{\infty} i(c_n - c_{-n}) \sin n\theta \\ &= \sum_{n=1}^{\infty} b_n \sin n\theta, \end{aligned}$$

where

$$\begin{aligned} a_n = c_n + c_{-n} &= \frac{1}{2\pi} \int_{-\pi}^{\pi} \psi(\phi)[e^{-in\phi} + e^{in\phi}]\, d\phi \\ &= \frac{1}{\pi} \int_{-\pi}^{\pi} \psi(\phi) \cos n\phi\, d\phi, \qquad n \geq 1; \\ b_n = i(c_n - c_{-n}) &= \frac{i}{2\pi} \int_{-\pi}^{\pi} \psi(\phi)(e^{-in\phi} - e^{in\phi})\, d\phi \\ &= \frac{1}{\pi} \int_{-\pi}^{\pi} \psi(\phi) \sin n\phi\, d\phi, \qquad n \geq 1. \end{aligned}$$

We write $c_0 = a_0/2$ so that a_0 will be defined by the same integral formula as a_n, with $n = 0$.

These formal calculations are not trivial to justify, but we are not going to make any use of the sine-cosine form.

Now, instead of writing the Poisson integral of a function ψ, we can get the same harmonic function by calculating the coefficients c_n (or a_n and b_n), writing out the Fourier series of ψ, and hence obtaining the solution of the Dirichlet problem with boundary values $\psi(\theta)$.

EXERCISE 21.1 Let $\psi(\theta) = 0$ for $0 < \theta < \pi$ and 1 for $\pi < \theta < 2\pi$. Write a series for the function $u(r, \theta)$ that is harmonic in the unit disk with these boundary values.[1] Use the series to compute $u(1/2, \pi/2)$ numerically.

EXERCISE 21.2 Obtain an integral formula for the partial sums of a Fourier series. (Compare Exercise 4.17.)

EXERCISE 21.3 If a harmonic function is represented for $|z| < 1$ by the series

$$u(r, \theta) = \sum_{-\infty}^{\infty} c_n r^{|n|} e^{in\theta},$$

show that a conjugate harmonic function is given by the series

$$-i \sum (\operatorname{sgn} n) c_n r^{|n|} e^{in\theta}.$$

(Use the Cauchy–Riemann equations in polar coordinates, Exercise 2.4.)

The question of whether a Fourier series actually converges to the function from which we got it is something that would take us too far from our subject; however, here are some of the facts.[2] The Fourier series of ψ converges to $\psi(x)$ for almost all x if $|\psi|^2$ is integrable, and in particular if ψ is continuous; but it does not necessarily converge everywhere. If $|\psi|$ is merely integrable, the Fourier series might not converge anywhere at all. The Fourier series of an integrable function is not only Abel summable at points of continuity, but even $(C, 1)$ summable, to the value of the function. If ψ is of bounded variation, for example if the graph of ψ is composed of a finite number of bounded monotonic pieces ("Dirichlet's condition"), the series converges everywhere, with sum $\frac{1}{2}[\psi(\theta+) + \psi(\theta-)]$.

Supplementary exercises

The following exercises give boundary values $\psi(\theta)$ for Dirichlet problems in the unit disk. Write series solutions for these problems.

1. $\psi(\theta) = 0, \; -\pi < \theta < 0$
$\quad 1, \; 0 < \theta < \pi/2$
$\quad 0, \; \pi/2 < \theta < \pi$

2. $\psi(\theta) = 0, \; -\pi < \theta < 0$
$\quad \theta, \; 0 < \theta < \pi$

3. $\psi(\theta) = \sin\theta,\ 0 < \theta < 2\pi$

4. $\psi(\theta) = 0,\ -\pi < \theta < 0$
$\qquad\quad \sin\theta,\ 0 < \theta < \pi$

5. $\psi(\theta) = e^{i\omega\theta},\ -\pi < \theta < \pi$, ω not an integer

6. $\psi(\theta) = |\theta|$ on $(-\pi, \pi)$

7. $\psi(\theta) = (\pi - \theta)/2$ on $(0, \pi)$; ψ odd on $(-\pi, \pi)$.

21B. Fourier series from Laplace's equation

It is interesting to see how the Fourier series for the solution of a Dirichlet problem for a disk can be obtained directly from Laplace's equation without going through the Poisson integral. We want to solve $u_{11} + u_{22} = 0$; in polar coordinates, this equation reads

$$\frac{1}{r}\frac{\partial}{\partial r}\left(r\frac{\partial u}{\partial r}\right) + \frac{1}{r^2}\frac{\partial^2 u}{\partial\theta^2} = 0.$$

We want a solution that satisfies the boundary condition $u(1, \theta) = \psi(\theta)$. We start from the far from obvious idea of trying to piece together a solution from functions of the form $R(r)\Theta(\theta)$, which satisfy Laplace's equation but do not necessarily satisfy the boundary conditions. A solution of this form would have to satisfy

$$\frac{1}{r}\frac{d}{dr}\left(r\frac{dR}{dr}\right)\cdot\Theta + \frac{1}{r^2}R\frac{d^2\Theta}{d\theta^2} = 0,$$

that is,

$$\frac{r}{R}\frac{d}{dr}\left(r\frac{dR}{dr}\right) = -\frac{1}{\Theta}\frac{d^2\Theta}{d\theta^2}.$$

Since the left-hand side depends only on r and the right-hand side depends only on θ, both sides must be constant, say

$$\frac{r}{R}\frac{d}{dr}\left(r\frac{dR}{dr}\right) = k^2 = -\frac{1}{\Theta}\frac{d^2\Theta}{d\theta^2}.$$

The θ equation is $\Theta'' + k^2\Theta = 0$, whose solutions are $e^{\pm ik\theta}$. Since the solution must be a function (of θ) of period 2π in order to be single-valued in the disk, k must be an integer. (If the constant that we denoted by k^2 had been negative, Θ would not be periodic.)

Now R satisfies

$$r\frac{d}{dr}\left(r\frac{dR}{dr}\right) = k^2 R,$$

and a little thought suggests $R = r^k$ as a solution. Since the solution must be finite at 0, we take $k \geq 0$. Thus we have found that $r^k e^{\pm ik\theta}$, $k = 0, 1, 2, \ldots,$

are acceptable solutions of Laplace's equation. Since the equation is linear, any linear combination of these solutions is a solution, and it seems plausible that an infinite linear combination

$$u(r, \theta) = c_0 + \sum_{\substack{-\infty \\ k \neq 0}}^{\infty} c_k r^{|k|} e^{ik\theta}$$

might be a solution. If it is to have the boundary values $\psi(\theta)$ as $r \to 1$, then (at least formally) we need

$$\psi(\theta) = c_0 + \sum_{\substack{k=-\infty \\ k \neq 0}}^{\infty} c_k e^{ik\theta}.$$

Now how do we find the coefficients c_n? This requires a further inspiration, namely, that the functions $e^{ik\theta}$ are **orthogonal** to each other, that is,

$$\int_{-\pi}^{\pi} e^{ik\theta} e^{-il\theta}\, d\theta = 0, \qquad k \neq l,$$

and

$$\int_{-\pi}^{\pi} e^{ik\theta} e^{-ik\theta}\, d\theta = 2\pi.$$

Hence (again formally) if we multiply the series for $\psi(\theta)$ by $e^{-in\theta}$ and integrate, we get

$$\frac{1}{2\pi} \int_{-\pi}^{\pi} \psi(\theta) e^{-in\theta}\, d\theta = c_n,$$

and then

$$u(r, \theta) = \sum_{-\infty}^{\infty} c_n r^{|n|} e^{in\theta},$$

with these coefficients c_n.

With this approach we now have a proposed solution of the Dirichlet problem in the form of an infinite series; it remains to investigate whether it actually represents a solution. There is no problem with convergence inside the unit disk, because of the factors $r^{|n|}$. The difficult part is to show that the sum of the series has the right boundary values.

This method is an example of the **method of separation of variables**; it has many applications to other boundary value problems in two or more dimensions, where methods depending on complex analysis are not always available.

EXERCISE 21.4 Solve the Dirichlet problem for an annulus, centered at 0, with inner radius ρ and outer radius σ. You will need to include the solutions of

$$r\frac{d}{dr}\left(r\frac{dR}{dr}\right) = k^2 R$$

corresponding to $k \le 0$ that we discarded above because they were unbounded at 0; these are not only the functions r^k, $k < 0$, but also $\ln r$.

Supplementary exercises

Write series for solutions of the following Dirichlet problems for the annulus $1 \le |z| \le 2$ (see Exercise 21.4).

1. For $r = 2$, $\psi(\theta) = \cos 2\theta$; for $r = 1$, $\psi(\theta) = \sin \theta$.
2. For $r = 2$, $\psi(\theta) = 1$, $0 < \theta < \pi$; 0, $\pi < \theta < 2\pi$. For $r = 1$, $\psi(\theta) = 0$.
3. For $r = 2$, $\psi(\theta)$ as in Supplementary Exercise 2; for $r = 1$, $\psi(\theta) = 0$, $0 < \theta < \pi$; 1, $\pi < \theta < 2\pi$.
4. For $r = 2$, $\psi(\theta) = |\theta|$, $-\pi < \theta < \pi$. For $r = 1$, $\psi(\theta) = 100$.

Notes

1. The series can be summed in closed form; for another method of solving the same problem, which leads directly to the closed form, see Exercise 24.2.
2. It is interesting to notice how long the development took. Dirichlet's test for convergence of Fourier series goes back to 1829; Fejér's theorem on $(C, 1)$ summability did not appear until 1904; Carleson's theorem on the convergence of the Fourier series of square-integrable functions is as recent as 1966. Kolmogorov published the first construction of an everywhere-divergent Fourier series in 1926. The "bounded variation" test had, of course, to wait for the concept of bounded variation, which seems to have been introduced by C. Jordan in 1893.

22 *Conformal Mapping*

22A. Analytic functions as conformal maps

In order to make further progress with Dirichlet problems (and for other reasons), we need to discuss analytic functions $w = f(z)$ as mappings from a region D in a z plane to a region Δ in a w plane.

A mapping (or just a "map") is said to be **conformal** if it is one-to-one and preserves the magnitude and sense of angles between curves.[1] More precisely, a map is conformal at a point z_0 if it preserves angles between curves at z_0; it is conformal in a region D if it is conformal at every point of D and one-to-one in D. The essential facts are that:

If $w = f(z)$ and f is analytic at z_0 with $f'(z_0) \ne 0$, then the map is conformal at z_0; if $f'(z_0) = 0$, then the map is not conformal at z_0; and conversely, if $w = f(z) = \phi(x, y)$ is a conformal map of a region D onto a region Δ, then f is analytic in D and $f'(z) \ne 0$ in D.

We had better begin by saying just what the phrase "preserves angles" means. Suppose that we have a continuous curve C that passes through z_0, and let Γ be the image of C in the w plane. If z_1 is a nearby point on C, then $\arg(z_1 - z_0)$ is the angle that the chord through z_0 and z_1 makes with the positive real direction, and $\arg[f(z_1) - f(z_0)]$ is the corresponding angle in the w plane. Then

$$\theta = \arg \frac{f(z_1) - f(z_0)}{z_1 - z_0}$$

is the angle through which the chord has been rotated by the map. (More precisely, it is the angle between the directions of the old chord and the new chord; the image of the original chord is not necessarily a line segment, although it will nearly be one if z_1 is near z_0.) If $f'(z_0) \neq 0$, the limit of θ as $z_1 \to z_0$ is $\arg f'(z_0)$, which is independent of the curve C. Thus two curves that meet at an angle γ at z_0 are mapped to two curves that meet at the same angle (although both curves may have been turned away from their original directions).

On the other hand, if $f'(z_0) = 0$, we may assume that $f'(z) \not\equiv 0$, since otherwise f is a constant, and there is not much to say about a map that takes a region to a single point. Then in a neighborhood of z_0 we have $f(z) - f(z_0) = (z - z_0)^n g(z)$, where $n \geq 2$ is the order of the zero of $f(z) - f(z_0)$. If z_1 is a point near z_0, we have $\arg[f(z_1) - f(z_0)] = n \arg(z_1 - z_0) + \arg g(z_1)$. Therefore curves that meet at an angle γ at z_0 become curves that meet at an angle $n\gamma$ at $f(z_0)$. This is what we mean by saying that *angles at z_0 are multiplied by n.* An angle of more than π then becomes an angle of more than 2π, so that if we think of observing the images of larger and larger angles, we eventually have the image of a neighborhood of z_0 overlapping itself, and f does not have an inverse. This shows again that (as we saw in Sec. **13A**) a function f cannot be univalent in a neighborhood of a point where $f' = 0$. If we consider f only in an angle with vertex z_0 and of opening less than $2\pi/n$, we can define a branch of the inverse of f in the intersection of this angle with a neighborhood of z_0, and the inverse function will transform curves that meet at an angle γ into curves that meet at an angle γ/n. For example, if we consider $w = z^2$ only in the first quadrant, the positive real and imaginary axes are mapped into two curves (actually, half lines) that meet at angle π; conversely, the inverse function $z = \sqrt{w}$ carries the real axis to the sides of the quadrant. Notice particularly that the orientations of angles are preserved even though their sizes are changed. That is, if w follows the real axis through 0 from left to right, then $\sqrt{w}$ follows the boundary of the first quadrant down the imaginary axis and to the right on the positive real axis in the z plane.

Since conformal maps are one-to-one, the relative order of points is preserved by a conformal map. That is, if A, B, C are points of an arc, in that order, their images A', B', C' must appear on the image of the arc in the same order. This principle is worth mentioning because it can prevent you from looking for an impossible mapping.

22B. Conformal maps as analytic functions

It is true, but far from simple to prove, that if $w=f(z)$ is a homeomorphism (one-to-one, continuous, with a continuous inverse) from D to Δ, and if at every point z_0 of D

$$\arg\frac{f(z_1)-f(z_0)}{z_1-z_0}$$

has a finite limit as $z_1 \to z_0$, then[2] f is analytic in D. We shall prove this only when f has continuous first and second derivatives.

Let $f(z)=u(x, y)+iv(x, y)$ produce a conformal map at z_0. Let $z=z_0+re^{i\theta}$; then (since we are assuming the existence of continuous derivatives) we have

$$f(z)-f(z_0)=r(f_1\cos\theta+f_2\sin\theta)+o(r) \qquad \left(f_1=\frac{\partial f}{\partial x}\text{ at } z_0,\text{ etc.}\right)$$

Consequently, unless $f_1=f_2=0$ at z_0, we have, when z is near z_0,

$$\arg[f(z)-f(z_0)]=\arg T(\theta)+o(1),$$

where

$$T(\theta)=f_1(x_0, y_0)\cos\theta+f_2(x_0, y_0)\sin\theta.$$

Hence, if we consider two points z_1 and z_2, the angle between $z_1-z_0=re^{i\theta_1}$ and $z_2-z_0=re^{i\theta_2}$ is $\theta_1-\theta_2$, whereas that between $f(z_1)-f(z_0)$ and $f(z_2)-f(z_0)$ is $\arg T(\theta_1)-\arg T(\theta_2)+o(1)$. Since the map is conformal, it follows that

$$\arg T(\theta_1)-\arg T(\theta_2)=\theta_1-\theta_2 \qquad (\text{mod } 2\pi)$$

This, in turn, implies that $T(\theta)=e^{i\theta}$ times a constant, because if we take $\theta_1=\theta$ and $\theta_2=0$, it says that $\arg T(\theta)-\arg T(0)=\arg e^{i\theta}$, or $\arg T(\theta)=\theta+\text{constant}$, or $T(\theta)=Ke^{i\theta}$.

Thus, unless $f_1=f_2=0$ at z_0, we have

$$f_1\cos\theta+f_2\sin\theta=Ke^{i\theta}.$$

For $\theta=0$ we get $f_1=K$; for $\theta=\pi/2$ we get $f_2=Ki$; hence at z_0 we have $f_1=-if_2$, the complex form of the Cauchy–Riemann equations (p. 16). Consequently, the Cauchy–Riemann equations are satisfied at every point where the map is conformal, unless $f_1=f_2=0$; but in the latter case, the Cauchy–Riemann equations are satisfied automatically. Hence $f(z)$ is analytic, and we cannot have $f_1=f_2=0$ because if that happens at a point z_0, the map is not one-to-one near z_0, as we saw above.

22C. Conformal maps and harmonic functions

The property of conformal maps that makes them useful for Dirichlet problems is that they transform a solution of Laplace's equation into a solution of Laplace's equation; that is, they transform harmonic functions into harmonic functions. When a conformal mapping is continuous in the closure of a region, it automatically carries boundary values to boundary values.

To show that a conformal map preserves harmonic functions, let $w = f(z)$ map a region D in the z plane to the region Δ in the w plane, so that $w = u + iv$, $u = u(x, y)$, $v = v(x, y)$, and let $\phi(u, v)$ be harmonic in Δ. We want to show that $\phi[u(x, y), v(x, y)]$ satisfies Laplace's equation in x and y.

EXERCISE 22.1 Calculate the derivatives $\partial\phi/\partial x$, $\partial\phi/\partial y$, $\partial^2\phi/\partial x^2$, $\partial^2\phi/\partial y^2$ by applying the chain rule and see what happens, given that $\phi(u, v)$, $u(x, y)$, and $v(x, y)$ satisfy the Laplace equation and that the derivatives of u and v are connected by the Cauchy–Riemann equations.

22D. Correspondence of regions

In using conformal mappings, we need to be able to show that specified regions actually correspond to each other under a mapping. In simple cases it is easy to verify directly that the regions in question correspond to each other in a one-to-one way. In more complicated cases it may be easier to show that the boundaries correspond to each other in a one-to-one way and then appeal to Darboux's theorem (Sec. **13C**) in order to verify that the regions themselves correspond.

In practice we often want to use Darboux's theorem when one or both of D and S are unbounded. We consider only the case when D and S have exterior points. Let a be an exterior point of D; then the transformation $w = 1/(z - a)$ carries D to a bounded region D' in an obviously one-to-one conformal way. Similarly, if b is a point outside S, $\zeta = 1/(w - b)$ carries S to a bounded region S'. Now Darboux's theorem applies to D' and S', and hence by composition of mappings, to D and S.

The following mapping, which could easily enough be analyzed directly, will illustrate the use of Darboux's theorem; we shall meet this mapping again in Sec. **26**. Consider the mapping $w = z^2 = x^2 - y^2 + 2ixy$, and suppose that we are interested in knowing what maps to the w region $0 \le v \le 1$ (a horizontal strip). Since $v = 2xy$, points satisfying $2xy = 0$ or $2xy = 1$ map to $v = 0$ and to $v = 1$ (Fig. 22.1, p. 182). It is a good guess that a region D bounded by a branch of the hyperbola $2xy = 1$ and two semiaxes will be a conformal image of the the interior of the strip S: $0 < v < 1$ in the w plane. By Darboux's theorem, we need only know that the map is one-to-one between the boundaries. The following procedure is really unnecessarily elaborate for this simple example, but it suggests what to do in more complicated situations.

Let's consider the ray $x > 0$, $y = 0$. Then $u = x^2$, $v = 0$, and as x goes from 0 to ∞, u increases (strictly) from 0 to ∞, so that w describes the u axis from 0 to $+\infty$, just once. On the ray $x = 0$, $y > 0$, $u = -y^2$, $v = 0$, and as y goes from $+\infty$ to 0, u increases from $-\infty$ to 0. So the left-hand and lower boundary of D, traced from top to right via 0, corresponds univalently to the u axis from $-\infty$

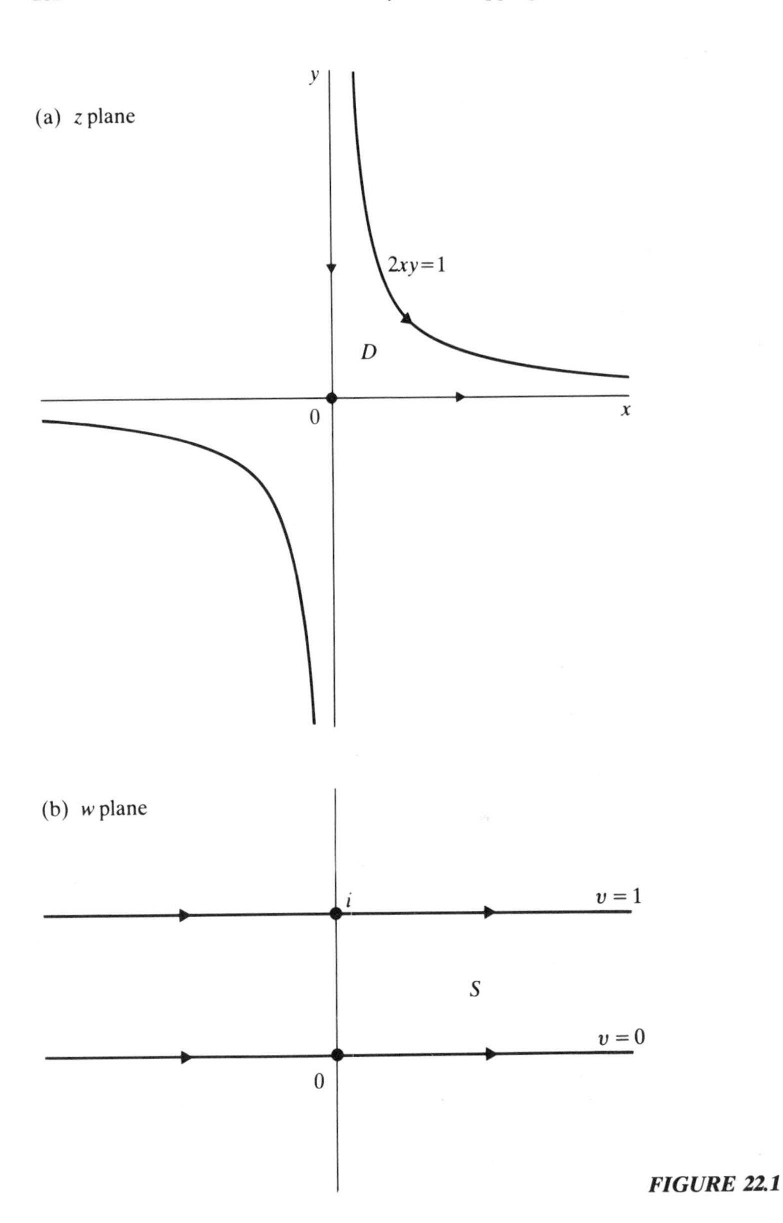

FIGURE 22.1

to $+\infty$. Now consider the upper branch of the hyperbola $2xy = 1$. Then $u = x^2 - 1/(4x^2)$, $v = 1$, and when x goes from 0 to $+\infty$, u goes from $-\infty$ to $+\infty$. Moreover, u always increases, because $du/dx = 2x + 1/(2x^3) > 0$. We now have the correspondence between the boundaries of D and S indicated in Fig. 22.1. It follows that the mapping is one-to-one between D and S. A similar argument applies to any hyperbola $xy = c$ or to the entire first quadrant (which maps to the upper half of the w plane).

Notes

1. Sometimes a map is called directly conformal if it is conformal in the sense defined in Sec. **22A**, and conformal if it just preserves magnitudes of angles.
2. Proved by D. Menshov [Menchoff], *Math. Ann.* *95* (1926): 641–670. A detailed discussion of this and more general results is given in Markushevich, *Selected Chapters*.

23 *Some Applications of Conformal Mapping to Physics*

We saw in Sec. **20A** that there are physical problems in which we need to solve Dirichlet problems with piecewise constant boundary values. For particular geometric configurations, these Dirichlet problems can sometimes be solved by inspection, or by using Fourier series. In more general situations, conformal mapping is useful because a conformal map carries harmonic functions to harmonic functions. Consequently, it may be possible to solve an apparently complicated problem by solving a simpler problem and then transforming the solution by conformal mapping. This process is used, for example, for solving problems about fluid flow, electrostatics, and heat. To explain how it is used, we need some physical terminology, which I present first for fluid flow as perhaps the most intuitive application. You should understand that the mathematical models of such phenomena are only highly simplified first approximations to real problems; nevertheless, they are physically helpful and suggest interesting mathematical problems.

23A. Terminology of fluid flows

Consider a fluid that is flowing in a channel or lake of uniform depth, and in horizontal layers, so that a horizontal cross section at any level represents the whole flow. We may then idealize the flow as two-dimensional, and think of it as if it were the flow of a plane sheet of two-dimensional fluid in some region. There is a velocity vector $\mathbf{v} = p\mathbf{i} + q\mathbf{j}$ that specifies the velocity at each point. We assume that the fluid is **incompressible** (which means what it sounds like), and **nonviscous**. The latter term means that there is no internal friction, and no tendency for the fluid to stick to the boundary of the region in which it is flowing. We also suppose that the flow is **irrotational**, which in mathematical language says that $\operatorname{curl} \mathbf{v} = 0$, and in physical language that there are no local whirlpools in the fluid: A small paddlewheel in the fluid has no tendency to revolve. Under these assumptions, it is shown in physics that the velocity vector is the gradient[1] of a **potential** ϕ, so that $p = \partial\phi/\partial x$ and $q = \partial\phi/\partial y$, where ϕ is harmonic. Then ϕ is called the **velocity potential**, and the level curves $\phi = \text{constant}$ are **equipotentials**. The conjugate harmonic function ψ (which exists if the flow is in a simply connected region, as we usually assume) is the **stream function**, and its level curves $\psi = \text{constant}$ are orthogonal to the equipotentials. Since the derivative of ϕ along an equipotential is evidently zero, the flow must be directed along the curves of constant ψ, which are the **streamlines**. The integral $\int_A (d\phi/d\mathbf{n})\, ds$ along an arc A of an equipotential is proportional to the amount of fluid crossing the arc A per unit time.

The analytic function $f = \phi + i\psi$ is the **complex potential**, f' is the **complex velocity**, and $|f'(z)|$ is the **speed** at z.

EXERCISE 23.1 Show that the vector velocity $\mathbf{v} = p\mathbf{i} + q\mathbf{j}$ has components $p = \operatorname{Re} f'$, $q = -\operatorname{Im} f'$.

Since the function $if(z)$ is also analytic, $-\psi$ and ϕ are the velocity potential and stream function of another flow, whose streamlines are the equipotentials of the original flow. One usually pays more attention to the streamlines than to the equipotentials, because if we draw the streamlines it is easy to visualize what the flow looks like.

Notice that if $f'(z_0) = 0$, the speed is zero at z_0; then z_0 is called a **stagnation point**. On the other hand, if $f'(z) \to \infty$ as $z \to z_0$, the mathematical model of the flow breaks down near z_0; however, we may conclude that the speed must be very large near z_0.

If the flow is taking place in a region whose boundary contains a curve C across which fluid cannot flow—that is, a rigid impermeable boundary—the velocity vector at points of C must be tangent to C. Therefore a rigid boundary is always a streamline. Hence an analytic function whose imaginary part is constant on each of one or more rigid boundaries (generally, different constants on different boundaries) is the complex potential of a flow that has these boundaries as streamlines. On the other hand, if we have a flow and insert a rigid boundary along part or all of a streamline, the appearance of the flow is not changed.

EXERCISE 23.2 Show that if C is a simple closed curve and the flow takes place inside C, with C as a single streamline, then the complex potential is constant inside C (so that nothing is flowing).

It follows that we can have an interesting flow inside a simple closed curve C only if there are singular points on C, points at which fluid is being supplied or removed.

For an unbounded region—for example, the strip between two parallel lines—different parts of the boundary can be different streamlines, with a singular point at ∞.

In practice, we are likely to want to solve the converse problem: Describe a flow with prescribed streamlines along the boundary (in other words, find a harmonic function with piecewise constant boundary values). That is, we want to solve a Dirichlet problem with piecewise constant boundary values. Such a problem does not have a unique solution until we make appropriate assumptions about the behavior of the solution at the singular points. (See Secs. **20A** and **24B**.)

23B. Other physical interpretations

Every flow problem can be interpreted as a problem about the flow of heat or about electrostatic potentials. The physical terminology is, of course, different. The heat problem is the problem of steady-state **temperatures** in a two-dimensional lamina with its flat faces insulated, or alternatively, a solid, homogeneous cylinder all of whose cross sections parallel to a given plane have the same (not necessarily circular) shape, and with boundary conditions imposed on the edges of the lamina or the surface of the cylinder. A simple and familiar example of this model is the door of a refrigerator, with given distributions of temperature on the inside and outside surfaces. The temperature T satisfies Laplace's equation; the curves $T = \text{constant}$ are **isothermals**. If U is the harmonic function conjugate to T, then the curves $U = \text{constant}$ are the lines along which heat flows. The quantity of heat crossing an arc A per unit time is proportional to $\int_A (\partial T/\partial \mathbf{n})\, ds$. When $\partial T/\partial \mathbf{n} = 0$ on A, no heat is flowing through A and we say that A is **insulated**.

EXERCISE 23.3 Show that a conformal map carries an insulated boundary to an insulated boundary.

We can interchange isothermals and lines of heat flow to get another heat problem, and we can interpret heat problems as fluid flow problems (in two ways).

Two-dimensional **electrostatic** problems are cross sections of three-dimensional problems, where all cross sections parallel to a specified plane are congruent. A **conductor** is the cross section of an actual conductor, and a **point charge** is the cross section of a uniform line charge perpendicular to the plane that we are considering (*not* an isolated point charge in 3-space). Then the potential ϕ is harmonic in regions that contain no charges; if $\phi + i\psi$ is an analytic function $f(z)$, then the force on a unit charge is proportional to $|f'(z)|$. Curves $\psi = \text{constant}$ are **lines of force**, and conductors are **equipotentials**, $\phi = \text{constant}$.

To illustrate, the harmonic function in the unit disk with boundary values a on the upper semicircular boundary and b on the lower can be interpreted in several ways: as describing a flow from a source at -1 to a sink at $+1$; as giving the temperatures in the disk when the upper boundary is held at temperature a and the lower boundary at temperature b; or as describing the electric field in the disk when the upper boundary is at potential a, the lower boundary is at potential b, and there are bits of insulation at ± 1. In the last case we have a **capacitor**.[2] We could also interchange the equipotentials and the flow or field lines, and have fluid or heat flowing from an extended source along the upper semicircle to the lower one, or point charges at ± 1 with current flowing between them. One point of view may be more appropriate than another for a given physical problem.

Notes

1. Some authors take the velocity to be the negative gradient of the potential. You will have to allow for this if you look up further applications in other books.
2. The old term "condenser" for a capacitor still survives in some mathematical writing.

24 *Some Special Flows*

We shall begin by constructing the flows generated by some simple analytic functions. We can then use conformal mapping to construct more complicated flows, using the principle that a conformal map carries the solution of a Dirichlet problem to the solution of another Dirichlet problem.

24A. Uniform flows

The simplest flow of all has complex potential z. (More generally, cz, $c > 0$; but we can suppose that the units have been chosen so that $c = 1$.) Then $dw/dz = 1$, so that the velocity is constant, parallel to the real axis, and from left to right. If we take the real axis as a rigid boundary, we have a uniform flow in the upper half plane; if we also take a line $y = k$ as a rigid boundary, we have a flow in the channel between two parallel lines. In any of these cases the flow is referred to as a **uniform flow**. The streamlines are the lines $y = \text{constant}$.

You may wonder about this uniform flow: Where does the fluid come from, and where does it go? We have to think of there being a source of fluid at ∞ and a sink, where fluid is being removed, also at ∞, so arranged that the fluid leaves the source, flows to the right, and then disappears at ∞. We need to have a mathematical description of sources and sinks; it will be simplest to consider a source and a sink separately at finite points before trying to combine them.

24B. Sources and sinks

Consider the complex potential $w = m \log z$. Since this is not a (single-valued) function, we have to consider it only in the plane cut from 0 to ∞, for example along the positive real axis, with $\operatorname{Im} \log z$ between 0 and 2π. The complex velocity is $m/z = (m/r)(\cos\theta - i \sin\theta)$. Then the speed is m/r (the fluid is flowing very fast as it leaves 0, and slowing down as it gets farther away). The flow is outward along rays from the origin ($\theta = \text{constant}$) because the velocity vector is $(m/r)(\mathbf{i}\cos\theta + \mathbf{j}\sin\theta)$. Hence our cut along the positive real axis is along a streamline, and we can disregard its presence. Now let the fluid have unit density and let us think how much fluid crosses a circle $|z| = r > 0$ in unit time; this is

$$\int_{|z|=r} \frac{\partial\phi}{\partial\mathbf{n}}\,ds = \int_0^{2\pi} \frac{\partial(m\log r)}{\partial r}\,r\,d\theta = \int_0^{2\pi} m\,d\theta = 2\pi m,$$

so that the same amount of fluid crosses (in unit time) each circle centered at 0. This is what we would intuitively expect from a point source that emits fluid at the rate $2\pi m$ per unit time. Hence we say that $m\log z$ is the complex potential of a **source** at 0, of **strength** m. Physically speaking, we can have a steady flow only if the source is balanced by a sink (or a distribution of sinks), so that the fluid has somewhere to go. A **sink** is defined as a negative source; its complex potential is $-m\log(z - z_0)$ if it is located at a finite point z_0.

If a flow has complex potential $g(z) + \log(z - z_0)$, where g is bounded in a neighborhood of z_0, we still say that there is a **source** at z_0, since near z_0 the logarithm overwhelms g. One says that the flow is the flow corresponding to g, but modified by the presence of the source. A simple example is the flow produced if a source is inserted into a uniform flow; it seems reasonable to expect that the streamlines resemble those for the source near z_0, but pushed to the right, whereas far from z_0 they will be much like those of the uniform flow. This expectation is justified in the following exercise.

EXERCISE 24.1 For the flow with complex potential $w = z + \log z$, draw some streamlines, particularly for $v = 0, 1, \pi/2, 2.5, \pi, 3.5, 4$. Suggest one or two physical problems that this flow models.

We can also insert a sink into the flow from a source (or vice versa); since the complex potential of each flow is analytic at the singular point of the other, we say that the flow is generated by the combination of an isolated source and an isolated sink. Suppose that the source is at $z = a$ and the sink is at $z = -a$; then the complex potential is $m\log[(z-a)/(z+a)]$, and the streamlines are the curves $\operatorname{Im} f(z) = \text{const}$. The stream function is $m[\arg(z-a) - \arg(z+a)]$, and saying that $\operatorname{Im} f(z) = \text{constant}$ means that the angle γ at z in Fig. 24.1

FIGURE 24.1

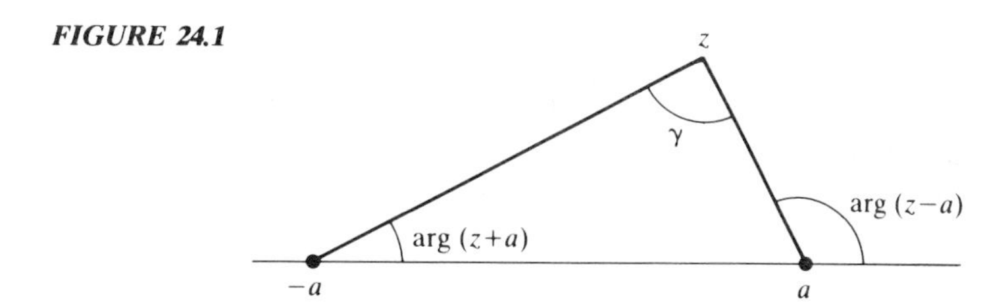

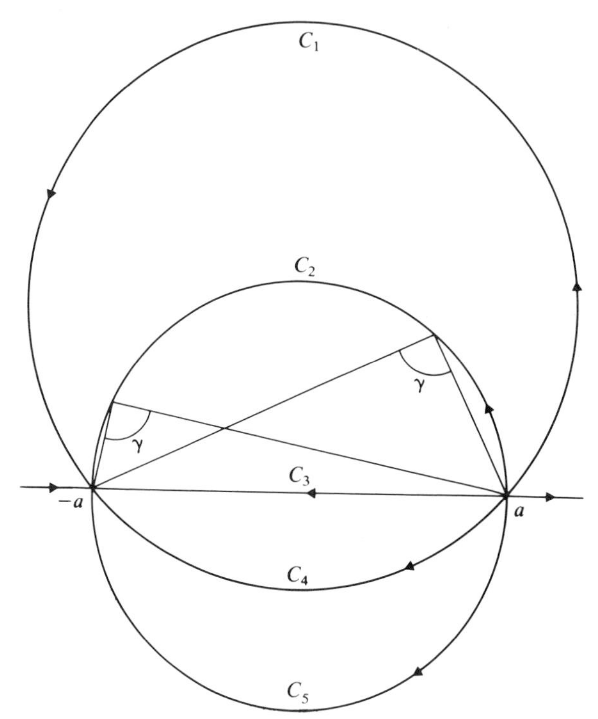

FIGURE 24.2

is constant. Consequently, the streamlines are portions of coaxial circles through a and $-a$; some examples are $C_1, \ldots, C_5$ in Fig. 24.2.

Figure 24.3 shows both the streamlines and the equipotentials.

EXERCISE 24.2 Show that the harmonic function given by Poisson's integral (Sec. **20B**) for $|z|<1$ with $\psi=+1$, $0<\theta<\pi$, $\psi=-1$ for $-\pi<\theta<0$, is a constant multiple of the stream function for the complex potential $\log[(1+z)/(1-z)]$, namely, $\tan^{-1}[2y/(1-x^2-y^2)]$.

Now suppose that we let $a\to\infty$ and replace m by am, so that the source and sink get stronger as they get farther apart (if we didn't do this, we wouldn't expect to have much effect on the finite part of the plane). Then if z is fixed, z/a is small and

$$ma\log\frac{a+z}{a-z}=ma\left[\frac{2z}{a}+O(a^{-2})\right],\qquad a\to\infty.$$

In the limit we get $2mz$, the complex potential of a uniform flow. This explains why we can regard a uniform flow as the flow from a source–sink pair at ∞. Such a pair is called a **doublet**.

The function $2mz$ has a simple pole at ∞. If we consider complex potentials

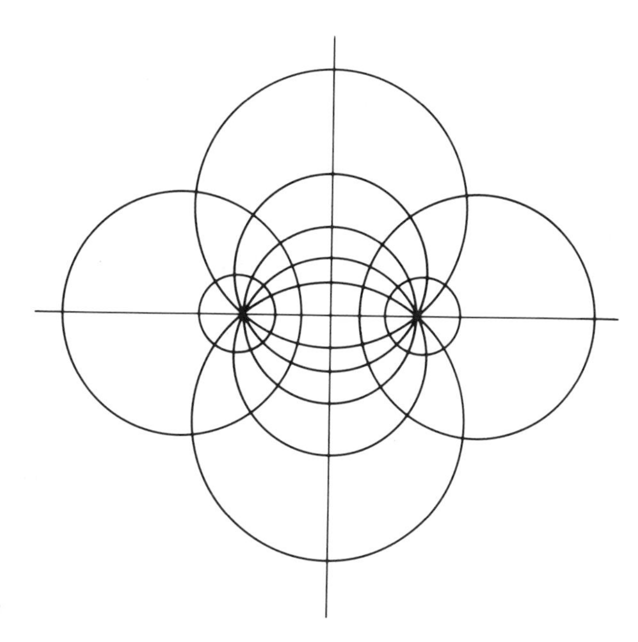

FIGURE 24.3

with poles of higher order at ∞, or even with essential singularities at ∞, we would get progressively more complicated flow patterns (see Exercise 24.4).

We can also have a doublet at a finite point, for example the origin. Since a conformal map preserves Laplace's equation, it transforms streamlines into streamlines. Since the complex potential of a doublet at ∞ is kz, it follows that k/z is the complex potential of a doublet at 0 (oriented so that the source is at the left; we can change the orientation by multiplying by a factor $e^{i\alpha}$). Similarly, $k/(z - z_0)$ is the complex potential of a doublet at z_0.

EXERCISE 24.3 Show that the streamlines for a doublet at z_0 are circles tangent to a horizontal line through z_0. Draw a figure showing the direction of flow.

To describe a uniform flow in a half plane, it would not be sufficient just to find a function that is analytic in the half plane (say, in $y > 0$) with imaginary part zero on the boundary. For example, $\operatorname{Im} z^2$ is harmonic in the upper half plane and zero on the real axis, but the complex potential z^2 belongs to a very different flow (Sec. **22D**). The example $\operatorname{Im} e^z = e^x \sin y$ generates a much more complicated flow.

EXERCISE 24.4 Sketch some streamlines for $\operatorname{Im} e^z$.

There is a sense in which we can say that $\operatorname{Im} z$ is the *smallest* function (not identically zero) that is harmonic in the upper half plane and zero on the real axis. More precisely, *if v is harmonic for $y > 0$, continuous for $y \geq 0$, and 0 for real z, and if $|v(x, y)| \leq Ay$, then $v(x, y) = cy$ with $-A \leq c \leq A$.*

To see this, continue v to the lower half plane by Schwarz's principle (Sec. **20F**) and then form an analytic function f of which v is the imaginary part. By Carathéodory's inequality (Sec. **16C**), applied to if, we have $f(z) = O(|z|)$. Then, by the general form of Liouville's theorem for harmonic functions (Exercise 16.9), $f(z) = az + b$. Since $\operatorname{Im} f = 0$ when $y = 0$, b is real. Taking $a = c + i\beta$, we get $\operatorname{Im} f = \operatorname{Im}[(c + i\beta)(x + iy)] = \beta x + cy$. Since $v = 0$ on the real axis, $\beta = 0$, and so $v = cy$.

In particular, if v grows so slowly that $v(y)/y \to 0$ as $y \to \infty$, v must be the constant 0.

Usually "the" flow in a region is understood to mean the flow obtained by a conformal map of a uniform flow.

EXERCISE 24.5 The function $\phi(z) = z + (1/z)$ is the potential for a uniform stream disturbed by a doublet at 0. Discuss the streamlines in the upper half plane and observe that one streamline contains the upper half of the circle $|z| = 1$; hence the flow outside the circle is the flow around a semicircular obstacle. Find the speed at the top of the circle and notice that this is the greatest speed on the corresponding streamline. This means, by Bernoulli's principle (pressure is inversely proportional to speed) that the fluid exerts an upward force on the obstacle. This suggests that a river tends to lift a log that lies on its bed.

Supplementary exercises

1. Write the complex potential for a flow obtained by inserting a sink into a uniform flow.
2. Write the complex potential for the flow from a pair of sources, one at $+1$ and one at -1. Sketch some streamlines.
3. If we consider two oppositely oriented doublets of equal strengths at $\pm a$, let $a \to 0$, and let the strengths of the doublets become infinite, we get what is called a quadrupole at 0. Find its complex potential.

   (One can also get the potential of a doublet in this way; the method used in the text for the doublet will also work here, but not quite as simply.)

24C. Reflection principles

Consider a distribution of point singularities (sources, sinks, doublets, etc.) at finite points in the open upper half plane. Let its complex potential be $f(z)$. The real axis cannot be expected to be a streamline of the corresponding flow, but it is easy to find a flow with the same singularities in the upper half plane for which the real axis *is* a streamline; we can think of the new flow as the original flow disturbed by the introduction of a wall along the real axis. In fact, the new complex potential is simply $f(z) + g(z)$, where $g(z) = \overline{f(\bar{z})}$. To see this, recall that (Exercise 2.5) g will be analytic in the lower half plane except for singularities at finite points that are the conjugates of the singular points of

f, and therefore all in the lower half plane. (Compare Schwarz's principle, Sec. **17C**.) On the real axis, $z = \bar{z}$ and $\overline{f(\bar{z})} = \overline{f(z)}$; therefore $f(z) + g(z)$ is real on the real axis, that is, its imaginary part is zero on the real axis. Consequently, $f + g$ has the real axis as a streamline, and has, in the upper half plane, the same sources, sinks, etc., as f.

For example, $(z - i)^{-1}$ is the complex potential of a doublet at i. Then $g(z) = (z + i)^{-1}$, and so $(z - i)^{-1} + (z + i)^{-1} = 2z/(z^2 + 1)$ is the complex potential of a flow with a doublet at i and a barrier along the real axis. The effect is the same as if we inserted a doublet of the same strength at the mirror image of the original doublet (similarly in the general case); in electrostatics this is called the **method of images**.

There is a similar method with reflection in a circle instead of a straight line; in hydrodynamics it is known as Milne-Thomson's **circle theorem**.[1] One could obtain it by a conformal transformation that sends the real axis to a circle, but it is simpler to guess the answer. If the original flow (which may now have a singularity at ∞) has only isolated point singularities anywhere in the plane, but none in the disk $|z| \leq a$, then inserting the circle C: $|z| = a$ as a barrier changes the complex potential from $f(z)$ to $f(z) + f^*(a^2/z)$, where f^* is obtained from f by conjugating the Maclaurin coefficients of f. The proof is simple: On C, $\bar{z} = a^2/z$, and so $w = f(z) + f^*(\bar{z})$ is real on C; $\operatorname{Im} w = 0$ on C, so C is a streamline. When z is outside C, a^2/z is inside, so w has the same singular points as f outside C, the point ∞ included. As an example, insert the circle $|z| = 1$ as a barrier into the uniform flow $w = z$. Then the complex potential is $w = z + (1/z)$, which happens also to be the potential of a uniform stream disturbed by a doublet (Exercise 24.5).

Note

1. For more on the circle theorem, see O. Popa, "Four New Expressions of the Circle Theorem," *Rev. Roumaine Math. Pures Appl.* *27*, no. 4, (1982): 489–494 (*Math. Rev.* *83k*: 30041).

25 *Möbius Transformations*

We have now seen a number of particular physical problems that can be solved by using particular analytic functions. However, we have not yet made much use of conformal mapping. We shall see in Sec. **26** how we can use conformal mapping to construct solutions of some less transparent problems. It will be convenient first to become acquainted with a special class of mappings that are useful in many contexts, both pure and applied. These are the Möbius transformations.

25A. Definitions; simple properties

A **Möbius transformation**[1] is a function of the form

$$w = \frac{az + b}{cz + d},$$

where a, b, c, d are complex numbers and $ad - bc \neq 0$. (If $ad = bc$, then w is a constant and we dismiss the transformation as being uninteresting.) We consider Möbius transformations in the extended plane; formulas involving ∞ are to be interpreted as the corresponding limits.

A few facts are obvious. A Möbius transformation is a one-to-one map of the extended z plane to the extended w plane, with $w = 0$ when $az + b = 0$ and $w = \infty$ when $cz + d = 0$. When $z = 0$, $w = b/d$ if $d \neq 0$, and $w = \infty$ if $d = 0$.

A Möbius transformation is a composition of magnifications, rotations, translations, and a reciprocal transformation $w = 1/z$. This is immediate if $c = 0$. If $c \neq 0$, write $w_1 = cz + d$ (magnification and rotation, followed by translation); $w_2 = 1/w_1$; $w_3 = [b - (ad)/c]w_2$ (another magnification and rotation); $w_4 = w_3 + (a/c)$ ("magnification" includes "minification").

The Möbius transformations form a group. If we identify a Möbius transformation with the matrix

$$\begin{pmatrix} a & b \\ c & d \end{pmatrix},$$

you can easily see that the composition of two transformations corresponds to the product of their matrices, and the inverse transformation corresponds to the inverse matrix.

EXERCISE 25.1 A Möbius transformation that carries the real axis to the real axis can be written with real coefficients a, b, c, d.

It is obvious that there is a Möbius transformation that carries any three distinct points to any three prescribed distinct points.

25B. Cross ratio[2]

The **cross ratio** of four distinct points is defined by

$$[z_1, z_2, z_3, z_4] = \frac{(z_1 - z_2)(z_3 - z_4)}{(z_1 - z_4)(z_3 - z_2)}.$$

If one of the four points is ∞, the cross ratio is to be interpreted by taking limits in the obvious way.

We are going to prove the not very intuitive fact that the cross ratio is invariant under Möbius transformations. We begin by introducing a transformation T, acting on points z, by $Tz = [z, z_2, z_3, z_4]$. Then T is clearly a Möbius transformation, and it carries z_2, z_3, and z_4 to 0, 1, and ∞, respectively. Furthermore, this T is the only Möbius transformation that carries z_2, z_3, and z_4 to 0, 1, ∞. Indeed, suppose that S has the same property. Then ST^{-1} is a Möbius transformation (because of the group property) which would have 0, 1, and ∞ as fixed points. {I use the notation in which ST is the transformation defined by $(ST)(z) = S[T(z)]$.}

EXERCISE 25.2 Show that a Möbius transformation that carries 0, 1, ∞ to 0, 1, ∞ (in that order) is the identity.

Now let S be any Möbius transformation, and form the cross ratio $[Sz_1, Sz_2, Sz_3, Sz_4]$. Let T be the transformation defined above; then TS^{-1} takes Sz_2, Sz_3, Sz_4 to 0, 1, ∞. Therefore, by the uniqueness property just mentioned, $[Sz_1, Sz_2, Sz_3, Sz_4]$ is the image of Sz_1 under TS^{-1}, which is to say that it is $TS^{-1}Sz_1 = Tz_1 = [z_1, z_2, z_3, z_4]$. We therefore have $[Sz_1, Sz_2, Sz_3, Sz_4] = [z_1, z_2, z_3, z_4]$, which is what we wanted to prove.

The invariance of the cross ratio gives us a way of writing down *the Möbius transformation that carries any three given distinct points to three prescribed points.* In fact, the Möbius transformation defined by $[w, w_2, w_3, w_4] = [z, z_2, z_3, z_4]$ is the Möbius transformation that carries z_2, z_3, z_4 (in that order) to w_2, w_3, w_4 (in that order).

This is an attractive general formula, but it does not always provide the most convenient way of getting a specific Möbius transformation, because of the algebra involved in solving for w in terms of z. In fact, when the specified points are simple numbers, it is often easier to work directly from the original definition of a Möbius transformation. Here is an illustration of the technique.

Example Find the Möbius transformation that carries the points 1, 0, i to i, 1, ∞ (in that order).

We start from

$$w = \frac{az + b}{cz + d}.$$

Since 0 goes to 1, we must have $b/d = 1$, so

$$w = \frac{az + b}{cz + b}.$$

Next, since i goes to ∞, $z = i$ must make the denominator zero. Hence we have $ci + b = 0$, $b = -ic$, and consequently,

$$w = \frac{az - ic}{cz - ic}.$$

Finally, since 1 goes to i,

$$i = \frac{a - ic}{c - ic}, \qquad \text{whence } a = c(1 + 2i).$$

The coefficient c is at our disposal (as long as we don't try to choose $c = 0$); let's take $c = 1$. Then

$$w = \frac{(1+2i)z - i}{z - i}.$$

This procedure can always be used; in the worst case, it leads to a system of three linear equations for the four numbers a, b, c, d, one of which can eventually be chosen arbitrarily. You must, however, not try to specialize the coefficients too soon. For example, if you are asked for a Möbius transformation that takes 0, 1, $1+i$ to 1, ∞, i, you must not decide in advance to take $a = 1$, because the transformation has to turn out to be $w = 1/(1-z)$.

Also see Sec. **25E** for other techniques for finding special Möbius transformations.

If the points z_1, z_2, z_3 are all real (one of them might be ∞), then the definition of the transformation T shows that Tz is real when z is real, and the same is of course true for T^{-1}. It follows that the image of the extended real axis (provided with *one* point ∞) is the extended real axis whenever T is real for real z.

25C. Transformation of circles

Next we show that *every Möbius transformation takes the real axis to either a circle or a straight line*; and conversely, every circle or straight line is the image of the real axis under some Möbius transformation.

When T is a magnification with rotation or a composition of several such transformations, it is evident that the image of the real axis is a straight line, and the image of a circle is a circle. Since every Möbius transformation is a composition of these transformations with the reciprocal transformation $w = 1/z$, we need only show that the composition of the latter transformation with translations and rotations takes the real axis to a circle or a straight line. For this it is enough to consider the transformation $Tz = 1/(z - \lambda)$. If z is real, $w = 1/(z-\lambda)$ says that $z = w^{-1} + \lambda$, and $\operatorname{Im}(w^{-1} + \lambda) = 0$. Since $\operatorname{Im}(1/w) = -v/|w|^2$, we therefore have either λ real, $v = 0$ (the real w axis), or $\operatorname{Im}\lambda = v/|w|^2$, that is, $(u^2 + v^2)\operatorname{Im}\lambda - v = 0$, the equation of a circle. Hence the image of the real axis is always a circle or a straight line.

We must also show that, conversely, *every circle or straight line is the image of the real axis under some Möbius transformation.* Evidently, every straight line is the image of the real axis under a translation (if the line is parallel to the real axis) or a rotation about the intersection of the line with the real axis. For a circle, translate it so that it passes through the origin and rotate it so that its center is at a positive real point p. Then it has an equation $|z - p| = p$ $(p \neq 0)$. Under $z = 1/w$, this becomes $|1 - pw| = p\,|w|$, or

$$(1 - pw)(1 - p\bar{w}) = p^2 w\bar{w},$$

which simplifies to $\operatorname{Re} w = 1/(2p)$, which is indeed the equation of a straight line. The original circle is the image of the real axis under the composition of the inverses of these transformations.

It is now easy to see that *every Möbius transformation of a straight line or circle is again a straight line or a circle* (although circles may be transformed into straight lines or vice versa). Let C be a circle or straight line and let Γ be another such figure. Since C is the image of the real axis under a Möbius transformation T, and Γ is the image of the real axis under a Möbius transformation S, then T^{-1} takes C to the real axis, and $S^{-1}T^{-1}$ takes C to Γ.

25D. Preservation of symmetry

Next we show that not only does a Möbius transformation transform a circle or straight line to a circle or straight line, but it also transforms a pair of points symmetric with respect to the original figure to a pair symmetric with respect to the second figure. In other words, *the property of being a pair of symmetric points is preserved by Möbius transformations.* This can be verified by straightforward computation, but the following argument is more elegant.

Since Möbius transformations preserve cross ratios, if z_2, z_3, and z_4 are three points on a given circle and T is any Möbius transformation, we have

$$[z, z_2, z_3, z_4] = [Tz, Tz_2, Tz_3, Tz_4].$$

We now use this principle repeatedly to transform the cross ratio $[z^*, z_2, z_3, z_4]$. First apply a translation to get

$$[z^*, z_2, z_3, z_4] = [z^* - a, z_2 - a, z_3 - a, z_4 - a],$$

where a is the center of the circle. By definition,

$$z^* - a = \frac{R^2}{\bar{z} - \bar{a}}.$$

Hence we have

$$[z^*, z_2, z_3, z_4] = \left[\frac{R^2}{\bar{z} - \bar{a}}, z_2 - a, z_3 - a, z_4 - a\right].$$

Now apply the transformation $Tz = R^2/z$ to get

$$\left[\bar{z} - \bar{a}, \frac{R^2}{z_2 - a}, \frac{R^2}{z_3 - a}, \frac{R^2}{z_4 - a}\right],$$

and then, since z_2, z_3, z_4 are on the circle, and therefore fixed under symmetry,

$$[\bar{z} - \bar{a}, \bar{z}_2 - \bar{a}, \bar{z}_3 - \bar{a}, \bar{z}_4 - \bar{a}].$$

The conjugate of a cross ratio is clearly the cross ratio of the conjugates of its entries, so we now have

$$[z-a, z_2-a, z_3-a, z_4-a] = \overline{[z, z_2, z_3, z_4]}.$$

$$\overline{[z-a, z_2-a, z_3-a, z_4-a]} = [\overline{z, z_2, z_3, z_4}].$$

Hence we see that a pair of points z, z^* symmetric with respect to the circle through z_2, z_3, z_4 satisfies

$$[z^*, z_2, z_3, z_4] = \overline{[z, z_2, z_3, z_4]}.$$

This equation could be taken as an alternative definition of symmetry of z and z^* with respect to the circle through z_2, z_3, z_4, since the preceding reasoning can be reversed step by step.

Now let T be any Möbius transformation; then, since T preserves cross ratios,

$$[z, z_2, z_3, z_4] = [Tz, Tz_2, Tz_3, Tz_4]$$

and

$$[z, z_2, z_3, z_4] = \overline{[z^*, z_2, z_3, z_4]} = \overline{[Tz^*, Tz_2, Tz_3, Tz_4]}.$$

Consequently, if z and z^* are symmetric with respect to C, then Tz and Tz^* are symmetric with respect to the circle through Tz_2, Tz_3, Tz_4.

Similarly, but more easily, we can show that the same equation characterizes symmetry of z and z^* with respect to a straight line that contains z_2, z_3, and z_4. It is enough to do this when the line is the real axis, because any other line can be carried to the real axis by a rigid motion, which preserves symmetry of points with respect to a line. For the real line, we evidently have

$$[\bar{z}, z_2, z_3, z_4] = \overline{[z, z_2, z_3, z_4]}$$

if z_2, z_3, z_4 are all real.

Now let T be any Möbius transformation and let z and z^* be symmetric with respect to a circle (or straight line) C. Since T preserves cross ratios, Tz and Tz_k $(k = 2, 3, 4)$ satisfy the same equation as z and z_k, so Tz and Tz^* are symmetric points.

25E. Construction of special Möbius transformations

The proposition that a Möbius transformation carries symmetric points to symmetric points is useful in finding Möbius transformations that have a desired effect. If we want a Möbius transformation that takes a given disk to a given disk (or half plane), the most direct approach would be to make three points on the boundary correspond to three points on the boundary [being careful about orientation: We have to be sure that the inside of the original set corresponds to the inside of the target set; we can do this by numbering the points so that regions inside are both on the right (or both on the left) as we go through the points]. This approach can lead to tedious algebra. It is often simpler to make one boundary point correspond to a boundary point, and an interior point and its symmetric point correspond to an interior point and its symmetric point.

Suppose, for example, that we want a Möbius transformation from the upper half plane to the unit disk. We can send 0 to 1 and the symmetric points i and $-i$ to the symmetric points 0 and ∞. Then, if $w=(az+b)/(cz+d)$, making $z=0$ correspond to $w=1$ gives us $b=d$. Then $z=i$ corresponds to $w=0$ and $z=-i$ corresponds to $w=\infty$, so $ai+b=0$ and $-ci+d=-ci+b=0$; therefore $a=ib$, $c=-ib$, and

$$w=\frac{ibz+b}{-ibz+b}=\frac{i-z}{i+z}.$$

EXERCISE 25.3 Show that the most general Möbius transformation from the upper half plane to the unit disk can be written in the form

$$w=e^{i\theta}\frac{z-\mu}{z-\bar{\mu}},$$

where θ is real and μ is any point in the upper half plane.

Now let us find *the most general Möbius transformation from the unit disk to the unit disk*. Here $z=0$ and $z=\infty$ correspond to symmetric points w and $1/\bar{w}$, and $w=0$ and $w=\infty$ correspond to symmetric points α and $1/\bar{\alpha}$.

EXERCISE 25.4 Carry out this plan to show that

$$w=e^{i\lambda}\frac{z-\alpha}{\bar{\alpha}z-1},\qquad |\alpha|<1,\ \lambda \text{ real}.$$

We can now understand the mysterious appearance of the symmetric point in the derivation of Poisson's formula in Sec. **20B**. For a function u, harmonic in the unit disk, we had (implicitly in Sec. **16A**, and explicitly in Sec. **20E**) the representation

$$u(0)=\frac{1}{2\pi}\int_{-\pi}^{\pi}u(re^{i\theta})\,d\theta,\qquad re^{i\theta}=\zeta,$$

for the value of u at the center. The transformation

$$w=T(\zeta)=\frac{\zeta+z}{\bar{z}\zeta+1}$$

maps $|\zeta|\le 1$ onto $|w|\le 1$ with $\zeta=0$ corresponding to $w=z$. Then $U(\zeta)=u[T(\zeta)]$ is harmonic in $|\zeta|<1$. Using the representation above for $U(0)$, we have

$$U(0)=u(z)=\frac{1}{2\pi}\int_{-\pi}^{\pi}u[T(\zeta)]\,d\arg\zeta.$$

But ζ is given by the inverse transformation

$$\zeta=\frac{w-z}{1-\bar{z}w}$$

and

$$d \arg \zeta = -i\frac{d\zeta}{\zeta}$$
$$= -i\left(\frac{1}{w-z} + \frac{\bar{z}}{1-\bar{z}w}\right) dw,$$

which is where we were at the beginning of the derivation on p. 167.

However, the pattern that we used in the last two examples of specific Möbius transformations is not always appropriate.

EXERCISE 25.5 Find the most general Möbius transformation from the upper half plane to the upper half plane. (*Hint*: This is very easy!) Then find the most general Möbius transformation from the right-hand half plane to the right-hand half plane.

We can use the ability of Möbius transformations to move points around in the plane in order to generalize theorems that originally depended on a particular point. For example, when $|f(z)| \leq 1$ in $|z| \leq 1$, and $f(0) = 0$, we showed in Exercise 16.6 (by using Schwarz's lemma) that $|f'(0)| \leq 1$.

EXERCISE 25.6 Show that if $|f(z)| \leq 1$ in $|z| \leq 1$, then $|f'(0)| \leq 1$ whether or not $f(0) = 0$.

EXERCISE 25.7 Show that if F is analytic and $\operatorname{Re} F < 0$ in the disk $|z| \leq 1$, then $|F'(0)| \leq -2 \operatorname{Re} F(0)$.

EXERCISE 25.8 Let f be analytic and $0 < |f(z)| < 1$ in $|z| \leq 1$. Show that $|f'(0)| \leq 2/e$.

An application of Möbius transformations to non-Euclidean geometry is given in Sec. **30**; you could read this now, since it uses none of the intermediate material.

Supplementary exercises

Find the Möbius transformations that take

1. $0, i, -1$ to $-1, 0, i$, respectively
2. $0, 1, \infty$ to $2, 0, 2i$
3. $0, 1, -1$ to $2, 1, \infty$
4. The unit disk to the unit disk, and $\frac{1}{2}i$ to 0.
5. Find the general Möbius transformation from the unit disk to the exterior of the unit disk.
6. Which of the pairs of configurations in Fig. S25.1 are equivalent under Möbius transformations?
7. Let f be analytic in $|z| \leq 1$, $|f(z)| \leq 1$, and $f(a) = 0$, where $|a| < 1$. How large can $|f'(a)|$ be? (If you get $|f'(\frac{1}{2})| \leq 2$, you haven't tried hard enough.)

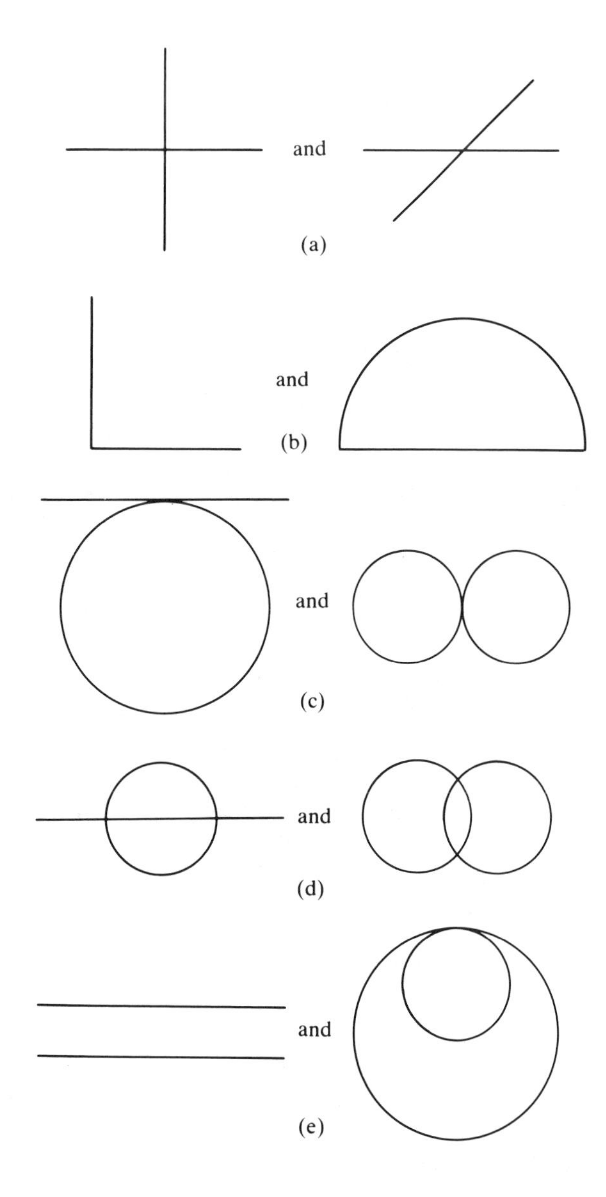

FIGURE S25.1

8. Let f be analytic in the upper half plane, $|f(z)| \leq 1, f(c) = 0$ (Im $c > 0$). How large can $|f'(c)|$ be?

Notes

1. Möbius transformations are also called fractional linear transformations, or linear fractional transformations, or bilinear transformations, or linear transformations, or homographic transformations.

2. Most of this section is adapted from Ahlfors's lectures; compare his book, chap. 3, sec. 3.

26 *Further Examples of Transformations and Flows*

26A. Inverse mappings

In Sec. **22D** we showed that the mapping $w = z^2$ provides a conformal map from the region D between the coordinate axes and the upper branch of the hyperbola $2xy = 1$ in the z plane to the strip $0 \leq v \leq 1$ in the w plane. Similarly, the lines $v = c$ are images of the upper branches of the curves $2xy = c$. Since the mapping is one-to-one, its inverse provides a map from the strip to the region D. In practice we are likely to emphasize the inverse mapping because it maps a uniform flow in the w plane to a less easily described flow in the z plane, and the inverse images of lines $v = \text{constant}$ are streamlines in the z plane.

Whenever we have a mapping $w = f(z)$, we can also study the inverse mapping provided that we select a single-valued branch of the inverse function. A branch of the inverse of $w = z^2$ is a branch of $z = \sqrt{w}$. However, in order to have a consistent notation in various problems, we shall write the inverse mapping as $w = \sqrt{z}$ and ask for the inverse images (in the z plane) of horizontal straight lines in the w plane, or of regions bounded by two of these lines.

In order to have a function $w = \sqrt{z}$, we let $z = re^{i\theta}$ and $w = r^{1/2}e^{i\theta/2}$, where $-\pi < \theta < \pi$, so that we are considering z in the plane cut along the negative real axis. Let us look at the inverse map of the boundary of the right-hand w half plane. If w starts at $u = 0$, v large and positive (informally, $w = i\infty$) and moves downward (v decreases), then z is real, negative, and of large absolute value. Since $z = u^2 - v^2 + 2iuv = -v^2$, $dz/dv = -2v$ is negative, and $z = x$ increases to 0 as v decreases. Then as w continues downward, z is negative, v is negative, and v decreases to $-\infty$. Hence the line $u = 0$ corresponds to the negative real axis described twice in opposite directions. This is not a situation to which Darboux's theorem applies directly, but we could apply the theorem to the z image of a line $u = c$ with c small and positive to show that the map is indeed one-to-one between the cut z plane and the w half plane.

Now let us see what curves in the z plane correspond to lines $u = c > 0$. If $u = c$, $z = c^2 - v^2 + 2ivc$, so $x = c^2 - v^2$, $y = 2vc$; that is, $y^2 = 4c^2(c^2 - x)$. This is the equation of a parabola opening to the left, with vertex at $(c^2, 0)$; as $c \rightarrow 0$, the parabolas collapse to the negative real axis described twice. See Fig. 26.1. Each parabola is a one-to-one image of a line $u = c$ under the inverse mapping.

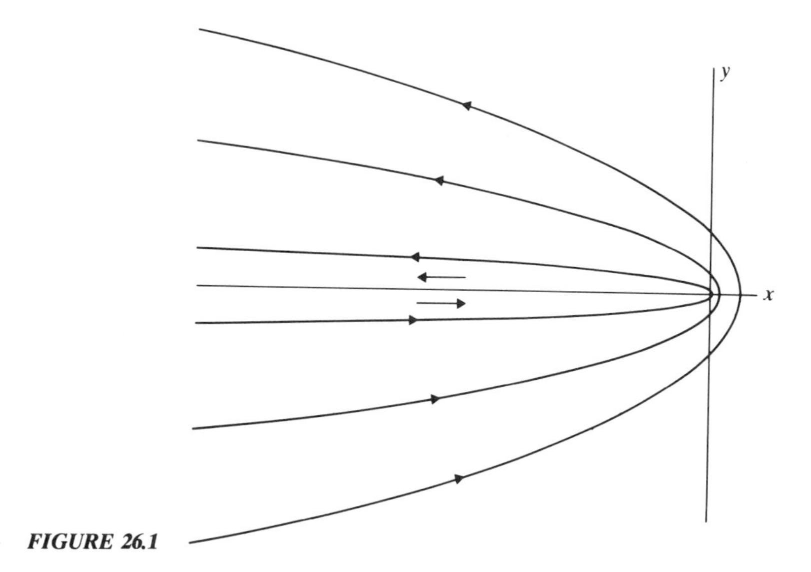

FIGURE 26.1

By taking the negative real axis and a single parabola as boundaries, we get a conformal map between the inside of a parabola with a barrier along part of its axis, and a vertical strip in the w plane. (We cannot get a map from the strip to the whole inside of the parabola in this way.[1]) We can interpret the corresponding flow as the flow around a linear breakwater in a parabolic bay; then the speed $|dw/dz|$ becomes infinite at $z = 0$. This kind of behavior is characteristic of flows in regions whose boundaries have sharp corners; when it occurs, it suggests that the mathematics does not model the physics too well. (However, physically speaking, there aren't any strictly one-dimensional barriers either.) It is reasonable to suppose that, in a real physical situation, the speed near the end of the barrier would be extremely large.

26B. Flows generated by the sine function and its inverse

An interesting example is the flow inside a semiinfinite strip, with the entire boundary as a single streamline, thought of as a conformal map of a uniform flow in a half plane. We can think of it informally as the half plane with its boundary folded up into the sides and base of the semiinfinite strip. It is not obvious what the complex potential should be, but we are going to show that the mapping $w = \sin z$ provides a conformal map of the half-strip $|x| < \pi/2$, $y > 0$, onto the upper half plane $v > 0$. In fact, we have

$$w = \sin z = \sin x \cosh y + i \cos x \sinh y,$$

so that $v = 0$ when $x = \pm\pi/2$ and when $y = 0$. This at least strongly suggests that the boundary of the half-strip D corresponds to the real axis in the w

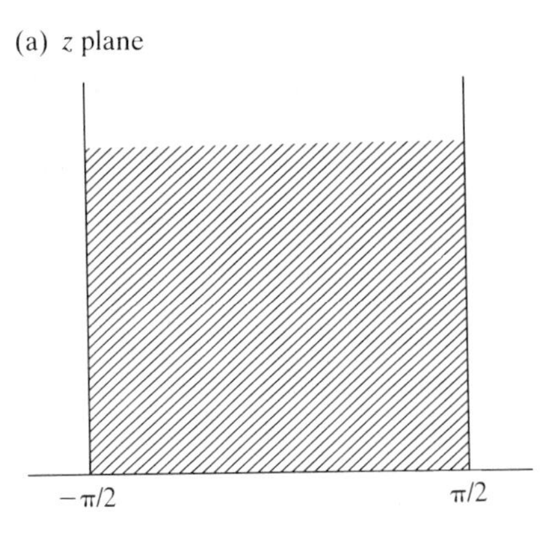

FIGURE 26.2

plane, but we really ought to prove it. So let z start at $y = \infty$ on the half-line $x = \pi/2$, and let y decrease. Then $w = \cosh y$ and w decreases until $y = 0$ at $w = 1$. At this point $dw/dz = 0$, and z must turn 90° to the right, from z's point of view; to our left as we look at Fig. 26.2. Now $y = 0$ and $w = \sin x$, so as x decreases, w decreases until $x = -\pi/2$, when $dw/dz = 0$ again. Now z turns 90° to the right again and $y \to +\infty$; as y increases, v decreases. Thus there is indeed a one-to-one correspondence between the boundary of D and the real axis in the w plane, as indicated in the figure. Hence there is also a one-to-one correspondence between the half-strip and the upper half plane (not the lower, because in each plane the region of interest is on the right as w or z traces the boundary).

Since the uniform flow in the w plane has complex potential w with streamlines $v = \text{constant}$, the flow in D has streamlines $\operatorname{Im} \sin z = \cos x \sinh y = \text{constant}$; a few are drawn in Fig. 26.3.

The flow in the w plane has a doublet at ∞, but the flow in the z plane cannot be that simple, since the lines $x = \pm\pi/2$ are not streamlines of a doublet at ∞. This difference corresponds to the circumstance that the complex

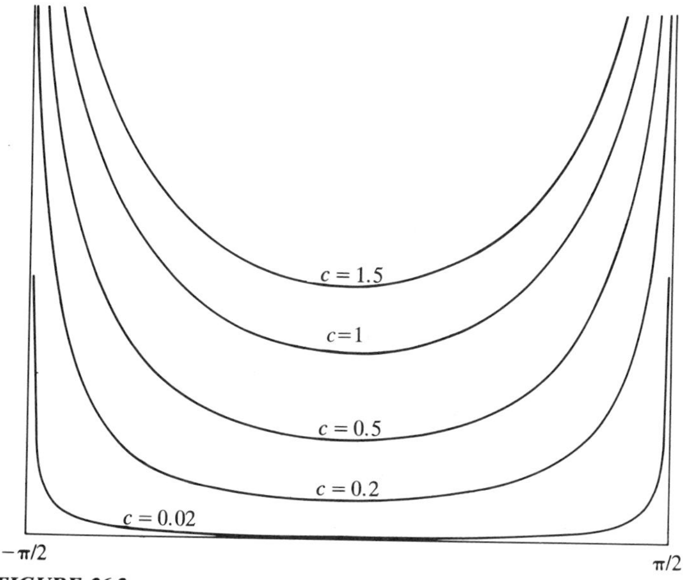

FIGURE 26.3

potential of the uniform flow has a simple pole at ∞, whereas $\sin z$ has an essential singularity at ∞ and all the terms of the power series of $\sin z$ contribute to it.

The inverse mapping (of a different region) also generates an interesting flow. Consider the images of the lines $x = c$, $|c| < \pi/2$, under $w = \sin z$. Since

$$u + iv = \sin x \cosh y + i \cos x \sinh y,$$

then, if $x = c$,

$$u = \sin c \cosh y, \qquad v = \cos c \sinh y,$$

$$\frac{u^2}{\sin^2 c} - \frac{v^2}{\cos^2 c} = 1.$$

The last equation is the equation of a hyperbola with foci at $(\pm 1, 0)$. If $0 < c < \pi/2$, then $\sin c > 0$ and the point $(\sin c, 0)$ is on the right-hand branch of the hyperbola; if $-\pi/2 < c < 0$, $(\sin c, 0)$ is on the left-hand branch. Thus the image of the line $x = c$ is the branch of the hyperbola that contains the vertex $(\sin c, 0)$. When $c \to \pm\pi/2$, the hyperbola degenerates to the pair of half lines $u < -1$ and $u > 1$, $v = 0$, each described twice. Taking these lines as boundaries, we have a flow from the upper half w plane to the lower half w plane through a hole in the u axis, as indicated in Fig. 26.4 (p. 204).

The complex velocity is $dz/dw = \sec z$, which becomes infinite as $z \to \pm\pi/2$ $(u \to \pm 1)$, so the speed becomes infinite at the edges of the hole. This may seem not unreasonable if you imagine the fluid as a crowd of people hurrying

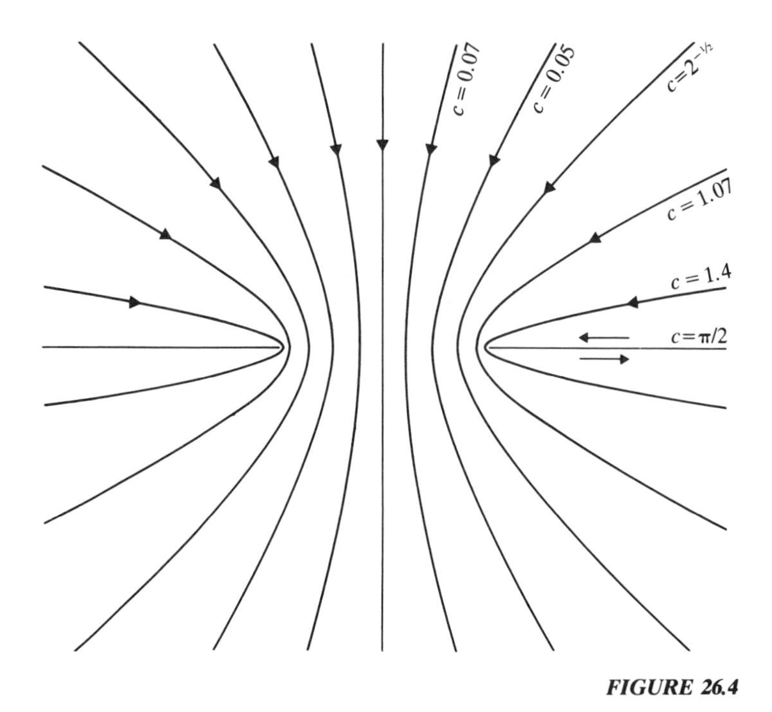

FIGURE 26.4

to get through a narrow doorway from a large theater into the outdoors. If you can look at such a situation from above, the streamlines will look surprisingly like the picture, considering how imperfect the fluid is in this case.

26C. Composition of mappings

Let us start with a simple problem. Suppose that we want the streamlines for a flow in the upper half plane with the negative and positive real axes as different streamlines $\psi = \pi$ and $\psi = 0$. There will have to be a singular point at 0, but we do not need to worry about what it will be like. If we can find a map $w = f(z)$ from the upper half z plane to the strip $0 < v < \pi$, such that $v = \pi$ corresponds to the negative real z axis and $v = 0$ corresponds to the positive real z axis, then $\psi = \operatorname{Im} f(z)$ will solve the problem.

EXERCISE 26.1 Show that $f(z) = \log z$ has the required properties.

Hence $\psi(x, y) = \theta = \tan^{-1}(y/x)$ solves our problem;[2] the streamlines are rays going out from the origin.

We saw in Sec. **24B** that the complex potential $w = \log[(z-1)/(z+1)]$ in the upper half plane has the real axis as a union of streamlines, as indicated in Fig. 26.5.

FIGURE 26.5

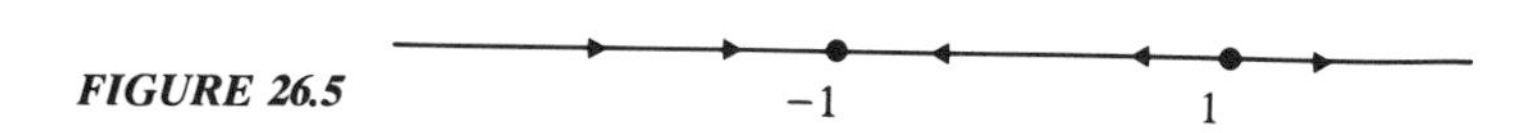

We could also have obtained this potential as the composition of $w = \log \zeta$ with $\zeta = (z-1)/(z+1)$.

Now consider the map that we had in Sec. **26B**, $z = \sin s$, $s = \sigma + it$, which maps the upper half z plane to the half strip $-\pi/2 < \sigma < \pi/2$, $t > 0$, in the s plane. We now consider a different flow in the half strip. The streamlines are the inverse images of the streamlines in the z plane under the mapping $z = \sin(\sigma + it) = \sin\sigma \cosh t + i \cos\sigma \sinh t$. But the streamlines in the z plane are the curves

$$\operatorname{Im} \log \frac{z-1}{z+1} = \text{const.}$$

Since

$$\frac{z-1}{z+1} = \frac{x^2+y^2-1}{(x+1)^2+y^2} + i\frac{2y}{(x+1)^2+y^2},$$

the imaginary part of the logarithm of this function is

$$\begin{aligned} \tan^{-1}\frac{2y}{x^2+y^2-1} &= \tan^{-1}\frac{2\cos\sigma\sinh t}{\sin^2\sigma\cosh^2 t + \cos^2\sigma\sinh^2 t - 1} \\ &= \tan^{-1}\frac{2\cos\sigma\sinh t}{\sinh^2 t - \cos^2\sigma}. \end{aligned}$$

Define θ by $\tan\theta = (\cos\sigma)/\sinh t$; then

$$\tan 2\theta = \frac{2\cos\sigma\sinh t}{\sinh^2 t - \cos^2\sigma} = \frac{2\tan\theta}{1-\tan^2\theta}$$

and our streamlines become

$$\begin{aligned} c = \tan^{-1}\frac{2\cos\sigma\sinh t}{\sinh^2 t - \cos^2\sigma} &= \tan^{-1}(\tan 2\theta) = 2\theta \\ &= 2\tan^{-1}\frac{\cos\sigma}{\sinh t}, \end{aligned}$$

or $t = \sinh^{-1}(\tau\cos\sigma)$ where $\tau = \cot(c/2)$. A few of them are shown in Fig. 26.6 (p. 206).

Möbius transformations can often be used effectively to obtain a conformal map between a simple configuration and a more complicated one. As an illustration, let us map the upper half z plane onto the half disk $|\zeta| < 1$, $\operatorname{Im}\zeta > 0$ (the upper half of the unit disk). We saw in Sec. **22D** that $w = \sqrt{z}$ maps the upper half z plane to the first quadrant of the w plane, so that our problem is reduced to mapping a quadrant to a half disk. Now the quadrant and the half disk are each bounded by two perpendicular arcs of "circles," in

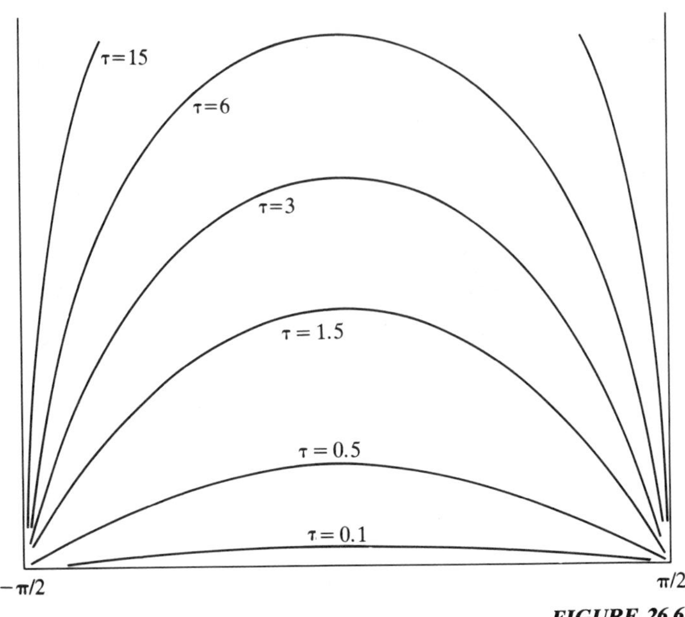

FIGURE 26.6

the extended sense of "circle or straight line," and therefore can be mapped onto each other by a Möbius transformation. Let us map the first quadrant of the w plane to the half disk $|\zeta|<1$, $\operatorname{Im}\zeta>0$. To find the transformation we note that a Möbius transformation that maps 1 to ∞ maps the semicircle to a half line, and one that maps -1 to 0 and 0 to 1 maps the diameter of a circle to the positive real axis. We saw in Sec. **25B** that this transformation can be written as a cross ratio; it turns out to be

$$w=\frac{\zeta+1}{1-\zeta} \quad \text{or} \quad \zeta=\frac{w-1}{1+w},$$

as is easily checked. Hence

$$\zeta=\frac{\sqrt{z}-1}{1+\sqrt{z}}$$

maps the z half plane to the ζ half disk.

EXERCISE 26.2 Find a conformal map from a half disk to a disk.

EXERCISE 26.3 Find a conformal map from the unit disk, cut along the radius from 0 to 1, to the whole disk.

EXERCISE 26.4 Map the complement of the line segment $[-1, 1]$ onto the unit disk.

One can concoct many examples of flows by making conformal transformations of examples that are already known, and if the results seem to be possibly useful, they can be collected in a reference book.[3] Problems for regions for which potentials are on record are easy to solve. It is much harder to start from a configuration of streamlines in which you are interested and find a sequence of mappings that will produce it from a uniform flow. If the boundary of the region is a polygon, the Schwarz–Christoffel transformation (not discussed in this book) may be helpful. Finally, there are modern computational methods for numerical conformal mapping that are probably more satisfactory for many practical purposes than complicated theoretical solutions.[4]

Notes

1. The transformation $w = \tan^2(\pi z^{1/2}/4)$ is given in tables as a map from the inside of a parabola to the unit disk in the w plane; the disk is easily mapped to a half plane (Sec. **25E**); and if the half plane is the upper half s plane, then $\zeta = \log s$ maps this half plane to the strip $0 < \operatorname{Im} \zeta < \pi$.
2. We could have found this more directly by noticing that $\theta = \arctan(y/x)$ is the imaginary part of $\log z$ in the upper half plane, hence harmonic, and is π or 0 on the negative and positive real axes.
3. For an extensive collection of conformal maps, see Kober.
4. See Henrici, chap. 16.

27 *Dirichlet Problems in General*

27A. Mappings between disks and other regions

We have seen that we can solve Dirichlet problems for a disk, and also for regions where closures are conformal images of disks.[1] In Sec. **28** we shall prove the **Riemann mapping theorem**, which says that *every simply connected region with at least two boundary points can be mapped conformally onto the unit disk.*

EXERCISE 27.1 What simply connected regions have fewer than two boundary points, and why can't we map them conformally on a disk?

It follows that we can say that every simply connected region with at least two boundary points can be mapped conformally onto every other such region, for example to a half plane.

If, then, we want a harmonic function with prescribed boundary values in a simply connected region D, we can (in principle) map D to the unit disk Δ, solve the corresponding Dirichlet problem in Δ, and map back. For Dirichlet problems of the simple kind that we have been using as illustrative examples,

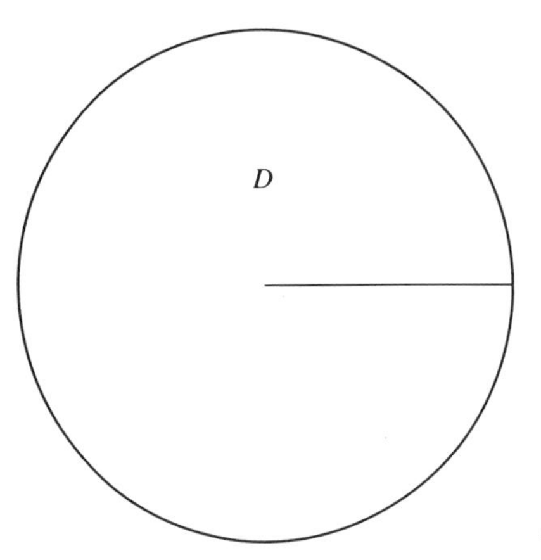

FIGURE 27.1

this procedure is effective. For continuous boundary values, we shall indicate in Secs. **27B** and **27C** how one can prove that the Dirichlet problem can be solved for simply connected regions. It can also be solved for mildly discontinuous boundary functions of the kind that occur in simple physical problems, provided that we do not expect convergence at discontinuities of the boundary function. A Dirichlet problem for a region like D in Fig. 27.1 makes reasonable sense if we think of the cut as the limit of sharp spines; here the opposite sides of the cut might be assigned different boundary values, and therefore have to be thought of (somehow) as different sets. Situations like this can be dealt with by the theory of *prime ends,*[2] which in this case provides a topology in which the points on opposite sides of the cut become different; and there is a one-to-one correspondence between the closure of the unit disk and the closure of D. This theory will also cope with much more complicated kinds of boundary, for example the one indicated in Fig. 27.2, in which the sinuous line condenses to an interval.

There are regions and boundary values for which the Dirichlet problem has no solution. One example is the punctured disk $0<|z|<1$, with $u(z)=0$ for $|z|=1$ and $u(0)=1$. If there were a solution u, it would be bounded, by the theorem on removable singularities (Sec. **20D**), and the mean value theorem

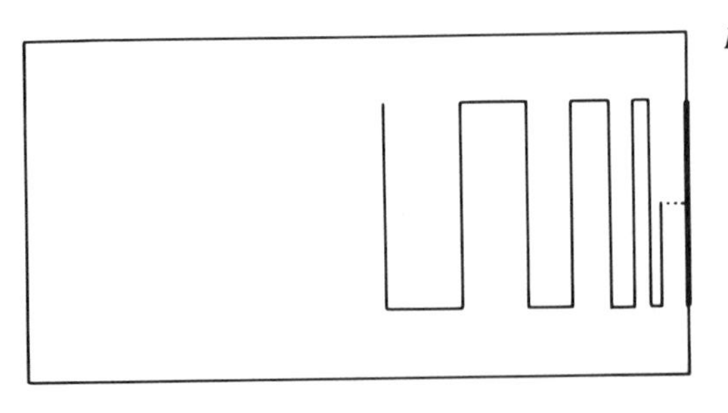

FIGURE 27.2

for harmonic functions would make it 0 at 0. It is the isolated boundary point that causes the trouble.

27B. The Green function

For the disk, Poisson's formula gives a more or less explicit solution of the Dirichlet problem. As I remarked in Sec. **20C**, one would hope for something similar for regions other than disks, and here we shall see what it is. Our discussion will be largely formal; I merely want to give you an idea of how the Poisson integral can be generalized.

The **Green function**[3] for D with "pole" at z_0 (note the new usage of "pole") is a function of the form $g(z; z_0) = -\log r + g_1(z)$, where $r = |z - z_0|$, g_1 is harmonic in D with the boundary values of $\log r$ on ∂D, the boundary ∂D is sufficiently smooth, and the normal derivative of g exists on ∂D. Thus g is a solution of the Dirichlet problem with zero boundary values, saved from being identically zero because it fails to be harmonic at z_0. If D is bounded, the Green function is unique if it exists, because the difference of two Green functions with the same pole would be harmonic in D, with boundary values 0.

Before trying to construct g, let us see what it would be good for if we did construct it. You need to know Stokes's theorem in the plane, otherwise called Green's theorem, which we met in Sec. **2C**. It says that for a simply connected region D with a sufficiently smooth boundary C,

$$\iint_D \left(\frac{\partial p}{\partial x} + \frac{\partial q}{\partial y}\right) dS = \int_C (p\,dy - q\,dx).$$

Let u and v be given functions, and take $p = u(\partial v/\partial x)$, $q = u(\partial v/\partial y)$. Then we have

$$\iint_D (u\,\nabla^2 v + \nabla u \cdot \nabla v)\,dS = \int_C u\left(\frac{\partial v}{\partial x}dy - \frac{\partial v}{\partial y}dx\right) = \int_C u\frac{\partial v}{\partial \mathbf{n}}\,ds,$$

where $\mathbf{n}$ is the outward normal to C. Interchange u and v and subtract, to get

$$\iint_D (u\,\nabla^2 v - v\,\nabla^2 u)\,dS = \int_C \left(u\frac{\partial v}{\partial \mathbf{n}} - v\frac{\partial u}{\partial \mathbf{n}}\right) ds.$$

Now let u be harmonic, $v = g(z; z_0)$, and apply the preceding equation to D with a small disk of radius ε and boundary C_ε around z_0 removed and a cut to make the new region simply connected: See Fig. 27.3 (p. 210). The left-hand side drops out, so does the integral back and forth along the cut, and we have

$$\int_C u\frac{\partial g}{\partial \mathbf{n}}\,ds = \int_{C_\varepsilon} \left(u\frac{\partial g}{\partial \mathbf{n}} - g\frac{\partial u}{\partial \mathbf{n}}\right) ds$$

(integral along C_ε now in the positive direction). Since $g(z; z_0) = -\log r +$

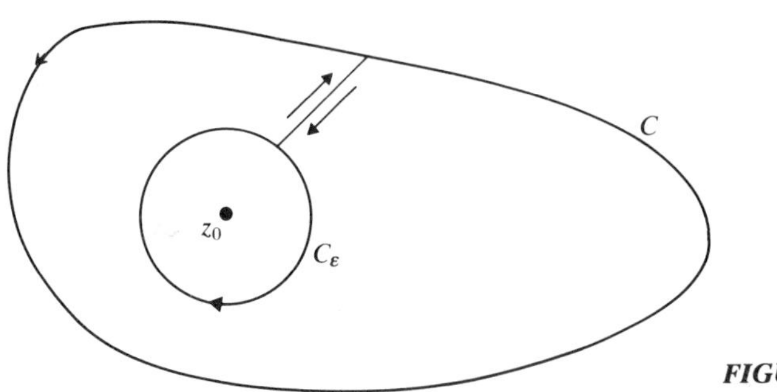

FIGURE 27.3

$g_1(z)$, where $r = |z - z_0|$, we have

$$\frac{\partial g}{\partial \mathbf{n}} = -\frac{1}{\varepsilon} + \frac{\partial g_1}{\partial r}, \qquad g = -\log \varepsilon + g_1,$$

$$\int_C u \frac{\partial g}{\partial \mathbf{n}}\, ds = \int_0^{2\pi} \varepsilon \left\{ u\left(-\frac{1}{\varepsilon} + \frac{\partial g_1}{\partial r} \right) + (\log \varepsilon - g_1) \frac{\partial u}{\partial r} \right\} d\theta.$$

Since $\partial g_1/\partial r$, g_1, and $\partial u/\partial r$ are all continuous at z_0, and $\varepsilon \log \varepsilon \to 0$ as $\varepsilon \to 0$, we obtain

$$\int_C u \frac{\partial g}{\partial \mathbf{n}}\, ds = -2\pi u(z_0).$$

Hence we have a representation for harmonic functions u inside C:

$$u(z_0) = -\frac{1}{2\pi} \int_C u \frac{\partial g}{\partial \mathbf{n}}\, ds,$$

where s is arclength on C. That is, the Green function provides us a representation for a harmonic function in D in terms of its values on the boundary of D (assuming that Green's theorem applies to D, which will certainly be the case if C is the kind of curve we consider elsewhere). This suggests, as in Sec. **20C**, that if we start with a boundary function, we ought to be able to construct the harmonic function with those boundary values.

Now that we see how a Green function can be used, let us see how to find it. Suppose that we know the function f that maps D to the unit disk with z_0 mapping to 0 (according to Riemann, such a function exists if D is simply connected and not the whole plane; see Sec. **28**). I claim that $-\log |f(z)|$ is the Green function g for D with pole at z_0, provided that $\partial g/\partial \mathbf{n}$ exists on ∂D.

To see this, let

$$f(z) = a_1(z - z_0) + a_2(z - z_0)^2 + \cdots$$

in a neighborhood of z_0; we have $a_1 \neq 0$ because the map is conformal. Since

$f(z) \neq 0$ for $z \neq z_0$ (the map is one-to-one), we have

$$-\log|f(z)| = -\log r + \text{harmonic function}.$$

As $z \to \partial D$, $|f(z)| \to 1$ and $\log|f(z)| \to 0$, so $-\log|f(z)|$ has the properties of a Green function and hence *is* the Green function.

Something that is harder to prove is that, conversely, if we know that the Green function g exists, we can construct the mapping function; then we can construct a harmonic conjugate h for g; and $w = \exp(-g - ih)$ is the mapping function. The existence of the Green function, and hence the existence of the mapping function, was originally considered obvious on physical grounds: The Green function is the electrostatic potential of a point charge at z_0 in the presence of a grounded conductor along C.

27C. The Poisson kernel from a Green function

You should have guessed by this time that the Poisson kernel is just $(2\pi)^{-1}$ times the negative of the normal derivative of the Green function for the unit disk. It follows from Exercise 1.22 or Exercise 25.4 that

$$w = \frac{z - z_0}{z\bar{z}_0 - 1}$$

maps the unit disk to the unit disk, with z_0 mapping to 0. Hence

$$g(z, z_0) = \operatorname{Re} \log \frac{\bar{z}_0 z - 1}{z - z_0}.$$

Write the logarithm as $\log(\bar{z}_0 z_0 - 1) - \log(z - z_0)$ and let $z = Re^{i\phi}$, $z_0 = re^{i\theta}$. Differentiate g with respect to R and let $R \to 1$:

$$\lim_{R\to 1}\left\{\frac{(\partial/\partial R)(\bar{z}_0\, Re^{i\phi} - 1)}{\bar{z}_0 z - 1} - \frac{(\partial/\partial R)(Re^{i\phi} - z_0)}{z - z_0}\right\} = \lim_{R\to 1} \frac{z_0 e^{i\phi}}{\bar{z}_0 z - 1} - \frac{e^{i\phi}}{z - z_0}$$

$$= \frac{re^{i(\phi-\theta)}}{re^{i(\phi-\theta)} - 1} - \frac{e^{i\phi}}{e^{i\phi} - re^{i\theta}}$$

$$= \frac{r}{r - e^{i(\theta-\phi)}} - \frac{1}{1 - re^{i(\theta-\phi)}}.$$

The real part is then

$$\frac{1}{2} r\left\{\frac{1}{r - e^{i(\theta-\phi)}} + \frac{1}{r - e^{-i(\theta-\phi)}}\right\} - \frac{1}{2}\left\{\frac{1}{1 - re^{i(\theta-\phi)}} + \frac{1}{1 - re^{-i(\theta-\phi)}}\right\}$$

$$= \frac{r^2 - 1}{1 + r^2 - 2r\cos(\theta - \phi)}.$$

Now it follows from what we know about conformal mappings that the

Green function of D_z is carried to the Green function of D_w by the map of D_z onto D_w that makes the poles correspond. Hence if we know the mapping function from the disk to D, and it is sufficiently smooth on $|z| = 1$, we can find the Green function of D; and we can solve the Dirichlet problem for D since we can solve the Dirichlet problem for the disk. This may help to justify the amount of time we have spent on Dirichlet problems for a disk.

EXERCISE 27.2 Find the Poisson kernel for the upper half plane.[4]

Notes

1. More precisely, we require regions that are conformal images of disks with the mapping continuous and one-to-one between the closed regions.
2. See, for example, Collingwood and Lohwater.
3. Most books still say "the Green's function," although they no longer say "Fourier's series." Compare Jackson, p. viii.
4. For a discussion along different lines, see R. Lange and R. A. Walsh, "A Heuristic for the Poisson Integral for the Half Plane and Some Caveats," *Amer. Math. Monthly 92* (1985): 356–358. For more about Dirichlet problems, see Henrici, chap. 15.

28 *The Riemann Mapping Theorem*

28A. Formulation and plan of proof

The Riemann mapping theorem states that if D is a simply connected region in the extended plane, with more than one point in its boundary, then there is a function f, analytic and univalent in D, that maps D onto the unit disk $\Delta: |w| < 1$.

This is not an easy theorem. The guiding idea is that since we require a function that is, in particular, analytic, univalent, and bounded in D, we start with the family of all such functions and try to cut it down until it contains just the function that we want. First we normalize the functions by the condition $f(c) = 0$, where c is a point of D, since some point of D has to map to 0. Each function maps D to a set that is contained in some disk centered at 0. We further normalize the functions by $f'(c) = 1$, and look for the function (if there is one) that maps D into the smallest possible disk. It should seem plausible that there must be a smallest disk, since otherwise we should have a limit function that maps D to the point 0, which, as we shall see, is impossible because $f'(c) \neq 0$. It will turn out, after a considerable amount of work, that the image $f(D)$ has to be the whole of the smallest disk. Once D is mapped onto some disk, it can (trivially) be mapped onto the unit disk.

We are going to need a number of preliminary definitions and theorems; these are also useful in other contexts.

28B. Normal families and equicontinuity

We say that a family $\mathscr{F}$ (not necessarily countable) of functions f is **normal** in a region D if *from each sequence of functions in* $\mathscr{F}$ *we can extract a subsequence that converges uniformly on each compact subset of* D (but not necessarily to an element of $\mathscr{F}$).[1]

In other contexts a normal family would be called sequentially compact, but the term "normal" is traditional in complex analysis. We shall be interested only in normal families of analytic functions. (Sometimes it is convenient to let "normal" cover also the possibility that the limit function may be the constant function ∞.)

We shall also need the definition of a property called equicontinuity. A condensed statement is that a family $\mathscr{F}$ of continuous functions on D is **equicontinuous** if the definition of continuity is satisfied uniformly over D and over the family. More precisely, *for every* $\varepsilon > 0$, *there exists* $\delta > 0$ *such that* $|f(z_1) - f(z_2)| < \varepsilon$ *whenever* $|z_1 - z_2| < \delta$, z_1 *and* z_2 *in* D, *and* f *is in* $\mathscr{F}$, *where* δ *is independent of* f *and of* z_1 *and* z_2.

The first theorem that we need is the

ARZELÀ-ASCOLI THEOREM *Equicontinuity and uniform boundedness of a family of continuous functions are sufficient for the family to be normal.*

Before we take up the proof of the Arzelà-Ascoli theorem, let us see why it is relevant to the proof of the Riemann mapping theorem. In that proof we need

MONTEL'S THEOREM *A family of functions that are analytic and uniformly bounded in a region* D *is necessarily normal in* D.

This will follow, in the light of the Arzelà-Ascoli theorem, if we show that such a family is equicontinuous on each compact subset D' of D.

When D' is a simple kind of set, for example, a closed disk, this equicontinuity theorem is an easy deduction from Cauchy's integral formula. When D' is a closed disk, let $|f(z)| \leq M$ for all f in $\mathscr{F}$ and $z \in D'$. The distance from D' to the boundary ∂D of D has a positive lower bound δ, because ∂D is closed and D' is compact. If z_1 and z_2 are in D' and L is the line segment joining them, then all points of L are at least δ away from ∂D. With integration along L we have

$$|f(z_1) - f(z_2)| = \left| \int_{z_1}^{z_2} f'(w)\, dw \right| \leq |z_1 - z_2| \sup_{w \in L} |f'(w)|\,.$$

By Cauchy's formula,

$$f'(w) = \frac{1}{2\pi i}\int_\gamma \frac{f(t)\,dt}{(t-w)^2},$$

where γ, of length λ, is the boundary of $D'' \supset D'$. Since the distance between the boundary of D and γ is at least some δ,

$$|f'(w)| \le \frac{M\lambda}{2\pi\delta^2}.$$

Consequently, $|f(z_1) - f(z_2)| \le M\lambda\,|z_1 - z_2|/(2\pi\delta^2)$, uniformly for z_1, z_2 in D'', and f in $\mathcal{F}$.

Now consider a general compact D'. Let σ be the shortest distance from points of D' to ∂D (again, σ exists because D' is compact and ∂D is closed). We can cover each point $z \in D'$ by a disk Δ (centered at z) of diameter $\sigma/2$ with $\Delta \subset D$. It is a general property of compact sets that there exists $\varepsilon > 0$ such that for each z_1 in D' there is a Δ such that every point of a disk d about z_1, of radius δ, is in Δ. If z_2 is in d, every point of the line segment L connecting z_1 and z_2 is in d. Repeat the previous argument.

At this point you can, if you like, go to Sec. **28C** and read the proof of the Riemann mapping theorem.

We now prove the Arzelà-Ascoli theorem. The first step is to show that if $\mathcal{F}$ is an infinite family of functions that are uniformly bounded on D and $\{z_n\}$ is a countable sequence of points of D, then there is a sequence $\{f_k\}$ of functions in $\mathcal{F}$ such that $\{f_k\}$ converges at each point of $\{z_n\}$.

Since $\{f(z_1)\}$ $f \in \mathcal{F}$ is a bounded set of complex numbers, it contains a convergent sequence $\{f_{n'}(z_1)\}$. A further subsequence $\{f_{n''}(z_2)\}$ converges [and of course $\{f_{n''}(z_1)\}$ converges too]. Continuing in this way, we have $\{f_{n^{(m)}}(z_k)\}$ convergent for $k = 1, 2, \ldots, m$, and every m. Now, by the Cantor diagonal process, we form the sequence $\{f_{n^{(n)}}(z)\} = \{f_k(z)\}$, which consists of the nth term of the nth sequence; this sequence converges at $z = z_m$ for every m.

Now let D' be a compact subset of D. We cover D' by a countable network of squares of arbitrarily small size, for example, by drawing parallels to the coordinate axes. The corners of these squares that are in D' form a countable set $\{z_n\}$. As we just showed, there is a sequence $\{f_k\}$ of elements of $\mathcal{F}$ that converges at all the points z_n. We are going to use the equicontinuity of $\mathcal{F}$ to show that $\{f_k\}$ also converges everywhere else in D'.

Let $\varepsilon > 0$ be given; from the definition of equicontinuity, there is a positive δ such that, for all f, $|f(z') - f(z'')| < \varepsilon$ if $|z' - z''| < \delta$ and z' and z'' are in D'.

Since D' is bounded, there is a finite set of squares from our network, each of side less than $\delta/\sqrt{2}$, that covers D'; let S be the set of corners of these squares. The sequence $\{f_m(z)\}$ converges uniformly on S, since S is a finite subset of $\{z_k\}$. That is, given an $\varepsilon > 0$, there is an N such that

$$|f_m(z_k) - f_{m'}(z_k)| < \varepsilon$$

if $m > N$ and $m' > N$, for all the z_k in S. But if z is any point in D', there is some $z_k \in S$ with $|z - z_k| < \delta$ and hence, by the definition of δ,

$$|f_m(z) - f_m(z_k)| < \varepsilon \qquad (\text{all } m).$$

Combining the two preceding inequalities, we have

$$|f_m(z) - f_{m'}(z)| \leq |f_m(z) - f_m(z_k)| + |f_m(z_k) - f_{m'}(z_k)| + |f_{m'}(z_k) - f_{m'}(z)| < 3\varepsilon.$$

Thus we have found a sequence $\{f_n\}$ that converges uniformly in D'.

EXERCISE 28.1 Show further that there is a single sequence $\{f_n\}$ that converges for every z in D, and uniformly in each D'.

28C. The proof

We now have the material needed for a proof of the Riemann mapping theorem. We define a class H, consisting of the functions that are analytic, bounded, and univalent in D. Then H is not empty. We can see this as follows. First, if the complement of D possesses a disk, let the center of the disk be p; then $1/(z-p)$ is bounded and univalent and therefore in H. Otherwise, by hypothesis, D contains at least two boundary points; one of these boundary points must be finite. If the other one were at ∞ (and there were no others), D would be a punctured plane, which is not simply connected. Hence we may suppose that there are two finite boundary points a and b. We can now choose an analytic branch $B(z)$ of $\sqrt{(z-a)/(z-b)}$ (this is where we use the simple connectivity of D). To define B, first find a branch of $\log(z-a)$, which is possible because D is simply connected and a is not in D; similarly, find a branch of $\log(z-b)$. Then $B(z) = \exp\{\frac{1}{2}[\log(z-a) - \log(z-b)]\}$.

Now B is univalent because the Möbius transformation $w = (z-a)/(z-b)$ is univalent. If $w_1 \neq 0$, ∞ is a point of the image D_1 of D by the mapping B, a neighborhood of w_1 is in D_1, and hence a neighborhood of $-w_1$ is not in D_1. Thus the complement of D_1 contains a disk centered at some point d, and so, by the case considered above, $1/(B-d)$ belongs to H.

Now let c be any point of D. Let M be the class of functions f that are analytic, bounded, and univalent in D, with $f(c) = 0$ and $f'(c) = 1$. The class M is not empty, since we know that there are functions g that are analytic, univalent, and bounded in D; since they are univalent, $g'(z) \neq 0$ in D; then $[g(z) - g(c)]/g'(c)$ belongs to M.

Each f in M maps D to some region D', with c mapping to 0; and $m_f = \sup_{z \in D} |f(z)|$ is finite. Let $s = \inf m_f$ for f in M. From the definition of infimum, there are functions f_n in M such that $\sup_{z \in D} |f_n(z)| \leq s + 1/n$. These functions f_n are uniformly bounded (by $s+1$) in D and so form a normal family. Some subsequence $\{f_{n_k}\}$ of $\{f_n\}$ therefore has a limit function f, which has $f(c)=0$. Since the convergence is uniform on compact subsets of D, we have $f'(c)=\lim_{k\to\infty} f'_{n_k}(c)=1$ (by the theorem on differentiating a uniformly convergent sequence, in Sec. **7D**). Since $|f_n(z)|\le s+1/n$, we have $|f(z)|\le s$. But $f(z)\not\equiv 0$, since $f'(c)\ne 0$. Hence $s>0$. Finally, f is univalent by Hurwitz's theorem (see Exercise 13.4). Consequently, f belongs to M and maps D to a subset of the disk S: $|z|<s$.

We are now going to show that f actually maps D onto the whole disk S. To do this, we suppose that the image of D is a proper subset of S, and construct a new function that maps D into a disk still smaller than S, thus contradicting the definition of s.

Suppose, then, that f maps D to a proper subset of S. The image of D has a boundary point b in S. We must have $b\ne 0$, since c, an interior point of D, maps to 0; hence its image is an interior point of the image of D, by the open mapping theorem (Sec. **12D**).

Now let $w=f(z)$, and construct

$$w_1=s\left(\frac{s(w-b)}{s^2-\bar{b}w}\right)^{1/2}.$$

Since $w\ne b$ when $z\in D$, and the symmetric point $s^2/\bar{b}$ is outside S, we have $|w_1|<s$, and we see, as for $B(z)$, above, that there is an analytic branch of the square root (we can choose either one). Since the Möbius transformation is a univalent function, w_1 is univalent, and the square root has the effect of making the image of S smaller. However, when $z=c$ and $w=0$, w_1 is not 0, so w_1 is not in M (when thought of as a function of z). We can bring c to 0 by subjecting w_1 to a Möbius transformation of S, so that we get

$$w_2=\frac{s^2[w_1-w_1(0)]}{s^2-\overline{w_1(0)}w_1}.$$

If dw_2/dz is not 1 at $z=c$, we divide by it and get

$$w_3=\frac{w_2}{(dw_2/dz)_{z=c}}.$$

Then w_3 does belong to M.

Now we compute dw_2/dz at $z=c$. We shall find that $|dw_2/dz|>1$ at $z=c$, and consequently, since $|w_2|\le s$, we will have $|w_3|<s$, and indeed $m_{w_3}<s$; this contradicts the definition of s. We conclude that f maps D onto S; and we already know how to map S onto Δ.

Here is the computation:

$$\frac{dw}{dz}=1 \text{ at } z=c, \quad \text{and} \quad w_1(0)=s^{1/2}(-b)^{1/2}.$$

Hence,

$$\left.\frac{dw_1}{dz}\right|_{z=c} = \left.\frac{dw_1}{dw}\right|_{w=0} = \frac{1}{2}s^{-3/2}(-b)^{-1/2}(s^2 - |b|^2),$$

$$\left.\frac{dw_2}{dw_1}\right|_{w=0} = s^2\frac{s^2 - |w_1(0)|^2}{(s^2 - |w_1(0)|)^2} = \frac{s^2}{s^2 - |b|\,s},$$

$$\left.\frac{dw_2}{dz}\right|_{w=0} = \left(\frac{dw_2}{dw_1}\cdot\frac{dw_1}{dw}\right)_{w=0}$$

$$= \frac{1}{2}s^{-3/2}(-b)^{-1/2}(s^2 - |b|^2)\cdot\frac{s^2}{s(s - |b|)}$$

$$= \frac{1}{2\sqrt{-b}}\frac{s + |b|}{s^{1/2}};$$

$$\left|\frac{dw_2}{dz}\right|_{w=0} = \frac{s + |b|}{2(s|b|)^{1/2}} > 1$$

because

$$s + |b| - 2(s|b|)^{1/2} = (s^{1/2} - |b|^{1/2})^2 > 0.$$

28D. Some applications of normal families

One can use Montel's theorem to prove many useful results. I present two of these here.

The first application is a precise formulation of the principle that convergence on a large enough subset of a region D induces convergence on all compact subsets of D. We prove

VITALI'S THEOREM *Let the functions f_n be analytic and uniformly bounded in a region D and let $\{f_n\}$ converge on a subset S of D that has a limit point p in D (inside, not on the boundary). Then $\{f_n\}$ converges uniformly on each compact subset*[1] *of D.*

Notice that S does not have to be very large: It merely needs to be an infinite subset of some compact subset of D.

The sequence $\{f_n\}$ is normal, by Montel's theorem, and therefore contains a subsequence that converges, uniformly on each compact subset of D, to an analytic function f (Exercise 28.1). I claim that in fact $f_n \to f$, uniformly on each compact subset of D. If this were not the case, there would be an infinite compact subset E of D on which $\{f_n\}$ does not converge uniformly. This would mean that for some positive δ there is a subsequence $\{g_n\}$ of $\{f_n\}$ such that $\sup_E |g_n - f| \geq \delta$ for all n. (Compare the second formulation of uniform convergence in Sec. **7D**.) However, $\{g_n\}$ is also a normal family, so a subsequence of $\{g_n\}$ converges on D to an analytic function g, and therefore $f = g$ on S. Since S has a limit point in D, we have $f = g$ everywhere in D (the zeros of the difference are isolated unless the difference is identically zero). (This is where the connectedness of D is essential.) This contradiction of the choice of $\{g_n\}$ shows that f_n does indeed converge[2] to f.

As a further application, we illustrate the general principle that (in informal terms) if f is analytic and bounded in D and $f(z)$ approaches a limit L along an arc A that tends to a boundary point p of D, then $f \to L$ as $z \to p$ along all other arcs that are sufficiently close to A. Specifically, *if D is a half strip, say $x > 0$, $a < y < b$, containing the positive real axis, and f is analytic and bounded in D and $f(x) \to L$ (finite) as $x \to \infty$, then $f(x + iy) \to L$ uniformly for $a + c < y < b - c$ $(0 < c < b - a)$.*

The proof depends on the idea that if n is a positive integer, then $f(z + n)$ has much the same properties as $f(z)$. Consider the functions f_n defined by $f_n(z) = f(z + n)$ in the region G: $a < y < b$, $0 < x < 2$. Then $f_n(z) \to L$ as $n \to \infty$ for every real z in G; by Vitali's theorem, $f_n(z) \to L$ uniformly for $a + c \le y \le b - c$ and $\frac{1}{2} \le x \le \frac{3}{2}$. If $\{z_k\} = \{x_k + iy_k\}$ is any sequence of points in $a + c \le y \le b - c$, then $x_k = m + x$, where m is an integer and $\frac{1}{2} \le x \le \frac{3}{2}$, so $f(x_k) \to L$.

We can transfer this result (also known as "Montel's theorem") by means of conformal mapping to any similar situation with sufficiently simple geometry. For example, D can be an angle $\alpha < |\arg z| < \beta$, with $f(z) \to L$ as $z \to \infty$ along a ray in this angle; or D can be a disk and $f(z) \to L$ as $z \to P \in \partial D$ in an angle inside D with vertex at L (see Fig. 28.1). For other results in this direction, see Sec. **33B**.

The theorem that a uniformly bounded family is normal is by no means the best possible. This theorem says that if f_n are analytic and all the f_n omit all values outside some disk, then $\{f_n\}$ is normal. The exterior of the disk can be transformed to the interior of a small disk, so that a family $\{f_n\}$ whose members take no values inside a given disk is normal. The ultimate extension of this idea is that a family of analytic functions whose members all omit two points of the finite plane (say, 0 and 1) is normal; a family of meromorphic functions whose members omit three points of the extended plane (the same

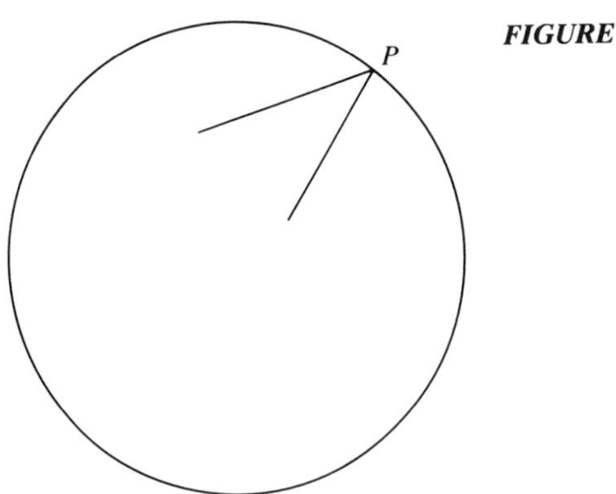

FIGURE 28.1

three points for every member of the family) is normal in the more general sense in which a normal family may have ∞ as a limit. This, however, is too difficult a theorem for us to prove here. It can be used to prove Picard's theorems, which say that an entire function (not a constant) can omit at most one finite value (as e^z does); a meromorphic function (not a constant) can omit at most two values (as $\tan z$ does: i and $-i$); more generally, an analytic function cannot omit more than two points in any neighborhood of an essential singular point—the ultimate generalization of the Casorati–Weierstrass theorem.[3]

28E. Extending a conformal map to the boundary

We have given a proof of the Riemann mapping theorem, but that theorem is not enough to let us solve Dirichlet problems. In Sec. **27B** I outlined a route from the mapping function to the Green function and hence to a generalization of the Poisson formula, but that demanded the use of Green's formula under conditions more general than you probably know it for, and more general than I want to present here. A more direct approach requires us to know that the mapping between the disk Δ and a region D can be extended to the boundaries of the two regions, something that is not always possible if the boundary of D is sufficiently complicated. In fact, *if D is a Jordan region* (that is, if its boundary is a Jordan curve), *it can be shown that the mapping function can be extended to a homeomorphism between the closures of the two regions,* but this is again a difficult theorem, and we shall consider only some special cases.[4]

First let us see why the extension lets us solve Dirichlet problems. Let $\phi_0(w)$ be a continuous boundary function on ∂D. The function ψ_0 determined by $\psi_0(z) = \phi_0[f^{-1}(z)]$ is defined and continuous on the boundary of Δ because a continuous univalent function has a continuous inverse. Then we can use the Poisson integral to find a function ψ that is harmonic in Δ with ψ_0 as boundary function. Hence the function ϕ defined by $\phi(w) = \psi[f(w)]$ approaches $\phi_0(w)$ as $w \to w_0$, and ϕ is harmonic in D.

In many applications we do have a conformal map that can be extended to the boundary. For example, in the flow around a corner (Sec. **22D**), the region between two branches of a hyperbola is mapped on a horizontal strip with continuous correspondence of the closures of the two regions. Thus we can solve the Dirichlet problem for the region between two hyperbolic curves if we can solve the Dirichlet problem for a strip.

We can also extend a conformal map to (and indeed past) a boundary if the geometry of the boundary is simple enough. For example, suppose that we have a map f from D to Δ and that part of the boundary of D is an open interval L on a straight line (which we may take to be the real axis, by subjecting D to a translation and rotation if necessary). Let us assume (as in Fig. 28.2, p. 220) that we can draw, about each point of L, a circle that meets the boundary of D only at points of L. This is to keep the rest of the

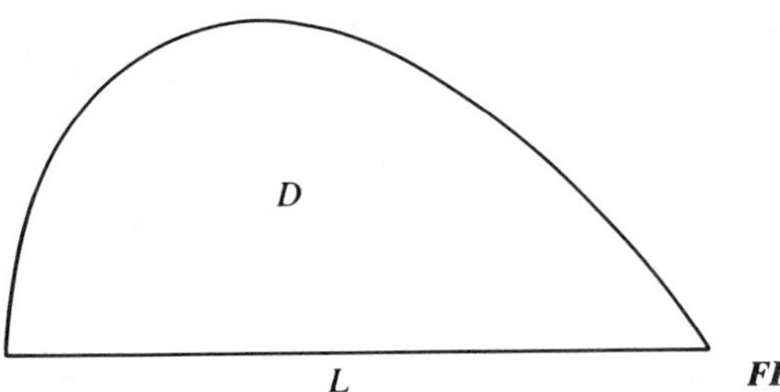

FIGURE 28.2

boundary from getting too close to L: We want to avoid a situation like that indicated in Fig. 28.3, where there are parts of the boundary that condense toward L, in which case the proof would not work. If the situation is like that in Fig. 28.2, a sufficiently small half disk with center w_0 on L and lying in D will not contain the inverse image of 0 under the mapping function $z=f(w)$. Hence there will be a branch of $\log f(w)$ in that half disk whose real part approaches 0 as $w \to w_0$, because $|f(w)| \to 1$ as $w \to w_0$ by the following exercise.

EXERCISE 28.2 If $z=f(w)$ maps D conformally on Δ, then if w approaches a boundary point of D, the values of $f(w)$ "approach the boundary of Δ"; that is, if $w_n \to w_0$, where w_0 is on the boundary of D, then every convergent subsequence of $\{f(w_n)\}$ approaches some point of $\partial\Delta$ (with the understanding that different subsequences might approach different points).

Hence $i \log f$ is real on the diameter of the half disk and can be continued analytically across L by the strong Schwarz reflection principle (Sec. **20F**); consequently, f can also be continued across L. In particular, f must be continuous on L.

The same kind of reasoning would work if L were an arc of a circle, or a curve of the same general sort.[5] This principle will take care of the kinds of mappings that arose in Sec. **26**.

Notes

1. Convergence that is uniform on each compact subset is often called either "compact" or "normal" convergence. Vitali's theorem was discovered independ-

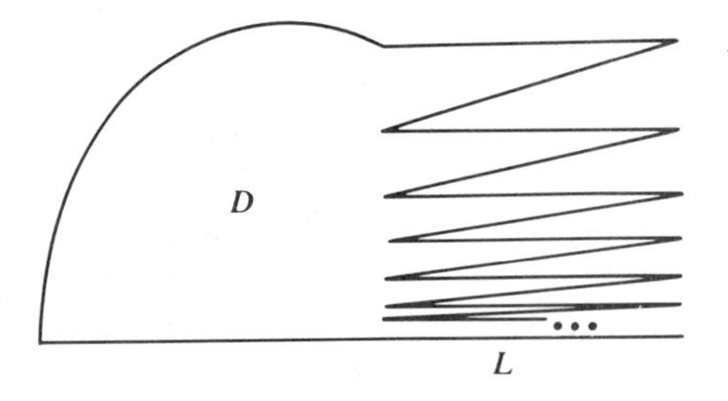

FIGURE 28.3

ently by M. B. Porter (the first discoverer of overconvergence), and is sometimes called the Vitali–Porter theorem.

2. This arrangement of the proof was suggested by G. B. Folland.

3. Actually, the story does not end here: Picard's theorems have given rise to an enormous amount of literature on the distribution of the values of analytic functions, and it continues to grow to this day.

4. For references on everything connected with the Riemann mapping theorem, see Burckel, chap. IX, and Henrici, chap. 16. The converse of the theorem stated here is also true: If the map extends continuously to the boundary, then the boundary is a Jordan curve.

5. The most general case is that when L is an analytic arc, that is, a simple arc defined by $z = z(t)$, where $z(t)$ is the restriction to the real axis of an analytic function, with $z'(t) \neq 0$.

29 *Intuitive Riemann Surfaces*

This section is an informal introduction to the idea of a Riemann surface, with illustrative examples. The objective is to clarify the geometry of conformal maps generated by functions that are not one-to-one in the whole plane (which are most of them: see Exercise 13.3).

We begin with the mapping $w = z^2$. This mapping was discussed in Sec. **22D**, where we showed that the first quadrant in the z plane corresponds to the upper half w plane. We could have shown in the same way that the second quadrant in the z plane corresponds to the lower half w plane. What, then, is going to correspond to the other z quadrants?

We can look at the same problem in a different way if we write[1] $z = \sqrt{w}$ and notice that in a neighborhood of each point $w \neq 0$ or ∞ we can define two analytic branches of $\sqrt{w}$, and that there are paths on which analytic continuation leads from one branch to another. However, a function is supposed to be single-valued; one way to avoid the difficulty is to introduce a cut in the w plane and consider a single-valued branch of $\sqrt{w}$. This procedure is apt to seem somewhat artificial, since there is not much reason to put the cut in any particular location.

What we do (following Riemann) is to introduce two w planes ("*sheets*"), one on top of the other, and define $\sqrt{w}$ as a (single-valued) function on the doubled plane. This is an example of a Riemann surface. Then, in the context of our original question, the image of the lower half z plane is going to be the second sheet of the Riemann surface. The sheets join at 0, and are connected across (say) the positive real axis in such a way that a path crossing this ray passes from one sheet to the other. Then, if $w = re^{i\theta}$, we have $z = r^{1/2}e^{i\theta/2}$ in sheet 1 from $\theta = 0$ to 2π, in sheet 2 from 2π to 4π; back in sheet 1 from 4π to 6π; and so on. On this surface, the mapping is one-to-one between the double

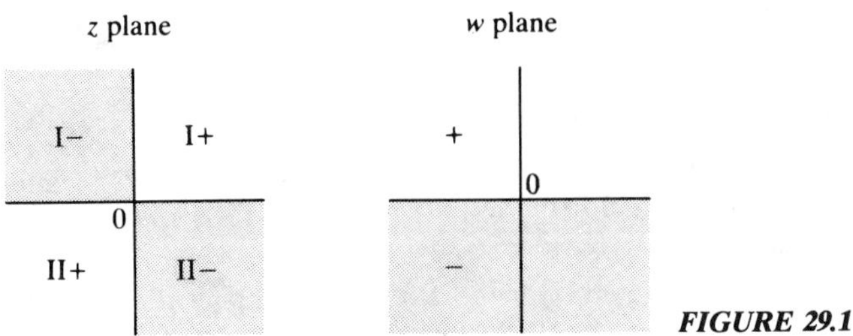

FIGURE 29.1

w plane and the single z plane, and it is easy to see what happens when we continue $\sqrt{w}$ analytically along a given path. It is clear that we cannot make a model of a surface like this in three-space, although we can realize it abstractly as a two-dimensional manifold.[2] For descriptions of how this is done, I refer you to books on Riemann surfaces; for our purpose, precise definitions are not necessary.

We saw in Sec. **22D** that the first quadrant in the z plane corresponds to the upper half w plane. If we treat the other quadrants in the same way, we obtain the correspondence indicated in Fig. 29.1, where + and − refer to upper and lower half w planes, I and II to the sheets, and the lower half w plane and its images in the z plane are shaded.

Fig. 29.2 is intended to help you visualize the connections between the sheets. Here you are to imagine that the line segments labeled 1 and 2 in sheets 1 and 2 have been rotated 90° in the third dimension to show how the sheets join each other across the real axis. Across the negative real axis, a moving point stays in the same sheet; across the positive real axis, it changes sheets. The join could equally well have been made across some other line, or even along a curve; the resulting surface would be abstractly the same.

The pairs of z quadrants, which are regions that correspond to complete w planes, are called **fundamental domains**.

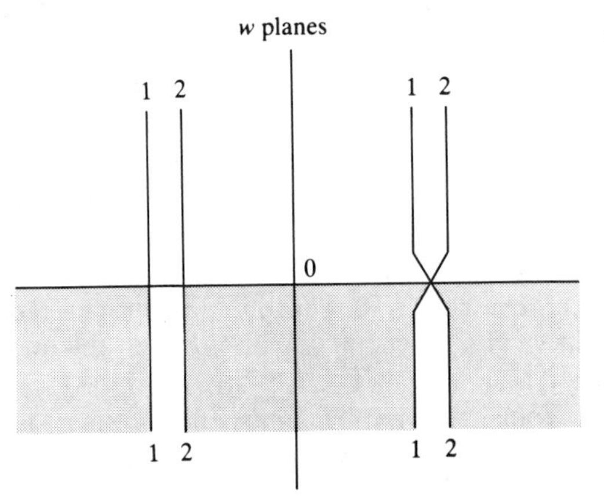

FIGURE 29.2

We can now think of $z = \sqrt{w}$ as a function on the Riemann surface. If w follows, for example, a circuit starting in the first quadrant and going counterclockwise around 0, it will return to its starting point in sheet 2; a second circuit will return it to its starting point in sheet 1. If we decide that $+1$ is the square root of 1 in the first sheet, we can see from the figure which sign to attach to any square root as a result of following a particular path.

EXERCISE 29.1 Construct a Riemann surface for $z = \log w$ (or $w = e^z$). What are the fundamental domains? Observe that the periodicity of e^z is reflected in the fact that the fundamental domains are simply translations of each other.

EXERCISE 29.2 Construct a Riemann surface for $w = \sin z$ (see Sec. **26B** for a start). The surface will be particularly tidy if you label the fundamental domains that join along $(\pi/2, 3\pi/2)$ with the same Roman numeral.

A less transparent example[3] is $w = z^3 - 3z$. There is no simple way to write the three values of z in terms of w, but if we construct a Riemann surface over the w plane, we can see how z behaves as w varies. (See Fig. 29.3.)

FIGURE 29.3

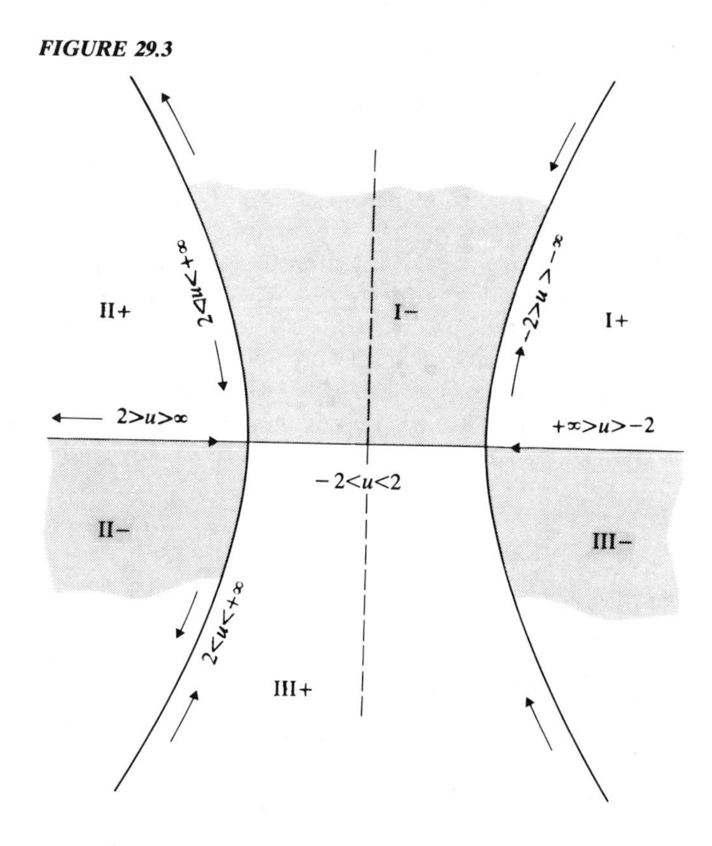

We have

$$w = (x+iy)^3 - 3(x+iy) = u + iv,$$
$$u = x^3 - 3xy^2 - 3x,$$
$$v = 3x^2y - y^3 - 3y = 3y\left(x^2 - \frac{1}{3}y^2 - 1\right).$$

Consider points on the real z axis; let w start at $+\infty$ on the real axis and move to the left. Then $u = x^3 - 3x$, $du/dx = 3(x^2-1)$. Therefore $du/dx > 0$ from $x = +\infty$ to $x = 1$, which corresponds to $u = -2$. Then we have a one-to-one correspondence between these parts of the real axes. Now let w continue to the left. Because $dw/dz = 0$ at $x = 1$, conformality fails at this point and z turns to the right; then $x^2 - \frac{1}{3}y^2 - 1 = 0$, since this is what happens when $v = 0$ and $y \neq 0$. Then z is on a branch of the hyperbola $3x^2 - y^2 = 3$ with $y > 0$. Since $u = x^3 - 3xy^2 - 3x$, we now have $\partial u/\partial y = -6xy$. When x and y are positive, y increases as u decreases, so z follows the upper half of the right-hand branch of the hyperbola. The region I+ between the real z axis and this branch must then correspond to the upper half w plane.

Now start w on the real axis at $u = -\infty$ and let z return along the same part of the hyperbola. As u increases, y decreases, and when w reaches -2, z is back at $+1$. Again, because $dw/dz = 0$ here, z must turn to the right and follow the real axis until z reaches -1, when $w = +2$, and then follow the upper part of the left-hand branch of the hyperbola as u goes from $+2$ to $+\infty$. The region labeled I − in Fig. 29.3 now must correspond to a lower half plane, since it is on the right as w goes from $-\infty$ to $+\infty$ along the real axis. Hence the upper half of the w plane must join the lower half across the common boundary of I+ and I−, that is, between $-\infty$ and -2.

Now start w at $u = +\infty$ with z on the upper part of the left-hand branch of the hyperbola, and let u decrease. When u reaches $+2$, $z = -1$, and z must turn and follow the real axis from $+2$ to $-\infty$ as w goes from $+2$ to $+\infty$. Now the z region labeled II+ must correspond to a different upper half w plane, which we take to be sheet 2. Then I− joins II+ from $w = +2$ to $+\infty$.

Continuing in this way, we get the correspondence indicated in Fig. 29.3, and connect the sheets as in Fig. 29.4. We can now read off from Fig. 29.4 how

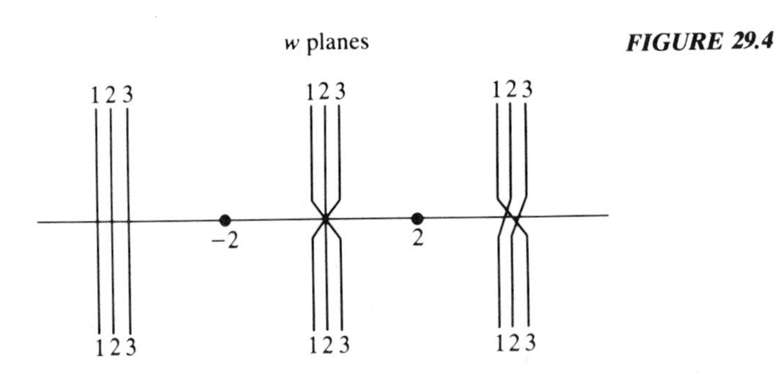

FIGURE 29.4

the inverse function of $w = z^3 - 3z$ behaves as a function on the Riemann surface. For example, if w starts in sheet 1 at a point like $3 + i$ and goes counterclockwise around 2 but not around -2, w will return to $3 + i$ having gone into sheet 3 and then back to sheet 1, and z returns to its original value. However, if w goes around both 2 and -2 and returns, it will remain in sheet 1 until it crosses the ray $(2, +\infty)$, and return in sheet 3, hence with a different value for z.

Supplementary exercises

Construct Riemann surfaces for

1. $w = z^3$

2. $w = \sqrt{(z-1)(z+1)}$

3. $w = z^3 - 3z^2$

4. $w = z + \dfrac{1}{z}$

Notes

1. It is more usual to see w and z interchanged, so that we would consider $w = \sqrt{z}$, but we have consistently been writing transformations the other way; a change in notation might be confusing. Ideally, you should be able to change notation without becoming confused, but there is no need to borrow trouble.

It is worth noticing that it is not always possible to continue one branch of a function into another. For example, the formula $w = 2^z$ determines infinitely many different functions.

2. Since Riemann surfaces cannot be drawn in ordinary space as coverings of the plane without self-intersections, they tend to seem mysterious, although the degree of abstraction required is not much more than that needed for acceptance of the complex numbers. One could play chess on a model of a Riemann surface as easily as on a torus or a Klein bottle, but I cannot imagine what Aldous Huxley was thinking of when he had characters in *Brave New World* playing "Riemann surface tennis."

3. There is a more detailed study of this surface in Osgood, pp. 388ff.

CHAPTER 5

Miscellaneous Topics

30 *A Non-Euclidean Geometry*

30A. Non-Euclidean geometry

One of the applications of Möbius transformations is the construction of a non-Euclidean geometry. You probably know that one of the axioms of Euclidean geometry is that through a point not on a straight line there is just one straight line parallel to the given line. (This is not Euclid's own formulation, but is equivalent to it.) A non-Euclidean geometry is one in which some axiom of Euclidean geometry is violated; usually it is this particular axiom that is replaced by something else. One of the great discoveries of the nineteenth century was that there *are* non-Euclidean geometries, that is, geometric systems in which the points and "straight lines" satisfy all of Euclid's axioms except the parallel axiom, but in which there are either no parallels to a line through a point not on it, or more than one. One can build such geometries axiomatically, just as Euclidean geometry is built, but to be really convinced one needs to see actual examples. For the case when there are no parallels, we can use the surface of a sphere, where the "straight lines" are great circles. It is harder to construct an example where there is more than one parallel through a point. We are going to construct a model (originated by Poincaré) of such a geometry. If you want to draw figures in this geometry, you need to be able to locate a point symmetric to a given point with respect to a circle; and to draw a circle through two points inside a circle, orthogonal to that circle. For details, see Sec. **30B**.

Our "plane" is going to be the open unit disk; the "straight lines" are to be circular arcs orthogonal to the unit circle. The circle itself is not part of the plane; arcs that meet on the circle are "parallel." See Fig. 30.1. Given a

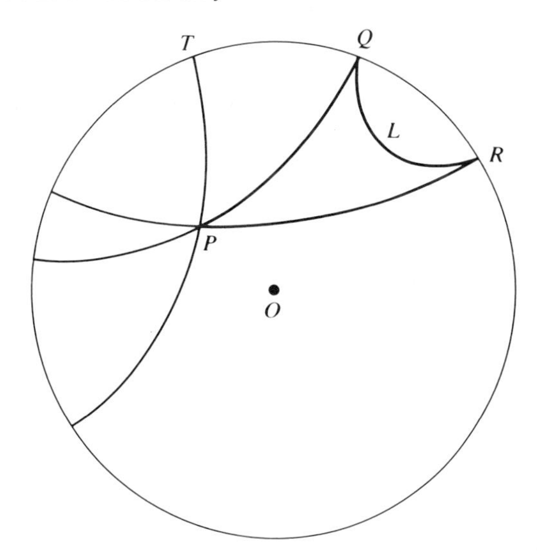

FIGURE 30.1

"straight line" L and a point P not on L, we can construct two arcs PQ and PR meeting L on the circle and therefore parallel to L; an arc like TP never meets L at all.

One can verify that the "straight lines" satisfy all the postulates of Euclidean geometry except the parallel postulate. If we take "rigid motions" to be the Möbius transformations of the unit disk into itself, we can call two figures congruent if they are equivalent under such a rigid motion. There is a "distance" between points that is invariant under rigid motions; we can discover a formula for it in terms of the points if we make reasonable assumptions about the properties that it ought to have. We require that distance is nonnegative and symmetric; invariant under rigid motions; additive for points on the same non-Euclidean straight line; and approximately Euclidean for short distances, that is, $\lim_{z\to 0} d(0, z)/|z| = 1$.

We begin by finding a formula for $d(r) = d(0, r)$. The transformation from z to $ze^{i\theta}$ is a non-Euclidean rigid motion, so $d(0, z) = d(0, |z|)$. A radius of the unit disk is a segment of a non-Euclidean straight line, so if $0 < r < r + h < 1$,

$$d(0, r+h) = d(0, r) + d(r, r+h).$$

The transformation $w = (z - r)/(1 - rz)$ is a non-Euclidean rigid motion that takes $z = r$ to $w = 0$ and $z = r + h$ to $w = h/(1 - hr - r^2)$. Since rigid motions are assumed to preserve distances,

$$d(r, r+h) = d\left(0, \frac{h}{1 - hr - r^2}\right).$$

We now have

$$d(0, r+h) - d(r) = d\left(\frac{h}{1 - hr - r^2}\right).$$

Divide by h and use the limit property that we assumed. We get $d'(r) = 1/(1-r^2)$, and so $d(r) = \frac{1}{2}\log[(1+r)/(1-r)]$. A non-Euclidean rigid motion carries z_1 and z_2 to $w = 0$ and $w = (z_2 - z_1)/(1 - \bar{z}_1 z_2)$. Since such a motion preserves distances,

$$d(z_1, z_2) = d\left(0, \left|\frac{z_1 - z_2}{1 - \bar{z}_1 z_2}\right|\right),$$

and when we use the formula for $d(0, r)$, we get

$$d(z_1, z_2) = \frac{1}{2}\log\frac{|1 - \bar{z}_1 z_2| + |z_1 - z_2|}{|1 - \bar{z}_1 z_2| - |z_1 - z_2|}.$$

We can now state the "invariant form" of Schwarz's lemma, discovered by G. Pick:

Let f be analytic in $|z| < 1$ and let $|f(z)| < 1$. Then either $d[f(z_1), f(z_2)] < d(z_1, z_2)$ whenever $z_1 \neq z_2$, $|z_1|$ and $|z_2| < 1$; or else $d[f(z_1), f(z_2)]$ always equals $d(z_1, z_2)$ and f is a Möbius transformation of the unit disk onto itself.

30B. Constructions

In order to draw respectable diagrams in our non-Euclidean geometry, we need to be able to carry out two geometric constructions: to construct a point symmetric to a given point with respect to a circle, and to construct a circle that passes through two given points and is orthogonal to the first circle.

First problem Circle C has center O, radius r; P is given. Find P' so that $OP \cdot OP' = r^2$. See Fig. 30.2.

Here OS is perpendicular to OP; $TO = OP$; SP' is perpendicular to TS; $OS = r$, and $OP'/r = r/OT (= r/OP)$ because triangles TOS and $OP'S$ are similar. Hence $OP \cdot OP' = r^2$.

Second problem Find P' and Q' symmetric to P and Q with respect to C. Draw a circle through any three of P, P', Q, Q'. See Fig. 30.3. [To do this, locate its center as the intersections of the perpendicular bisectors of (say) PP' and PQ'.]

I claim that this circle G is orthogonal to C and passes through the fourth point. This can be shown algebraically, but it is easier to simplify the problem by a Möbius transformation. Let the transformation T take C to a line L (see Fig. 30.4, p. 230). The images p, p' of P, P' are symmetric with respect to L, and so are the images q, q' of Q, Q'. The circle Γ through p, q, p', q' is easily constructed (its center is on L and it is evidently orthogonal to L). Under the inverse transformation T^{-1}, Γ becomes a circle through P, Q, P', Q' and is orthogonal to C because a Möbius transformation not only carries circles or straight lines to circles or straight lines, but is conformal and therefore

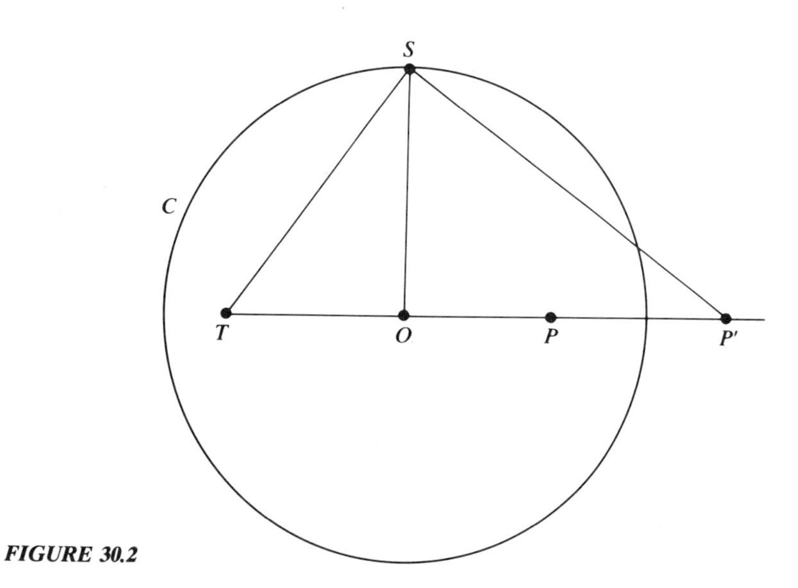

FIGURE 30.2

preserves the angles between them. Conversely, we see that if the circle G is orthogonal to C, and we pick points P' and Q' on G, their symmetric points with respect to C will be on G.

EXERCISE 30.1 Show how to construct a circle orthogonal to C and joining two points on C.

EXERCISE 30.2 Construct a non-Euclidean triangle with given vertices. How does the sum of the angles seem to relate to 180°?

FIGURE 30.3

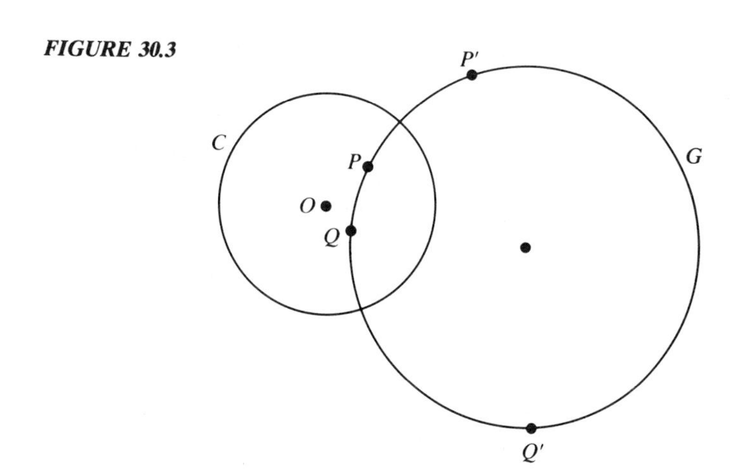

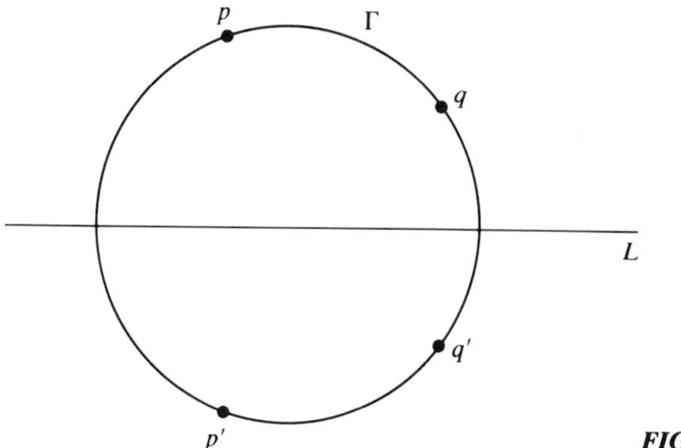

FIGURE 30.4

31 *Infinite Products*

Infinite products arise naturally if we try to generalize the factorization of polynomials to a similar factorization of transcendental (that is, nonpolynomial) entire functions. If we try to do this, we run into several difficulties almost at once. In the first place, an entire function might have no zeros at all (e^z). Second, a transcendental entire function usually has infinitely many zeros z_n (with a limit point at ∞), and an infinite product of terms $(z - z_n)$ does not look promising; even if we use factors $1-(z/z_n)$, we do not necessarily get a convergent product (as we shall see below). To make any progress, we need first to study the convergence of infinite products.

You might expect that a separate theory of infinite products would be unnecessary, since we can always convert products into sums by taking logarithms. However, we shall see that convergence theorems for products do not look quite like convergence theorems for series.

31A. Convergence of infinite products

It would seem reasonable to define a product $\prod_{n=1}^{\infty} z_n$ to mean the limit of the *partial products* $\prod_{n=1}^{N} z_n$. (We take an empty product to mean 1, since its logarithm would be an empty sum, which is usually understood to mean 0.) A little thought shows that if the partial products are to have a chance of approaching a limit other than 0, we had better have $z_n \to 1$. Hence the standard form of an infinite product is usually taken to be

$$\prod_{n=1}^{\infty} (1 + a_n),$$

where a_n are complex numbers. We say that the product converges if, first, the partial products approach a finite limit L; second, $a_n \neq -1$ for all sufficiently large n (say, for $n > N_0$); and third,

$$\lim_{N\to\infty} \prod_{N_0+1}^{N} (1+a_n) \neq 0.$$

The third condition forces $1+a_n \to 1$, and hence $a_n \to 0$ for a convergent product. Consequently, we may always suppose that $|a_n| < \frac{1}{2}$ for sufficiently large n, since otherwise the product necessarily diverges. Notice that a convergent product can perfectly well have the limit 0, but only when at least one factor *is* zero. When

$$\lim_{N\to\infty} \prod_{N_0}^{N} (1+a_n) = 0$$

for all N_0, we say (for reasons that will appear in a moment) that the product *diverges to* 0.

Let us now investigate the connection between convergence of a product and convergence of the corresponding series of logarithms. If $\prod_{N_0}^{N} (1+a_n) \to L \neq 0$, then we know from Exercise 1.20 that $\prod_{N_0}^{N} |1+a_n| \to |L|$ and that we can choose the angles of the terms and of L so that $\sum_{n=N_0}^{N} \arg(1+a_n) \to \arg L$. Consequently, $\sum_{n=N_0}^{\infty} \log(1+a_n)$ converges with some determination of each logarithm. Conversely, if $\sum_{N_0}^{\infty} \log(1+a_n)$ converges to S for some determination of each logarithm, then $\prod_{N_0}^{\infty} (1+a_n)$ converges to $e^S \neq 0$. This shows why we regard $\prod (1+a_n)$ as divergent if $\prod_{N_0}^{N} (1+a_n) \to 0$ for all N_0: because $\sum_{n=N_0}^{\infty} \log(1+a_n)$ would diverge to $-\infty$.

EXERCISE 31.1 Show that in fact if $\prod (1+a_n)$ converges (to a nonzero limit), then $\sum \operatorname{Log}(1+a_n)$ converges, where Log denotes the principal value of the logarithm.

It is convenient to have tests for convergence of infinite products, tests that are expressed directly in terms of the a_n.

EXERCISE 31.2 Show that when $|z| < \frac{1}{2}$, we have

$$|\log(1+z) - z| \leq |z|^2,$$

$$\left|\log(1+z) - z + \frac{1}{2}z^2\right| \leq |z|^3,$$

$$\left|\log(1+z) - z + \frac{1}{2}z^2 - \frac{1}{3}z^3\right| \leq |z|^4,$$

and so on.

We can now write down a number of simple sufficient conditions for the convergence of $\prod (1+a_n)$.

First, if $a_k \geq 0$, then $\prod (1+a_k)$ converges if and only if $\sum a_k$ converges. (This is not necessarily true if the a_k are real, but the a_k change sign; see Exercise 31.4).

To establish this, suppose that $\sum a_k$ converges and $a_k \geq 0$; then $\sum a_k^2$ converges, because eventually $0 \leq a_k < 1$ and then $a_k^2 \leq a_k$. By Exercise 31.2, $\sum |\log(1+a_k) - a_k|$ is dominated by $\sum a_k^2$ and so converges. Hence the series $\sum \log(1+a_k)$ converges and $\prod (1+a_k)$ converges.

Conversely, let $\prod (1+a_k)$ converge and $a_k \geq 0$. Then $\sum \log(1+a_k)$ converges (Exercise 31.1), and we may suppose that $a_k < \frac{1}{2}$. Now we have

$$\begin{aligned}\log(1+a_k) &= a_k - \frac{1}{2}a_k^2 + \frac{1}{3}a_k^3 - \cdots \\ &= a_k\left(1 - \frac{1}{2}a_k + \frac{1}{3}a_k^2 - \cdots\right) \\ &\geq a_k\left(1 - \frac{1}{2}a_k\right) \geq \frac{3}{4}a_k,\end{aligned}$$

since $a_k \leq \frac{1}{2}$. Hence $\sum a_k$ converges.

EXERCISE 31.3 Show that $\prod (1+a_k)$ converges if both $\sum a_k$ and $\sum |a_k|^2$ converge.

EXERCISE 31.4 Investigate the convergence of $\prod (1+a_k)$ in the following cases:

(*a*) $a_k = (-1)^k k^{-1/3}$
(*b*) $a_{2n+1} = (2n+1)^{-1/2}$, $a_{2n+2} = -(2n)^{-1/2}$, $n = 1, 2, 3, \ldots$
(*c*) $a_k = z/k$, $k = 1, 2, 3, \ldots$
(*d*) $a_n = -z^2/n^2$
(*e*) $a_{2n} = z/n$, $a_{2n-1} = -z/n$, $n = 1, 2, 3, \ldots$
(*f*) Consider Fig. 31.1.[1] If the construction is continued indefinitely, does the figure remain bounded?

A product $\prod [1 + a_n(z)]$ is said to *converge uniformly* if the corresponding series of logarithms converges uniformly.

EXERCISE 31.5 Show that $\prod_{n=1}^{\infty} (1 - z^2/n^2)$ converges uniformly on each compact set that excludes all the points $\pm n$.

We can now see why we cannot usually represent an entire function with infinitely many zeros z_n in the form

$$\prod_{n=1}^{\infty} (z - z_n),$$

or even

$$z^m e^{g(z)} \prod_{n=1}^{\infty} \left(1 - \frac{z}{z_n}\right),$$

where g is entire. If, for example, the z_n were all positive, then when z is real

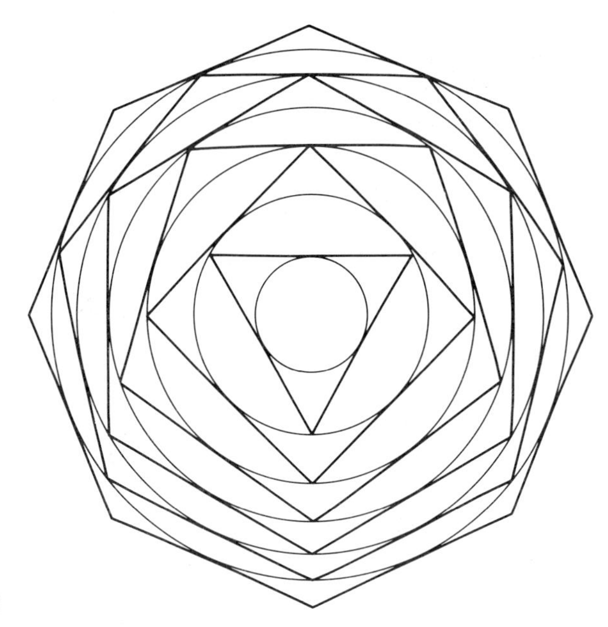

FIGURE 31.1

and negative, the product could converge only if $\sum 1/z_n < \infty$ (by the test given just after Exercise 31.2). This condition says that the $|z_n|$ must increase rather fast. For example, $z_n = n$ would not do, and therefore $1/\Gamma(-z)$ cannot be represented by a product of this kind. The function $\sin \pi z$ has zeros at the integers; we could not have $(\pi z)^{-1} \sin \pi z = \prod_{n=-\infty}^{\infty} (1 - z/n)$, $n \neq 0$. However, $\prod (1 - z^2/n^2)$ does converge, and in fact we are going to show in Sec. **31B** that

$$\sin \pi z = \pi z \prod_{n=1}^{\infty} \left(1 - \frac{z^2}{n^2}\right).$$

This is **Euler's product** for the sine function.

Supplementary exercises

Investigate the convergence of the infinite products in Supplementary Exercises 1–4.

1. $\prod_{n=1}^{\infty} [1 + (-1)^n n^{-1/2}]$

2. $\prod_{n=1}^{\infty} [1 + (-1)^n n^{-2/3}]$

3. $\prod_{n=1}^{\infty} \left(1 - \frac{z}{n}\right) e^{z/n}$

4. $\prod_{n=1}^{\infty} \left(1 - \frac{z}{n^{1/2}}\right) \exp[n^{-1/2} z + (2n)^{-1} z^2]$

5. Let $f(x)=\prod_{n=1}^{\infty}(1+x^{2^n})$, $0<x<1$. It is trivial that the product converges. To what function does it converge?
6. Show that $\prod_{n=1}^{\infty}(1+z^{n^2})$ converges for $|z|<1$ to a function $f(z)$. Can f be continued analytically through the point $z=1$?

31B. The product representation[2] of $\sin \pi z$

Rather than showing directly that Euler's product represents $\sin \pi z$, we are going to proceed indirectly, via residue calculus, to find a representation for $\sin \pi z$, which will turn out to be Euler's formula. It is somewhat easier to do this in a general situation and then specialize it to get what we want.

We consider the integral

$$I_n=\frac{1}{2\pi i}\int_{C_n}\frac{f(w)\,dw}{w(w-z)},$$

where f is a function, meromorphic in the finite plane, with simple poles a_n, of residue 1, and f is analytic at 0. Furthermore, we suppose that C_n are simple closed curves, tending to ∞ as $n\to\infty$ and of length $O(n)$, on which $|f(w)|$ is bounded. Of course this would be a rather useless discussion unless we know that functions f and curves C_n with all these properties actually exist. In fact, they do.

EXERCISE 31.6 Show that $f(w)=\cot w-w^{-1}$ has these properties if C_n is the rectangle with vertices $\pm(n+\frac{1}{2})\pi\pm n\pi i$. Also show that $I_n\to 0$.

Let us then see what the residue theorem says about I_n. This integral is equal to the sum of the residues of its integrand at all the poles inside C_n, which are at (1) 0, (2) z, and (3) the poles a_n of f. The corresponding residues are

1. $\dfrac{-f(0)}{z}$,
2. $\dfrac{f(z)}{z}$,
3. $\dfrac{1}{a_n(a_n-z)}=\left(-\dfrac{1}{a_n}+\dfrac{1}{a_n-z}\right)\dfrac{1}{z}$.

(Remember that the residues of f at a_n are all 1.) Hence, since $I_n\to 0$ as $n\to\infty$, we have

$$f(z)=f(0)+\sum_m\left(\frac{1}{a_m}-\frac{1}{a_m-z}\right),$$

where we sum over all a_m that are inside C_n and then let $n\to\infty$.

In particular, this gives us a series representation for $\cot \pi z$; but that is not what we are interested in at this point.

We now define $g(z) = z^{-1} \sin z$ and take $f(w) = g'(w)/g(w) = (d/dw) \log g(w) = \cot w - w^{-1}$. Then $a_n = n\pi (n \neq 0)$.

EXERCISE 31.7 Show that $f(0) = 0$.

Now we have

$$\frac{d}{dz} \log \frac{\sin z}{z} = \sum_{\substack{n=-\infty \\ n \neq 0}}^{\infty} \left(\frac{1}{n\pi} - \frac{1}{n\pi - z} \right),$$

where the sum is the limit of sums over $-N < n < N$ $(n \neq 0)$ as $N \to \infty$, and converges uniformly on compact sets that avoid the points $n\pi$.

Next we integrate over a path from 0 to z; with some choice of each logarithm we have

$$\log \frac{\sin z}{z} = \sum_{\substack{n=-\infty \\ n \neq 0}}^{\infty} \left[\frac{z}{n\pi} + \log(n\pi - z) - \log n\pi \right].$$

How much indetermination is introduced by the logarithms? To answer this question, let us consider real values x of z. Then the logarithmic terms are

$$\log(n\pi - x) - \log n\pi = \log \frac{n\pi - x}{n\pi} = \log\left(1 - \frac{x}{n\pi}\right);$$

or, if we combine the terms with positive and negative n,

$$\log\left(1 - \frac{x}{n\pi}\right) + \log\left(1 + \frac{x}{n\pi}\right) = \log\left(1 - \frac{x^2}{n^2\pi^2}\right),$$

where, since $(1 - x^2/(n^2\pi^2))$ is either a positive or negative real number, each logarithm is determined up to an additive $k\pi i$ (k = integer). Hence we obtain

$$\frac{\sin x}{x} = \prod_{n=1}^{\infty} \left(1 - \frac{x^2}{n^2\pi^2}\right),$$

up to a factor $e^{ik\pi} = \pm 1$. But when $x = 0$ the product must be 1, and hence k is an even integer. Thus we have

$$\sin x = x \prod_{n=1}^{\infty} \left(1 - \frac{x^2}{n^2\pi^2}\right)$$

for real x. Since the product converges uniformly on each compact set that excludes the points $n\pi$ (Exercise 31.5) the product represents an analytic function, and the representation holds with x replaced by any $z \neq n\pi$. We have therefore verified Euler's product for the sine function.

Notice that the separate products

$$\prod_{n>0} \left(1 - \frac{z}{n\pi}\right) \quad \text{and} \quad \prod_{n<0} \left(1 - \frac{z}{n\pi}\right)$$

do not converge.

31C. Wallis's formula

Here we derive a famous formula by taking $z = \pi/2$ in Euler's product. We find

$$1 = \frac{\pi}{2} \prod_{n=1}^{\infty} \left(1 - \frac{1}{4n^2}\right),$$

$$\frac{2}{\pi} = \lim_{N\to\infty} \left[\left(1 - \frac{1}{2}\right)\left(1 + \frac{1}{2}\right)\right]\left[\left(1 - \frac{1}{4}\right)\left(1 + \frac{1}{4}\right)\right] \cdots \left[\left(1 - \frac{1}{2N}\right)\left(1 + \frac{1}{2N}\right)\right],$$

that is,

$$\frac{2}{\pi} = \frac{1}{2} \cdot \frac{3}{2} \cdot \frac{3}{4} \cdot \frac{5}{4} \cdot \frac{5}{6} \cdot \frac{7}{6} \cdots \frac{2N-1}{2N} \cdot \frac{2N+1}{2N} \cdots$$

$$= \lim_{N\to\infty} \frac{[(2N)!]^2(2N+1)}{2^{4N}(N!)^4},$$

or

$$\frac{\pi}{2} = \lim_{N\to\infty} \frac{2^2 \cdot 4^2 \cdot 6^2 \cdots (2N)^2}{1^2 \cdot 3^2 \cdot 5^2 \cdots (2N-1)^2(2N+1)}.$$

This is **Wallis's product**. It follows that

$$\sqrt{\pi} = \lim_{N\to\infty} \frac{2 \cdot 4 \cdot 6 \cdots (2N)}{1 \cdot 3 \cdots (2N-1)N^{1/2}}$$

31D. Products with convergence factors

Since when $|z/z_n| < 1$,

$$\log\left(1 - \frac{z}{z_n}\right) = -\frac{z}{z_n} - \frac{1}{2}\frac{z^2}{z_n^2} - \cdots,$$

then when $|z_n|$ is large, a product

$$\prod \left(1 - \frac{z}{z_n}\right) e^{z/z_n}$$

ought to be more likely to converge than $\prod (1 - z/z_n)$; a product

$$\prod \left(1 - \frac{z}{z_n}\right) \exp\left(\frac{z}{z_n} + \frac{1}{2}\frac{z^2}{z_n^2}\right)$$

ought to do even better; and so on. In fact, Weierstrass showed that every entire function f can be represented in the form

$$z^m e^{g(z)} \prod \left(1 - \frac{z}{z_n}\right) \exp\left(\frac{z}{z_n} + \frac{1}{2}\frac{z^2}{z_n^2} + \cdots + \frac{1}{k_n}\frac{z^{k_n}}{z_n^{k_n}}\right),$$

where $k_n \to \infty$ sufficiently fast and g is an entire function [so that $e^{g(z)}$ has no zeros], and the z_n are taken in order of increasing modulus. This representation is so general that it is not very useful.

There is a more useful product representation for entire functions that are of **finite order** in the following sense. Let $M(r) = \max_{|z| \le r} |f(z)|$; then f is said to be of order ρ if ρ is the greatest lower bound of numbers s for which $\limsup_{r\to\infty} r^{-s} \log M(r) < \infty$. [Roughly speaking, this means that $M(r)$ is not much larger than $\exp(r^\rho)$, and sometimes about that large.] Then **Hadamard's product representation** says:

If f is of order ρ with zeros at (possibly) 0 and z_n, then

$$f(z) = z^m e^{P(z)} \prod_{n=1}^{\infty} \left(1 - \frac{z}{z_n}\right) \exp\left[\frac{z}{z_n} + \frac{1}{2}\left(\frac{z}{z_n}\right)^2 + \cdots + \frac{1}{p}\left(\frac{z}{z_n}\right)^p\right]$$

where $p \le \rho$ and $P(z)$ is a polynomial of degree at most ρ.

For example, $\sin \pi z$ is of order 1; Euler's product is a collapsed form of the Hadamard product with $p = 1$. The reciprocal of the gamma function is also of order 1; its Hadamard product is

$$\frac{1}{\Gamma(z)} = z e^{\gamma z} \prod_{n=1}^{\infty} \left(1 + \frac{z}{n}\right) e^{-z/n},$$

where γ is Euler's constant,

$$\gamma = \lim_{n\to\infty} \left(1 + \frac{1}{2} + \frac{1}{3} + \cdots + \frac{1}{n} - \log n\right).$$

(Although γ has been computed to thousands of decimal places, it is not known whether γ is rational or irrational.)

We shall not prove the product formula for the reciprocal gamma function, but we may notice that, combined with Euler's product for the sine, it shows again that $\Gamma(z)\Gamma(1 - z) = \pi/\sin \pi z$.

Notes

1. This figure, drawn by David B. Singmaster, appeared in *London Math. Soc. Newsletter,* no. 122 (October 1985); it is reproduced here with Dr. Singmaster's permission. Dr. John W. Wrench, Jr., has provided me with several references to similar figures with the polygons differently oriented; the oldest is *Mathematics and the Imagination,* by E. Kasner and J. Newman (Simon & Schuster, New York, 1940), p. 312. The figure is sometimes attributed to Kepler, who apparently considered the dual figure (where the polygons are successively inscribed instead of circumscribed), but only the first few stages.

2. There are many different and fairly elementary proofs of Euler's formula. See, for example, W. F. Eberlein, "On Euler's Infinite Product for the Sine," *J. Math. Anal. Appl. 58* (1977): 147–151. Also see *Math. Rev. 83e*:01032.

32 *Rate of Growth Versus Number of Zeros*

32A. A general principle

An entire function of large order, that is, one whose maximum modulus $M(r)$ increases rapidly as $r \to \infty$, may have very few zeros or even none at all [think of $\exp(z^n)$ or $\exp(e^z)$]. On the other hand, if the function has too many zeros z_n in a disk of radius r, the product $\prod (1 - z/z_n)$ will not converge (think of $z_n = n$ as contrasted with $z_n = n^2$). Hadamard's theorem quoted in Sec. **31D** suggests that the more zeros there are, the larger the order of $M(r)$ has to be. This may seem paradoxical, since you might expect a function with many zeros to be small in modulus. The actual situation is quite different: There is a principle that says that when a function has many zeros, its modulus must frequently be large in order to compensate; if the modulus is not large, an analytic function with many zeros is in danger of vanishing identically. We already know an extreme case when the principle can be quantified: If $M(r) = O(r^n)$ as $r \to \infty$, then by Exercise 7.8 we know that the function is a polynomial of degree at most n and therefore has at most n zeros. We are now going to take up some more general results based on the same principle.

32B. Jensen's formula

We consider a function f that is analytic in a disk $|z| \leq R$, but not identically zero; we also assume (to keep the formulas simple) that $f(0) \neq 0$. Let f have zeros $\{z_k\}$, $k = 1, \ldots, n$, in $|z| < R$, and suppose that $f(z) \neq 0$ on $|z| = R$. Then **Jensen's theorem** says that

$$\frac{1}{2\pi} \int_0^{2\pi} \log |f(Re^{i\theta})| \, d\theta - \log |f(0)| = \sum_{j=1}^{n} \log \frac{R}{|z_j|}.$$

Notice that if f has no zeros, then $\log |f|$ is harmonic and Jensen's theorem reduces to the mean value theorem for harmonic functions (Secs. **16A**, **20E**). In the general case the formula shows that if $|f(Re^{i\theta})|$ is not large, there cannot be too many zeros in $|z| < R$, in accordance with our general principle.

The right-hand side of Jensen's formula can be written more neatly in terms of $n(t)$, the counting function of the zeros (that is, the number of zeros of modulus less than t): It is $\int_0^R t^{-1} n(t) \, dt$. This is because $n(t)$ is a step function that jumps by k when t crosses a value for which the circumference $|z| = t$ contains k zeros (a multiple zero is counted according to its multiplicity). The necessary calculation is easiest to follow when no circle $|z| = t$ contains more than one (simple) zero; the general case can be carried out in the same way. We have $n(0) = 0$, since $f(0) \neq 0$, and $n(t) = 0$ for $0 < t < |z_1|$. Then $n(t) = 1$ from $|z_1|$ to $|z_2|$, $n(t) = 2$ from $|z_2|$ to $|z_3|$, and so on. We therefore have

$$\begin{aligned}\int_0^R \frac{n(t)}{t}\,dt &= \int_{|z_1|}^{|z_2|} \frac{dt}{t} + 2\int_{|z_2|}^{|z_3|} \frac{dt}{t} + 3\int_{|z_3|}^{|z_4|} \frac{dt}{t} + \cdots + N\int_{|z_N|}^{R} \frac{dt}{t}\\ &= \log|z_2| - \log|z_1| + 2(\log|z_3| - \log|z_2|) + \cdots + N(\log R - \log|z_N|)\\ &= -\log|z_1| - \log|z_2| - \cdots - \log|z_N| + N\log R\\ &= \log\frac{R}{|z_1|} + \log\frac{R}{|z_2|} + \cdots + \log\frac{R}{|z_N|}.\end{aligned}$$

We now prove Jensen's theorem. We saw in Exercise 12.1 that, when z_0 is a zero of f, the function f'/f has a simple pole at z_0 with residue 1. It follows that, when g is analytic in $|z| \le R$, we have

$$\frac{1}{2\pi i}\int_{|z|=R} g(z)\frac{f'(z)}{f(z)}\,dz = \sum g(z_n),$$

where z_n are the zeros of f inside $|z| = R$ [assuming that $f(z) \neq 0$ on $|z| = R$]. This suggests that the right-hand side of Jensen's formula ought to be connected with

$$J = \frac{1}{2\pi i}\int_C (\log z)\frac{f'(z)}{f(z)}\,dz,$$

but $g(z) = \log z$ is not analytic (since it is not single-valued). Nevertheless, we can consider the same integral when C is the circle $|z| = R$ with a cut along the positive real axis from 0 to R. If f happens to have zeros on the cut, we replace $f(z)$ by $f(ze^{i\lambda})$ with λ chosen so that $f(ze^{i\lambda})$ has no zeros on the cut. There is such a λ, since there are only finitely many zeros to avoid; and the change does not affect the moduli of the zeros. The function $\log z$ is determined in the cut disk by taking $\log(-1) = i\pi$. Then the integrand in J is analytic except for simple poles at the zeros of f inside C, and we have

$$J = \sum_{j=1}^{N} \log z_j.$$

On the other hand, we can write J as the sum of three integrals, one along the upper side of the cut, one along the lower side, and one along the circle. As in Sec. **11**, the integral along the upper side is

$$\frac{1}{2\pi i}\int_0^R (\log x)\frac{f'(x)}{f(x)}\,dx;$$

the integral along the lower side is

$$-\frac{1}{2\pi i}\int_0^R (\log x + 2\pi i)\frac{f'(x)}{f(x)}\,dx;$$

and their sum is

$$-\int_0^R \frac{f'(x)}{f(x)}\,dx = \log f(0) - \log f(R).$$

The integral around the circle can be written as

$$\frac{1}{2\pi i}\int_{|z|=R} (\log z)[\log f(z)]'\,dz.$$

After integration by parts, this becomes

$$\frac{1}{2\pi i}(\log z)\log f(z)\Big|_{|z|=R} - \frac{1}{2\pi i}\int_{|z|=R}\log f(z)\frac{dz}{z}.$$

Putting the three integrals together and taking real parts, we have

$$\sum_{j=1}^{N}\log|z_j| = \operatorname{Re} J = \log|f(0)| - \log|f(R)| + \operatorname{Re}\left\{\frac{1}{2\pi i}(\log z)\log f(z)|_{|z|=R}\right\}$$
$$-\frac{1}{2\pi}\int_0^{2\pi}\log f(Re^{i\theta})\,d\theta.$$

We still have to compute the change in $(\log z)\log f(z)$ around $|z| = R$.

When z goes around the circle $|z| = R$, $\log z$ starts as $\log R$ and ends as $\log R + 2\pi i$, whereas $\log f(z)$ starts as $\log f(R)$ and ends (by the argument principle) as $\log f(R) + 2\pi i N$, if there are N zeros. Hence the change in $[\log f(z)](\log z)$ around $|z| = R$ is

$$(\log R + 2\pi i)[\log f(R) + 2\pi i N] - (\log R)[\log f(R)]$$
$$= 2\pi i[\log|f(R)| + N(\log R + 2\pi i)].$$

If we divide by $2\pi i$ and take real parts, we get

$$\operatorname{Re}\frac{1}{2\pi i}(\log z)\log f(z)\Big|_{|z|=R} = \log|f(R)| + N\log R.$$

Hence the net result of computing J in two ways is

$$\sum_{j=1}^{N}\log|z_j| = \log|f(0)| + N\log R - \frac{1}{2\pi}\int_0^{2\pi}\log|f(Re^{i\theta})|\,d\theta,$$

that is,

$$\frac{1}{2\pi}\int_0^{2\pi}\log|f(Re^{i\theta})|\,d\theta - \log|f(0)| = \sum_{j=1}^{N}\log\frac{R}{|z_j|},$$

the conclusion of Jensen's theorem.

The proof is easily modified to show that when f is meromorphic, with zeros z_n and poles p_n, then

$$\frac{1}{2\pi}\int_0^{2\pi}\log|f(Re^{i\theta})|\,d\theta-\log|f(0)|=\sum_{j=1}^{N}\log\frac{R}{|z_j|}-\sum_{k=1}^{M}\log\frac{R}{|p_k|}.$$

32C. Some consequences of Jensen's theorem

First notice that if $|f(0)|=1$ and $|f(z)|\le M$ for $|z|\le R$, then

$$\int_0^R\frac{n(t)}{t}\,dt\le\max_{|z|=R}\log|f(z)|.$$

This inequality puts a restriction on the number of zeros that such a function can have in a disk $|z|<r<R$. For example,

$$n\left(\frac{R}{2}\right)\log 2=n\left(\frac{R}{2}\right)\int_{R/2}^{R}\frac{dt}{t}\le\int_{R/2}^{R}\frac{n(t)}{t}\,dt\le\log M;$$

and still more precise statements can be made.

EXERCISE 32.1 Let f be an entire function that satisfies $|f(z)|\le Ae^{B|z|}$ (A and B constants, $B<1$). Let f have zeros z_n with $|z_n|=n$ ($n=1,2,3,\ldots$). Show that $f(z)\equiv 0$.

EXERCISE 32.2 Let $f(z)$ be an entire function with $|f(z)|\le A\exp(B|z|^2)$. If f has at least n zeros on each circle $|z|=n$ ($n=1,2,3,\ldots$), then $f(z)\equiv 0$ if $B<\frac{1}{4}$.

Suppose now that f is analytic and bounded in the disk $|z|<1$, with $f(0)\ne 0$. It is quite possible—in fact, rather likely—that f has infinitely many zeros in the unit disk. Let $|f(z)|\le M$, and number the zeros in order of increasing modulus. Then Jensen's theorem says that, when $R<1$,

$$\sum_{|z_j|<R}\log\frac{R}{|z_j|}\le\Big|\log|f(0)|\Big|+\frac{1}{2\pi}\int_0^{2\pi}\log|f(Re^{i\theta})|\,d\theta$$

$$\le\Big|\log|f(0)|\Big|+\log M.$$

Since the right-hand side is independent of R, we can let $R\to 1$ and obtain the inequality

$$\sum_{j=1}^{\infty}\log\frac{1}{|z_j|}\le\big|\log|f(0)|\big|+\log M.$$

Since $|z_j|\to 1$ as $j\to\infty$ (otherwise there would be a limit point of zeros inside the unit disk), this inequality shows that the moduli of the zeros cannot approach 1 so slowly that the series would diverge. A more transparent way of saying this is to say that $\sum_{j=1}^{\infty}(1-|z_j|)$ must converge. This follows because

$$\log\frac{1}{|z_j|}=-\log[1-(1-|z_j|)]=(1-|z_j|)+\frac{1}{2}(1-|z_j|)^2+\cdots\ge 1-|z_j|.$$

We did not use the full force of the condition that $|f(z)|$ is bounded; it would have been enough to assume that

$$\int_0^{2\pi} \log |f(re^{i\theta})|\, d\theta$$

is bounded above.

Notice that Jensen's theorem lets us bound $\int_0^R t^{-1}n(t)\,dt$ in terms of $\max_{|z|=R} \log |f(z)|$, but not the other way around. For example, f might not have any zeros at all, or $|f(z)|$ might have many small values that would make $\log |f(z)|$ negative and of large modulus.

Supplementary exercises

1. Let f be an entire function of order less than 2, that is, $|f(z)| \le A \exp(B\,|z|^{2-\varepsilon})$, $\varepsilon > 0$, where A and B are constants. Suppose that $f(m+in) = 0$ for all integral m and n. Show that $f(z) \equiv 0$.

2. Let f be analytic and bounded in the unit disk and let $f[\exp(-1/n)] = 0$, $n = 1, 2, \ldots$. Show that $f(z) \equiv 0$.

3. Let f be an entire function such that $|f(z)| \le Ae^{c|z|}$, $c < 4$. Suppose that $f(n) = f(in) = 0$, $n = 0, \pm 1, \pm 2, \ldots$. Show that $f(z) \equiv 0$.

33 *Generalizations of the Maximum Principle*

33A. Phragmén–Lindelöf theorems

Phragmén and Lindelöf discovered that the maximum principle continues to hold in the presence of an exceptional point if the function is subjected to an extra condition that in itself does not look strong enough to guarantee the conclusion. In other words, if $|f(z)| \le M$ on the boundary of D except for one point (frequently ∞), then either $|f(z)| \le M$ in D or else $|f|$ must grow extraordinarily fast as z approaches the exceptional point.

Let us illustrate the basic idea of a Phragmén–Lindelöf theorem by proving the simplest theorem of this kind:

> *If* lim sup $|f(z)| \le M$ *as z approaches each point of the boundary of a simply connected region D, except for one point z_0, and if $|f(z)| \le N$ in D, then actually $|f(z)| \le M$ in D. (Naturally we are to suppose that $N > M$.)*

The idea of the proof is that we first find a family of functions $g_\varepsilon(z)$ such that $g_\varepsilon(z)$ is analytic in D for each ε, $|g_\varepsilon(z)| \le 1$, $g_\varepsilon(z) \to 0$ as $z \to z_0$, and $|g_\varepsilon(z)| \to 1$ as $\varepsilon \to 0$, $z \ne z_0$. If we have such a family $g_\varepsilon(z)$, we will have $\limsup |f(z)g_\varepsilon(z)| \le M$ as z approaches each boundary point of D, because

$|f(z)g_\varepsilon(z)| \to 0$ as $z \to z_0$. Then by Sec. **16B** we have $|f(z)g_\varepsilon(z)| \le M$ for $z \in D$. Since the right-hand side is independent of ε, we can let $\varepsilon \to 0$, and it follows that $|f(z)| \le M$ for $z \in D$ because $|g_\varepsilon(z)| \to 1$ for each $z \ne z_0$.

All that remains is to discover a suitable family g_ε. For a bounded D, this is easy. Let K be the maximum of $|z - z_0|$ for z in the closure of D. Then $g_\varepsilon(z) = [(z - z_0)/K]^\varepsilon$, $\varepsilon > 0$, has the desired properties; we may choose any branch of the power.

EXERCISE 33.1 Prove the same theorem when D is unbounded and has an exterior point z_1, and $z_0 = \infty$.

The corresponding theorem for harmonic functions could be proved in a similar way; but it is easier to argue that if u is harmonic in the simply connected region D, then $u = \operatorname{Re} f$ with f analytic (Sec. **19A**) and $|e^f| = e^u$ is bounded if and only if u is bounded above.

Instead of a single exceptional boundary point, we could allow a finite number of exceptional boundary points.

We now have a more general version of the maximum principle than we had before. However, the hypothesis that f is bounded is stronger than necessary if we also take the shape of D into account. The next two theorems may seem rather special, but they have interesting applications.

Having observed that e^z is bounded on the imaginary axis but not in the right-hand half plane, we might be led to ask what kind of condition we could impose on a function f that is analytic in the right-hand half plane and bounded on the imaginary axis in order to force it to be bounded in the right-hand half-plane. The absolute value of e^z grows like e^x in the right-hand half plane; a function $\exp(z^\delta)$, $0 < \delta < 1$, grows more slowly but is unbounded on the imaginary axis. Perhaps, then, growth at this lower rate is incompatible with boundedness on the imaginary axis. This turns out to be the case.

It is more useful, and more illuminating, to consider functions that are analytic in an angle $|\arg z| \le \alpha$, where $\alpha < \pi$. We will assume that f is known not to grow "too fast" as $|z| \to \infty$ inside the angle, and hope to infer that if f is bounded on the sides of the angle, then in fact f must have been bounded inside the angle all along.

Theorems for angles that are not symmetric with respect to the positive real axis can be reduced to the symmetric case by a change of variable (really, a rotation of the plane).

We now prove the **main Phragmén–Lindelöf theorem**:

If f is analytic for $|\arg z| < \alpha$, where $\alpha < \pi$, and continuous in the closed angle; if $|f(z)| \le M$ on the sides of the angle; and if $|f(z)| \le A \exp(B\,|z|^\mu)$, where $0 < \mu < \pi/(2\alpha)$ and A and B are positive constants; then $|f(z)| \le M$ inside the angle.

Here the essential condition is that μ is strictly less than $\pi/(2\alpha)$; the numbers

A and B are included mainly to avoid complications near $z = 0$. The condition $\alpha < \pi$ prevents the angle from overlapping itself.

For the proof, we need a function like the g_ε of the preceding theorem. That is, we need to have g_ε analytic in the interior of the angle, $|g_\varepsilon(z)| \leq 1$ on the sides of the angle, $g_\varepsilon(z) \to 0$ as $z \to \infty$, and $|g_\varepsilon(z)| \to 1$ as $\varepsilon \to 0$ for each z inside the angle. It is probably not obvious how to find such a function, but once we have thought of it, it is easy to see that it does the job. We take $g_\varepsilon(z) = \exp(-\varepsilon z^\sigma)$, where $\mu < \sigma < \pi/(2\alpha)$, and z^σ has its principal value. In fact, if $\arg z = \theta$,

$$z^\sigma = \exp[\sigma(\ln|z| + i \arg z)] = |z|^\sigma e^{i\sigma\theta},$$

$$g_\varepsilon(z) = \exp[-\varepsilon\,|z|^\sigma(\cos\sigma\theta + i\sin\sigma\theta],$$

$$|g_\varepsilon(z)| = \exp(-\varepsilon\,|z|^\sigma \cos\sigma\theta).$$

On the rays $\theta = \pm\alpha$ we have $|\sigma\theta| < \frac{1}{2}\pi$, and $\cos\sigma\theta$ is positive. Hence $|g_\varepsilon(z)| \leq 1$ on the sides of the angle and $|f(z)g_\varepsilon(z)| \leq M$ there. On an arc $|z| = R$, $|\theta| < \alpha$, on the other hand,

$$\begin{aligned} |f(z)g_\varepsilon(z)| &\leq A\exp[BR^\mu - \varepsilon R^\sigma \cos(\sigma\theta)] \\ &\leq A\exp[BR^\mu - \varepsilon R^\sigma \cos(\sigma\alpha)], \end{aligned}$$

because $\cos\sigma\theta$ cannot be smaller than $\cos\sigma\alpha$. Now, since $\sigma > \mu$, the exponent is negative for large R and $|f(z)g_\varepsilon(z)| \to 0$ as $R \to \infty$, uniformly in θ. Consequently, $f(z)g_\varepsilon(z)$ is bounded in the angle, and it follows that $|f(z)g_\varepsilon(z)| \leq M$ in the whole angle. Now let $\varepsilon \to 0$, and the conclusion of the theorem follows.

Notice that the proof would not work if $\mu = \pi/(2\alpha)$, since it depends on our having σ strictly between μ and $\pi/(2\alpha)$. However, we can obtain a theorem with $\mu = \pi/(2\alpha)$ if we strengthen the other hypotheses. To keep things simple, I give the corresponding theorem only for the half plane, $\alpha = \pi/2$. Here we assume that $|f(z)| \leq M$ on the imaginary axis, and that $|f(z)| \leq A\exp(\delta\,|z|)$ for each positive δ (only small values of δ are relevant), where the number A may depend on δ. With these hypotheses, it again follows that $|f(z)|$ is bounded by M in the half plane $\operatorname{Re} z \geq 0$.

Let us consider $g(z) = f(z)e^{-\delta z}$ in each of the first and fourth quadrants. Since $|g|$ is bounded by M on the imaginary axis and by A on the positive real axis, we can apply the preceding theorem to each quadrant and deduce that $|f(z)e^{-\delta z}| \leq \max(M, A)$ in each quadrant. Now we assumed that $f(z) = O(e^{\delta|z|})$ for *every* positive δ which of course also means that $f(z) = O(e^{\delta|z|/2})$ for every positive δ. Hence, in particular, $f(x) = O(e^{\delta x/2})$ as $x \to +\infty$, and so $f(x)e^{-\delta x} \to 0$ as $x \to \infty$. Hence we must have $A \leq M$, because otherwise $|f(z)e^{-\delta z}|$ would be maximized in the right-hand half plane at some finite interior point. Consequently, $|f(z)e^{-\delta z}| \leq M$ in the right-hand half plane, with M independent of δ. Letting $\delta \to 0$, we obtain $|f(z)| \leq M$ in the right-hand half plane.

In this and similar theorems it is always enough to have the upper bound on $|f|$ inside the angle holding only on a sequence of arcs that tend to ∞, since we can apply the maximum principle to the regions bounded by pairs of successive arcs and portions of the sides of the angle.

33B. Some applications of Phragmén–Lindelöf theorems

Our first application is related to the second theorem in Sec. **28D**. *Let f be analytic and bounded in the closed angle between the rays* $\arg z = \alpha$ *and* $\arg z = \beta$, $|\alpha - \beta| < 2\pi$, *and suppose that* $f(z) \to L$ *as* $r \to \infty$ *along both rays. Then* $f(z) \to L$ *uniformly in the angle as* $|z| \to \infty$.

We may simplify the problem by supposing that $L = 0$ [otherwise consider $f(z) - L$]; and, to begin with, we suppose that $|\alpha - \beta| < \pi$. (This assumption will be removed later.) We further simplify the problem by rotating the plane so that $\beta = -\alpha$, $0 < \alpha < \pi/2$. Now consider

$$g(z) = \frac{z}{z+K} f(z), \qquad K > 0.$$

For large $|z|$ we have

$$|g(z)| \le \frac{|z|}{|z+K|} |f(z)| \le \frac{r}{(r^2 + K^2)^{1/2}} |f(z)|.$$

We also have $|f(z)| \le M$, for some M, by hypothesis, in the angle; and $|f(z)| < \varepsilon$ if $r > r_1 = r_1(\varepsilon)$ on the sides of the angle. Then, when $|z| > r_1$, we have $|g(z)| \le |f(z)| < \varepsilon$ for $\arg z = \pm\alpha$, independently of K. Now choose $K = r_1 M/\varepsilon$. When $|z| < r_1$, we have

$$|g(z)| \le \frac{rM}{(r^2 + r_1^2 M^2/\varepsilon^2)^{1/2}} = \frac{r\varepsilon M}{(r^2\varepsilon^2 + r_1^2 M^2)^{1/2}} \le \frac{r\varepsilon M}{r_1 M} < \varepsilon.$$

Thus $|g(z)| < \varepsilon$ throughout the angle,

$$|f(z)| = |g(z)| \left| \frac{z+K}{z} \right| < \varepsilon \frac{|z| + K}{|z|},$$

and if $|z| > K$, this says that $|f(z)| < 2\varepsilon$. In other words, $f(z) \to 0$ uniformly in the angle.

If $|\beta - \alpha| > \pi$, we can consider $h(w) = f(w^p)$ with a sufficiently large integer p; then h is analytic for w in an angle of opening $|\beta - \alpha|^{1/p} < \pi$ if p is large enough.

We can use the preceding proposition to prove that if f is analytic and bounded in an angle $\alpha \le \arg z \le \beta$ and $f(z) \to L$ as $z \to \infty$ along one side of the angle, $f(z) \to M$ along the other side, then $L = M$ [and $f(z) \to L$ uniformly in the angle].[1]

To do this, we consider $g(z) = [f(z) - L][f(z) - M]$, which is also analytic and bounded in the angle and approaches 0 as $z \to \infty$ on both sides of the angle. By the preceding proposition, $g(z) \to 0$ uniformly in the angle, which is to say that

$$[f(z) - L][f(z) - M] \to 0$$

uniformly in the angle.

From this we can deduce both that $L = M$ and that $f(z) - L \to 0$. Let $\varepsilon^2 > 0$ be given; then for sufficiently large $|z|$ we have simultaneously $|f(z) - L| < \varepsilon$ on $\arg z = \alpha$, $|f(z) - M| < \varepsilon$ on $\arg z = \beta$, and $|f(z) - L|\,|f(z) - M| < \varepsilon^2$ for $\alpha \le \arg z \le \beta$. For any particular z, either $|f(z) - L| < \varepsilon$ or $|f(z) - M| < \varepsilon$ (or both). If $|f(z) - L| < \varepsilon$ for all rays in the angle, we would have a contradiction (if $M \ne L$) for small ε when $\arg z = \beta$. If $|f(z) - L|$ is sometimes greater than ε, there must be some θ $(\alpha \le \theta \le \beta)$ for which both inequalities are satisfied. [Consider the supremum of θ's for which $|f(z) - L| < \varepsilon$.] The same argument applies if $|f(z) - M| < \varepsilon$ near α. Then, by the triangle inequality,

$$|f(z) - M| + |f(z) - L| \le 2\varepsilon \quad (\arg z = \theta)$$

for sufficiently large $|z|$; since ε is arbitrary, $M = L$ and the conclusion follows.

We have seen that if a function f has its absolute value "not too large" in a half plane and is bounded on the boundary of the half plane, then it is actually bounded in the half plane. We might expect that we could let $|f|$ be larger in the half plane if we compensated by having $|f|$ decrease rapidly on the boundary. What actually happens is that $|f|$ cannot decrease too rapidly on the boundary unless $f \equiv 0$. More precisely, an analytic function (not identically 0) cannot grow more slowly than an exponential in a half plane and at the same time approach zero exponentially fast on the boundary. The following theorem of F. Carlson quantifies this principle: *Let f be analytic for* $\operatorname{Re} z > 0$, *continuous in* $\operatorname{Re} z \ge 0$, *and let f satisfy the conditions* $|f(z)| \le Ae^{k|z|}$ *in* $|\arg z| < \pi/2$ *and* $|f(iy)| \le Be^{-\varepsilon|y|}$, *where A, B, ε, k are positive numbers* (ε *usually small, k large*); *then* $f(z) \equiv 0$. The idea behind the theorem is that the exponential decrease on the boundary will be propagated to rays that are sufficiently close to the positive and negative imaginary axes. We will then have $|f|$ bounded in an angle of opening of less than π, by the main Phragmén–Lindelöf theorem, and hence bounded in the half plane; and once we know that, it will be easy to see that $f(z) \equiv 0$.

The details of the proof are extraordinarily delicate. Notice that we cannot get anywhere by applying the main Phragmén–Lindelöf theorem directly to f, or even to $f(z)e^{-kz}$, because the angle $|\arg z| < \pi/2$, is not quite small enough. What we do is to divide the half plane into smaller angles, considering appropriate auxiliary functions for each one. To begin with, consider $g(z) = f(z)e^{-kz}e^{-i\varepsilon z}$ in the first quadrant, with $\varepsilon > 0$. We have $|g(x)| \le A$ and $|g(iy)| \le B$, and it follows from the main Phragmén–Lindelöf theorem for an angle of opening $\pi/2$ [this requires only that $|f(z)| < P\exp(q\,|z|^2)$] that $|g(z)| \le C = \max(A, B)$ in the first quadrant. Then on a ray $\arg z = \theta$,

$0<\theta<\pi/2$, we have

$$|f(z)| \le C \exp(kr \cos\theta - \varepsilon r \sin\theta).$$

The right-hand side is bounded by C if $k\cos\theta = \varepsilon \sin\theta$, that is, if $\tan\theta = k/\varepsilon$. Arguing similarly in the lower half plane, we have $|f(z)| \le C$ for $z = re^{i\theta}$ with $\tan\theta = -k/\varepsilon$. Now we have $|f|$ bounded by C on two rays making an angle of $2\tan^{-1}(k/\varepsilon) < \pi$, and hence inside that angle. Also, $|f(z)| \le C$ on the imaginary axis and on the rays, and hence (by the Phragmén–Lindelöf theorem again, this time for the angles $\pi/2 > \arg z > \theta$ and $-\pi/2 < \arg z < -\theta$) in the rest of the half plane.

We now have $|f(z)| \le C$ in the right-hand half plane and $|f(iy)| < Be^{-\varepsilon|y|}$. Consider $g(z) = f(z)e^{pz}$, where p is a large positive number. Then $|g(iy)| = |f(iy)| \le Be^{-\varepsilon|y|}$, and for $x > 0$ we have $|g(z)| \le Ae^{k|z|+px} < Ae^{(k+p)|z|} = Ae^{k'|z|}$. Thus g satisfies the same hypotheses as f except for the size of the exponent k', which does not appear in $C = \max(A, B)$. Hence what we have done up to this point implies that $|g(z)| \le C$ in the half plane, that is, $|f(z)| \le Ce^{-px}$ with C independent of p. Letting $p \to \infty$, we see that $f(z) \equiv 0$.

Finally, we use the theorem just proved to prove an even more surprising result, generally known as **Carlson's theorem**. This says in informal language that $\sin \pi z$ is the smallest function that is analytic in the right-hand half plane, vanishes at the positive integers, and is not identically zero. More precisely, *if $f(z)$ is analytic for $x \ge 0$, if $|f(z)| \le Ae^{\tau|z|}$ with $\tau < \pi$, and if $f(n) = 0$ for $n = 0, 1, 2, \ldots$, then $f(z) \equiv 0$.* The example of $f(z) = \sin \pi z$ shows that the condition $\tau < \pi$ is essential.

It is natural to start by looking at the function $g(z) = f(z)/\sin \pi z$, since g will (after we remove its removable singular points) be analytic in the right-hand half plane. Since $|\sin \pi iy| = |\sinh \pi y|$, we have $g(iy) = O[e^{(\tau-\pi)|y|}]$ as $|y| \to \infty$. In order to apply the preceding theorem, we need an inequality for $|g(z)|$ of the form $|g(z)| \le ae^{k|z|}$ in the right-hand half plane. To obtain this inequality, we need lower bounds for $|\sin \pi z|$. By Exercise 4.14,

$$|\sin \pi z| = \{\sinh^2 \pi y + \sin^2 \pi x\}^{1/2} \ge \begin{cases} \sinh \pi|y| \\ |\sin \pi x| \end{cases}.$$

If $|y| \ge 1$, the first inequality gives us $|\sin \pi z| \ge \sinh \pi$, and so $|g(z)| \le A_1 e^{\tau|z|}$ with $A_1 = A/\sinh \pi$, when $|\text{Im } z| \ge 1$.

If $|y| < 1$, we get an upper bound for $|g(z)|$ by considering regions S_n defined by $|x - n| \le \frac{1}{2}$, $|y| \le 1$, $n = 1, 2, \ldots$. We already know that $|f(z)| \le Ae^{\tau|z|}$, and therefore on the vertical sides we have $|f(z)| \le A \exp[\tau(x^2 + y^2)^{1/2}] \le A \exp[\tau(n+1)]$. Since $|\sin \pi z| \ge |\sin \pi x|$, we have $|\sin \pi z| \ge 1$ when $z = x + iy$, $x = n + \frac{1}{2}$, and therefore

$$|g(z)| \le A \exp[\tau(n+1)]$$

on the boundary of S_n, and hence, by the maximum principle, inside S_n. But

$n \leq |z|+1$ for z in S_n, and consequently

$$|g(z)| \leq A \exp\{\tau(|z|+2)\} = Ae^{2\tau}e^{\tau|z|}.$$

The rectangle $0 \leq x \leq \frac{1}{2}$ has to be treated separately. Here g is analytic and has an upper bound B; in the preceding inequality, replace $Ae^{2\tau}$ by the larger of $Ae^{2\tau}$ and B. Hence g satisfies the hypotheses of the preceding theorem, and it follows that $g(z) \equiv 0$; therefore $f(z) \equiv 0$.

Supplementary exercise

1. Suppose that the entire function f satisfies $|f(z)| \leq A(\varepsilon)e^{\varepsilon|z|}$ for every $\varepsilon > 0$. Show that if f is bounded on the real axis, then f is a constant.

Note

1. This result is often called the Lindelöf principle. The hypotheses can be substantially weakened: See the paper by Sindalovskiĭ cited in Note 3 to Sec. **2**.

34 *Asymptotic Series*

34A. An example

The function

$$E(x) = \int_x^\infty t^{-1}e^{-t}\,dt, \qquad x > 0,$$

is one of several functions known collectively as exponential integrals. Its values are, of course, very small when x is very large, but since it is likely to occur multiplied by large factors, one may want to know its values rather accurately. By repeated integration by parts, we have

$$\begin{aligned} E(x) = -\int_x^\infty \frac{1}{t}\,d(e^{-t}) &= \frac{e^{-x}}{x} - \int_x^\infty \frac{e^{-t}}{t^2}\,dt \\ &= \frac{e^{-x}}{x} - \frac{e^{-x}}{x^2} + 2\int_x^\infty \frac{e^{-t}}{t^3}\,dt = \cdots \\ &= e^{-x}\left[\sum_{k=1}^{n} \frac{(k-1)!(-1)^{k-1}}{x^k}\right] + (-1)^n n! \int_x^\infty \frac{e^{-t}}{t^{n+1}}\,dt. \end{aligned}$$

Thus

$$e^x E(x) = \sum_{k=1}^{n} \frac{(-1)^{k-1}(k-1)!}{x^k} + R_n, \qquad R_n = (-1)^n e^x n! \int_x^\infty \frac{e^{-t}}{t^{n+1}}\,dt.$$

Although the series

$$\sum_{k=1}^{\infty} \frac{(-1)^{k-1}(k-1)!}{x^k}$$

diverges for every x, the remainder R_n after n terms satisfies

$$|R_n| \leq n!x^{-n-1}.$$

When $x \to \infty$, R_n is of smaller order than the nth term of the series. For example, if we take four terms of the series and $x = 10$, we get the value of $e^{10}E(10)$ with error less than 2.4×10^{-4}; when $x = 100$, the same number of terms gives $e^{100}E(100)$ with an error less than 2.4×10^{-9}.

34B. Properties of asymptotic series

A series that produces the kind of approximation we have just discussed is known as an asymptotic series (more precisely, an asymptotic power series). In a convergent power series we get a better approximation to a function by taking more and more terms for a given x. In an asymptotic series, on the other hand, we get a better approximation to the function by taking larger and larger values of x for a given (usually small) number of terms. However, if we take too many terms, we usually get a *poorer* approximation.

The formal definition is that $\sum a_n z^{-n}$ *is an asymptotic series for a given analytic function* f *in an angle* $\alpha < \arg z < \beta$ *if, for each* n,

$$\lim_{|z| \to \infty} z^n \left\{ f(z) - \sum_{k=0}^{n} a_k z^{-k} \right\} = 0,$$

uniformly in each interior angle; that is,

$$f(z) - \sum_{k=0}^{n} a_k z^{-k}$$

is uniformly $o(z^{-n})$ as $|z| \to \infty$ in the angle. We write $f(z) \sim \sum a_k z^{-k}$.

More generally,

$$f(z) \sim \phi(z) \sum a_k z^{-k}$$

is an asymptotic representation of $f(z)$ if

$$\frac{f(z)}{\phi(z)} \sim \sum a_k z^{-k}.$$

You should be warned that terminology varies a good deal from one book to another. In our definition the representation is to hold *uniformly in an angle* and f is required to be *analytic*.

EXERCISE 34.1 Show that if a (single-valued) f has, in a given angle, two asymptotic power series, then the series have the same coefficients.

However, it is possible for a function to have different asymptotic expansions in *different* angles. This is known as the Stokes phenomenon.

The calculus of asymptotic series is much like the calculus of convergent power series. However, the interesting asymptotic series are usually divergent, and provide an asymptotic representation of a function only in the angle between two rays $\arg z = \alpha$, $\arg z = \beta$. Also, in contrast to convergent power series, the same asymptotic series can represent more than one function. For example, if $E(z)$ is the function defined in Sec. **34A**, then $e^z E(z) - e^{-z}$ has the same asymptotic expansion as $e^z E(z)$ in an angle $\varepsilon - (\pi/2) < \theta < -\varepsilon + (\pi/2)$, $\varepsilon > 0$, because $z^n e^{-z} \to 0$ for each n, uniformly in each interior angle. That is, the function determines a series, but the series does not necessarily determine the function.

If the angle in which an asymptotic series represents an analytic function is the whole plane and the approximation is uniform in the plane, taking $n = 0$ in the definition shows at once that f has a removable singular point at ∞ and that f is represented by a convergent power series $\sum b_n z^{-n}$ outside some disk. Hence asymptotic series for single-valued functions are interesting only in angles of opening less than 2π.

Term-by-term addition of asymptotic series is clearly admissible.

EXERCISE 34.2 Show that if (in the same angle) $f(z) \sim \sum a_n z^{-n}$ and $g(z) \sim \sum b_n z^{-n}$, then $f(z)g(z) \sim \sum c_n z^{-n}$, where $c_n = \sum_{k=0}^{n} a_k b_{n-k}$ (the same formula as for the product of two convergent power series (Secs. **4B** and **15B**).

We may always integrate an asymptotic series term by term, in the sense that if $f(z) \sim \sum_{k=0}^{\infty} a_k z^{-k}$ *in an angle, then*

$$\int_z^{\infty} [f(w) - a_0 - a_1 w^{-1}]\, dw \sim \sum_{k=2}^{\infty} a_k (1-k)^{-1} z^{-k+1}$$

in the same angle, with integration along a ray.

We have

$$f(w) - \sum_{k=0}^{n} a_k w^{-k} = g(w) w^{-n},$$

where $g(w) \to 0$ uniformly as $w \to \infty$ in each interior angle. Then

$$\begin{aligned}\int_z^{\infty} [f(w) - a_0 - a_1 w^{-1}]\, dw &= \int_z^{\infty} \sum_{k=2}^{n} a_k w^{-k}\, dw + \int_z^{\infty} g(w) w^{-n}\, dw \\ &= -\sum_{k=2}^{n} \frac{a_k}{k-1} z^{-k+1} + \int_z^{\infty} g(w) w^{-n}\, dw.\end{aligned}$$

We now need to show that the last integral is $o(z^{-n+1})$. With $w = te^{i\lambda}$ and

$z = re^{i\lambda}$, the integral becomes

$$e^{-i(n-1)\lambda} \int_r^\infty g(te^{i\lambda}) t^{-n} \, dt$$

and its modulus does not exceed

$$\int_r^\infty |g(te^{i\lambda})| \, t^{-n} \, dt < \varepsilon \int_r^\infty t^{-n} \, dt = \varepsilon \frac{r^{-n+1}}{n-1},$$

for any positive ε, provided that r is large enough.

The situation for differentiation of asymptotic series is somewhat different.

EXERCISE 34.3 With $E(z)$ as in Sec. **34A**, show that $g(x) = e^x E(x) + e^{-x} \sin(e^{2x})$ has an asymptotic series *for positive real* x, but that its derivative does not; also show that the series is not an asymptotic series in our sense in any angle of positive opening.

This explains why some books say that asymptotic series cannot always be differentiated term by term. However, *if f is analytic in an angle A and has an asymptotic series (according to our definition) in A, then f' has an asymptotic series, obtained by formal differentiation.* That is, if

$$f(z) = \sum_{k=0}^{n} a_k z^{-k} + o(z^{-n})$$

then

$$f'(z) = \sum_{k=0}^{n} (-k) a_k z^{-k-1} + o(z^{-n-1}).$$

We have

$$f(w) = \sum_{k=0}^{n} a_k w^{-k} + o(w^{-n}),$$

uniformly in any interior angle A_1. By Cauchy's formula,

$$f'(z) = \frac{1}{2\pi i} \int_C \frac{f(w) \, dw}{(w-z)^2},$$

where C is a circle, centered at z, of radius $\lambda |z|$, with λ so small that C is inside A_1. Then if $\varepsilon > 0$ is given and $|z|$ is large enough,

$$f'(z) = \frac{1}{2\pi i} \int_C (w-z)^{-2} \left\{ \sum_{k=0}^{n} a_k w^{-k} + \phi(w) \right\} dw, \qquad |\phi(w)| < \varepsilon / |w|^n.$$

Consequently,

$$\begin{aligned}\left|f'(z)-\sum_{k=1}^{n} ka_k z^{-k-1}\right| &= \frac{1}{2\pi}\left|\int_C \frac{\phi(w)\,dw}{(w-z)^2}\right| \\ &= \frac{1}{2\pi}\left|\int_0^{2\pi} \frac{\phi(z+\lambda\,|z|\,e^{i\theta})}{\lambda\,|z|\,e^{i\theta}}\,d\theta\right| \\ &\le \frac{1}{\lambda\,|z|}\{|z|\,(1+\lambda)\}^{-n}\varepsilon \\ &= \varepsilon O(|z|^{-n-1}),\end{aligned}$$

uniformly in A_1. This shows that f' has the asymptotic series obtained by differentiating the series for f.

It is also true that the composition of a convergent power series with an asymptotic power series can yield the asymptotic series of the composition of the two functions; more precisely, if g is analytic in a disk and f has an asymptotic expansion with $f(\infty)$ in the disk of convergence of g, then the asymptotic expansion of $g[f(z)]$ is obtained by formal substitution.[1]

Supplementary exercises

1. Find (as in Sec. **34A**) an asymptotic power series for $\pi^{1/2}xe^{x^2}\operatorname{erfc}(x)$, where

$$\operatorname{erfc}(x) = 2\pi^{-1/2}\int_x^\infty e^{-t^2}\,dt.$$

2. Find an asymptotic power series for $x^{1-n}e^x\Gamma(n, x)$, where

$$\Gamma(n, x) = \int_x^\infty t^{n-1}e^{-t}\,dt$$

(the incomplete gamma function).

3. Let

$$f(x) = \int_0^\infty \frac{e^{-t}}{1+xt}\,dt.$$

Obtain an asymptotic power series for $f(x^{-1})$.

34C. Asymptotic series for Laplace transforms

One very useful method of finding asymptotic series provides an asymptotic series for functions that are represented by Laplace transforms; many functions can be represented in this way. The procedure (known as **Watson's lemma**) is to represent the function F of interest as

$$F(z) = \int_0^\infty e^{-zw}f(w)\,dw,$$

where $\operatorname{Re} z > 0$, the integral converges for $\operatorname{Re} z >$ some x_0, f is analytic at $w = 0$ and on the whole positive real axis, and $f(0) \neq 0$. Then expand f in its Maclaurin series and integrate term by term. The remarkable feature of this method is that it works, even though the Maclaurin series of f will usually converge only in a bounded disk. The formal calculation yields

$$F(z) \sim \sum_{k=0}^{\infty} \frac{k!a_k}{z^{k+1}},$$

where a_k are the Maclaurin coefficients of f. The proof that this actually produces an asymptotic series is rather complicated, and I omit it.[2]

EXERCISE 34.4 Use this method to find an asymptotic series for

$$J_0(z) = \int_0^{\infty} e^{-zt}(t^2 + 1)^{-1/2}\, dt$$

(J_0 is the Bessel function of order 0).

34D. Stirling's formula

Stirling's formula is the first term of an asymptotic expansion of the gamma function. It is useful in dealing with expressions that involve factorials, in, for example, probability theory, statistical mechanics, and even in discussing the convergence of the kinds of infinite series that appear in calculus courses (if we are willing to let the undergraduates know about it).

Stirling's formula says that, for $-\pi + \varepsilon < \arg z < \pi - \varepsilon$ $(\varepsilon > 0)$,

$$\Gamma(z + 1) \sim z^{z+(1/2)} e^{-z} (2\pi)^{1/2},$$

where the $\sim$ sign now means that the ratio of the two sides tends to 1 (read it as "is asymptotic to"). We shall prove the formula only when z is a real positive integer. Notice that the difference between the two sides of the formula actually increases with $|z|$, although their ratio tends to 1.

The formula can be made more precise, even for small n, by a slight modification:[3]

$$\Gamma(n + 1) \sim n^n e^{-n} \left[2\pi\left(n + \frac{1}{6}\right)\right]^{1/2},$$

which is easy to remember. The approximation becomes even closer if one more term of the asymptotic series is included. The series begins[4]

$$\Gamma(z + 1) \sim z^{z+(1/2)} e^{-z} (2\pi)^{1/2} \left\{1 + \frac{1}{12z} + \frac{1}{288z^2} - \cdots\right\}.$$

We start from the formula

$$\sum_{k=1}^{n} f(k) = \int_1^n f(t)\, dt + \frac{1}{2}[f(1) + f(n)] - \int_1^n f'(t) P(t)\, dt,$$

where f' is continuous and $P(t) = [t] - t + \frac{1}{2}$, a periodic function of period 1 with average 0 over a period. ($[t]$ means, as usual, the greatest integer less than or equal to t.) This is the simplest version of the **Euler–Maclaurin formula** connecting integrals and sums. It is easily verified by writing the integral as the sum of integrals over $(1, 2)$, $(2, 3)$, . . . , $((n-1), n)$, and integrating by parts.

EXERCISE 34.5 Verify the formula.

When $f(t) = \log t$, this gives

$$\begin{aligned}\log n! &= \int_1^n \log t\, dt + \frac{1}{2} \log n + \int_1^n t^{-1} P(t)\, dt \\ &= n \log n - n + 1 + \frac{1}{2} \log n + \int_1^n t^{-1} P(t)\, dt,\end{aligned}$$

and hence

$$n! \sim n^{n+(1/2)} e^{-n} C(n),$$

where $\lim_{n\to\infty} C(n) = C$ exists, by the reasoning that is often used to show the convergence of the integral

$$\int_0^\infty \frac{\sin x}{x}\, dx.$$

EXERCISE 34.6 Verify the existence of C.

This is Stirling's formula except that we still have to show that $C = (2\pi)^{1/2}$. In many applications the precise value of C is irrelevant.

There are several ways of calculating C, none of them very easy. The one that I shall present depends on showing first that

$$\lim_{n\to\infty} \frac{n^{1/2}\Gamma(n + \frac{1}{2})}{n!} = 1.$$

It is somewhat easier to prove that

$$\frac{n^{3/2}\Gamma(n + \frac{1}{2})}{\Gamma(n+2)} \to 1$$

and use the functional equation for the gamma function to reduce this to what we want.

From Sec. **7H** we have[5]

$$\begin{aligned}\frac{\Gamma(n + \frac{1}{2})\Gamma(\frac{3}{2})}{\Gamma(n+2)} &= B\left(n + \frac{1}{2}, \frac{3}{2}\right) \\ &= \int_0^1 (1-x)^{1/2} x^{n-1/2}\, dx\end{aligned}$$

$$= \int_0^\infty (1 - e^{-t})^{1/2} e^{-[n+(1/2)]t} \, dt$$

$$= \int_0^\infty t^{1/2} e^{-[n+(1/2)]t} \, dt - \int_0^\infty [t^{1/2} - (1 - e^{-t})^{1/2}] e^{-[n+(1/2)]t} \, dt$$

$$= \Gamma\left(\frac{3}{2}\right)\left(n + \frac{1}{2}\right)^{-3/2} - \int_0^1 [t^{1/2} - (1 - e^{-t})^{1/2}] e^{-[n+(1/2)]t} \, dt$$

$$- \int_1^\infty [t^{1/2} - (1 - e^{-t})^{1/2}] e^{-[n+(1/2)]t} \, dt.$$

The first term is $\Gamma(\frac{3}{2})n^{-3/2}[1 + o(1)]$, so we have to show that the other two terms are $o(n^{-3/2})$. We can actually show that they are $O(n^{-5/2})$. To do this we need some elementary inequalities.

EXERCISE 34.7 Show that

(*a*) $1 - e^{-t} < t, \ 0 < t < \infty$;

(*b*) $1 - e^{-t} > t - \frac{1}{2}t^2$ for $0 < t < 1$;

(*c*) $(1 - \frac{1}{2}t)^{1/2} > 1 - \frac{1}{2}t$ for $0 < t < 1$.

Now we have to get upper bounds for each of two integrals. The first one is

$$\int_0^1 [t^{1/2} - (1 - e^{-t})^{1/2}] e^{-[n+(1/2)]t} \, dt.$$

The expression in brackets is, by Exercise 34.7*b*, less than

$$\left[t^{1/2} - \left(t - \frac{1}{2}t^2\right)^{1/2}\right] = t^{1/2}\left[1 - \left(1 - \frac{1}{2}t\right)^{1/2}\right].$$

This, in turn, by Exercise 34.7*c*, is less than $t^{3/2}$. Hence the first integral does not exceed

$$\int_0^1 t^{3/2} e^{-[n+(1/2)]t} \, dt < \int_0^\infty t^{3/2} e^{-[n+(1/2)]t} \, dt = \Gamma\left(\frac{5}{2}\right)\left(n + \frac{1}{2}\right)^{-5/2}.$$

In the second integral (with limits 1 and ∞) we use Exercise 34.7*a* for the expression in brackets to get (because $t > 1$)

$$\int_1^\infty [t^{1/2} - (1 - e^{-t})^{1/2}] e^{-[n+(1/2)]t} \, dt < \int_1^\infty t^{1/2} e^{-[n+(1/2)]t} \, dt < \int_1^\infty t^{3/2} e^{-[n+(1/2)]t} \, dt$$

$$< \int_0^\infty t^{3/2} e^{-[n+(1/2)]t} \, dt = \Gamma\left(\frac{5}{2}\right)\left(n + \frac{1}{2}\right)^{-5/2}.$$

Hence both integrals are $O(n^{-5/2})$ and we indeed have

$$\frac{n^{1/2}\Gamma(n + \frac{1}{2})}{n!} \to 1.$$

Now we calculate $\Gamma(n+\frac{1}{2})$ explicitly. By using the functional equation of the gamma function repeatedly, we get

$$\begin{aligned}\Gamma\left(n+\frac{1}{2}\right) &= \left(n-\frac{1}{2}\right)\Gamma\left(n-\frac{1}{2}\right) = \left(n-\frac{1}{2}\right)\left(n-\frac{3}{2}\right)\Gamma\left(n-\frac{3}{2}\right)\\ &= \cdots = \left(n-\frac{1}{2}\right)\left(n-\frac{3}{2}\right)\cdots\frac{1}{2}\Gamma\left(\frac{1}{2}\right)\\ &= \frac{2n-1}{2}\cdot\frac{2n-3}{2}\cdots\frac{1}{2}\Gamma\left(\frac{1}{2}\right)\\ &= \frac{(2n)!\Gamma(\frac{1}{2})}{2^{2n}n!}\end{aligned}$$

and therefore

$$\frac{n^{1/2}(2n)!\,\Gamma(\frac{1}{2})}{(n!)^2 2^{2n}} \to 1.$$

Now apply Stirling's formula (with the unknown factor C) to the factorials in the preceding formula. We obtain

$$\frac{n^{1/2}(2n)^{2n}e^{-2n}(2n)^{1/2}\Gamma(\frac{1}{2})C}{n^{2n}e^{-2n}n\cdot 2^{2n}C^2} \to 1$$

and hence $2^{1/2}\Gamma(\frac{1}{2})/C = 1$. Therefore $C = (2\pi)^{1/2}$.

EXERCISE 34.8 Calculate C independently by using Wallis's formula (Sec. **31C**).

Supplementary exercises

1. In connection with Exercise 34.6, show that $\int_0^\infty \lambda(x)\sin x\,dx$ converges if $\lambda(x)$ is bounded and decreases to 0 as $x \to \infty$.
2. Find $n!$ exactly for as large as n as your calculator permits, and compare what you get from Stirling's formula.
3. Find the radius of convergence of $\sum z^n\Gamma(n+\frac{1}{2})/n!$.
4. Use the Euler–Maclaurin formula (Exercise 34.5) to find how many terms of $\sum_{n=1}^\infty 1/n^2$ would be needed to get the sum of the series with an error of less than 5×10^{-10} (roughly, nine decimal places). *Hint*: You can just as well write the Euler–Maclaurin formula with lower limit m and upper limit ∞. *Remark*: This is not a good way to calculate the sum of the series, which is actually $\pi^2/6$.
5. Euler's constant γ is defined as

$$\lim_{n\to\infty}\left(1+\frac{1}{2}+\frac{1}{3}+\cdots+\frac{1}{n}-\log n\right).$$

Use the Euler–Maclaurin formula to estimate how many terms of the harmonic series would be needed to find γ with an error of at most 5×10^{-10}. Assume that the necessary logarithms are available.

Notes

1. E. R. Love, "Functions of Asymptotic Expansions," *Bull. Austral. Math. Soc. 6* (1972): 307–312.
2. For a proof of Watson's lemma, see, for example, Copson, p. 218.
3. N. D. Mermin, "Improving an Improved Approximation to $n!$," *Amer. J. Phys. 51* (1983): 776.
4. The next few terms (quoted from Sibagaki, p. 71) are

$$-\frac{139}{51840z^3}-\frac{571}{2488320z^4}+\frac{163879}{209018880z^5}+\frac{5246819}{75246796800z^6}-\frac{534703531}{902961561600z^7}$$

5. This calculation is imitated from Titchmarsh, p. 58 (Sec. **1.87**).

35 *More About Univalent Functions*

Let us consider functions f that are analytic and univalent in the unit disk, in other words, functions that map the unit disk conformally on some region D. Since D can be any simply connected region with at least two boundary points, it might seem that there is not much to say, but it turns out that one can say quite a lot both about the power series of f and about the images of disks $|z| \le r < 1$, provided that we normalize f by assuming that $f(0) = 0$ and $f'(0) = 1$. We can do this by adding a constant to f, which merely translates D, and then multiplying f by a constant, which is at most a change of scale and a rotation. We shall therefore suppose that

$$f(z) = z + a_2 z^2 + a_3 z^3 + \dots, \qquad |z| < 1.$$

The class of univalent functions, normalized in this way, is usually called S (for "schlicht"). It was discovered, early in the history of S, that it is easier to start by considering the class[1] U of functions $F(z) = 1/f(1/z)$, where $f \in S$. These are precisely the functions that are analytic for $|z| > 1$ except for a simple pole at ∞ with residue 1, and are also univalent. We have

$$1\Big/\left(\frac{1}{z}+\frac{a_2}{z^2}+\frac{a_3}{z^3}+\cdots\right) = z\Big/\left(1+\frac{a_2}{z}+\frac{a_3}{z^2}+\cdots\right) = z + \alpha_0 + \alpha_1 z^{-1} + \alpha_2 z^{-2} + \cdots.$$

For the class U we can prove **Gronwall's area theorem**:

$$\sum_{n=1}^{\infty} n\, |\alpha_n|^2 \le 1.$$

To prove this, take $r > 1$ and compute the area of the bounded region whose boundary is the image of $|z|=r$ by $w=F(z)$. This boundary is a simple closed curve (because F is univalent); we write its equation as $w=Re^{i\phi}$, and then the area is $\frac{1}{2}\int R^2\,d\phi$ by elementary calculus. If we use the Cauchy–Riemann equations in polar coordinates we can write the area as

$$\begin{aligned}
\frac{1}{2}\int R^2\,d\phi &= \frac{1}{2}\int R^2\frac{\partial\phi}{\partial\theta}\,d\theta = \frac{1}{2}\int R^2\frac{r}{R}\frac{\partial R}{\partial r}\,d\theta \\
&= \frac{1}{2}r\int_0^{2\pi} R\frac{\partial R}{\partial r}\,d\theta = \frac{1}{4}\int_0^{2\pi} r\frac{\partial(R^2)}{\partial r}\,d\theta \\
&= \frac{1}{4}r\frac{\partial}{\partial r}\int_0^{2\pi}|F(re^{i\theta})|^2\,d\theta \\
&= \frac{1}{4}r\frac{\partial}{\partial r}\int_0^{2\pi}|re^{i\theta}+\alpha_0+\alpha_1 r^{-1}e^{-i\theta}+\cdots|^2\,d\theta \\
&= \frac{\pi}{2}r\frac{\partial}{\partial r}(r^2+|\alpha_0|^2+|\alpha_1|^2r^{-2}+\cdots) \\
&= \frac{\pi}{2}r(2r-2r^{-3}|\alpha_1|^2-4r^{-5}|\alpha_2|^2-\cdots) \\
&= \pi(r^2-r^{-2}|\alpha_1|^2-\cdots).
\end{aligned}$$

The key observation is that the area cannot be negative, so we have

$$r^2 \ge \sum_{n=1}^{\infty} n\,|\alpha_n|^2 r^{-2n}.$$

Letting $r\to 1$, we obtain[2]

$$\sum_{n=1}^{\infty} n\,|\alpha_n|^2 \le 1.$$

We can use this result to prove **Bieberbach's coefficient theorem**: *If $f\in S$, then $|a_2|\le 2$.*

There is equality in Bieberbach's theorem when $f(z)=z/(1-z)^2$.

For the proof, we begin by noticing that

$$\begin{aligned}
f(z^2) &= z^2+a_2z^4+a_3z^6+\cdots \\
&= z^2(1+a_2z^2+a_3z^4+\cdots).
\end{aligned}$$

We can define $g(z)$ by $[f(z^2)]^{1/2}=z(1+a_2z^2+\cdots)^{1/2}$, taking the branch of the square root that is $+1$ at $z=0$.

EXERCISE 35.1 Show that $g(z)$ belongs to S.

Hence $1/g(1/z)$ is in U. We can get its Laurent series by

$$g\left(\frac{1}{z}\right)=\frac{1}{z}\left(1+\frac{1}{2}a_2z^{-2}+\cdots\right),$$

$$\frac{1}{g(1/z)} = z\left(1 + \frac{1}{2}a_2 z^{-2} + \cdots\right)^{-1},$$

$$= z\left(1 - \frac{1}{2}a_2 z^{-2} + \cdots\right)$$

$$= z - \frac{1}{2}a_2 z^{-1} + \cdots.$$

Now, by the area theorem, $\frac{1}{4}|a_2|^2 \leq 1$, and $|a_2| \leq 2$.

As an application of Bieberbach's theorem we can establish a geometric property of the map of the unit disk under $f \in S$. Let c be a point outside the image of the unit disk. Then

$$\frac{cf(z)}{c - f(z)} = z + (a_2 + c^{-1})z^2 + \cdots$$

is also in S. Hence

$$\left|a_2 + \frac{1}{c}\right| \leq 2, \qquad \left|\frac{1}{c}\right| \leq 4, \qquad |c| \geq \frac{1}{4}.$$

Consequently, if $f \in S$, then f takes all values in the open disk with center 0 and radius $\frac{1}{4}$; in other words, the map of the unit disk by f must cover this disk. This is the most that can be said, because when $f(z) = z/(1+z)^2$, the point $\frac{1}{4}$ is in the image of $|z| = 1$. The number $\frac{1}{4}$ is *Koebe's constant* (because Koebe proved the existence of some such number).

Bieberbach conjectured in 1916 that if $f \in S$, then $|a_n| \leq n$. The conjecture was finally proved in 1984 by L. de Branges.[3]

In the special case when the a_n are all real, the proof that $|a_n| \leq n$ is rather easy. In fact, this inequality holds for functions f that have $a_0 = 0$, $a_1 = 1$, and take real values only on the real axis; such functions are called **typically real**.

Let $f(z) = u + iv$, with f real on the real axis and only there. Then, since f is real on the real axis it has real Maclaurin coefficients, so that we can write

$$f(z) = \sum_{n=0}^{\infty} a_n z^n, \qquad a_n \text{ real}.$$

Hence

$$v(re^{i\theta}) = \operatorname{Im}\left\{\sum_{n=0}^{\infty} a_n r^n(\cos n\theta + i \sin n\theta)\right\}$$

$$= \sum_{n=1}^{\infty} a_n r^n \sin n\theta,$$

an odd function of θ. For $0 < \theta < \pi$, v has constant sign, since if v changed sign at some θ and r, f would be real at $re^{i\theta}$, which we assumed doesn't happen. Consequently,

$$|a_n r^n| = \frac{2}{\pi}\left|\int_0^\pi v(re^{i\theta}) \sin n\theta \, d\theta\right| \le \frac{2}{\pi}\left|\int_0^\pi v(re^{i\theta}) n \sin \theta \, d\theta\right|,$$

because $|\sin n\theta| \le n\,|\sin \theta|$, as we can see inductively:

$$\begin{aligned}|\sin n\theta| &= |\sin[(n-1)\theta + \theta]| = |\sin(n-1)\theta \cos \theta + \cos(n-1)\theta \sin \theta| \\ &\le (n-1)\,|\sin \theta| + |\sin \theta|.\end{aligned}$$

Therefore, since v does not change sign,

$$|a_n r^n| \le \frac{2n}{\pi}\left|\int_0^\pi v(re^{i\theta}) \sin \theta \, d\theta\right| = n\,|a_1 r| \le nr,$$

and it follows (if we let $r \to 1$) that $|a_n| \le n$.

Although I said that Koebe's constant cannot be improved, one can actually say a little more.[4] Suppose that c_1 and c_2 are values omitted by f that lie on a line segment through 0 and on opposite sides of 0. Then one of c_1 and c_2 has absolute value greater than $\frac{1}{2}$.

To prove this, suppose that f does not take the value c_1. The function $c_2 f(z)/[c_2 - f(z)]$ is (as we saw above) in S, and therefore cannot omit any complex number of modulus less than $\frac{1}{4}$. But if f does not take the value c_1, then $c_2 f/(c_2 - f)$ does not take the value $c_1 c_2/(c_2 - c_1)$. Hence we have $|c_1 c_2|/(c_2 - c_1)| \ge \frac{1}{4}$, and so

$$\left|\frac{1}{c_1} - \frac{1}{c_2}\right| = \frac{1}{|c_1|} + \frac{1}{|c_2|} \le 4$$

(it is here that we use the assumption that c_1 and c_2 are on the opposite sides of 0, on a line segment through 0).

EXERCISE 35.2 If f and g are in S, do the following functions belong to S?

(a) $\frac{1}{2}(f+g)$ (b) $zf'(z)$ (c) $\displaystyle\frac{2}{z}\int_0^z f(w)\,dw$

The area theorem has been used to prove the isoperimetric property:[5] the area inside a simple closed curve of length L is at most equal to $L^2/(4\pi)$.

Supplementary exercises

1. Show that a limit (uniformly on compact sets) of elements of S belongs to S.

2. Show that there is a number K such that the closure of the image of the unit disk by an element of S must contain a triangle of area at least K, and that the largest value of K is attained for some element of S.

3. Assuming the validity of the Bieberbach conjecture, is there an element of S for which $f'(\frac{1}{2}) = 13$?

4. If $f \in S$ is $zf'(\frac{1}{2}z)$ necessarily in S?

The supplementary exercises for Sec. **13B** are also relevant here.

Notes

1. The class U is often denoted by Σ.

2. The preceding proof can be criticized on the grounds that we have not been careful to see that the area was computed with the correct sign. To see that in fact it is correct, we could deduce the integral formula that we used from a carefully stated version of Green's (or Stokes's) theorem (for example, Sagan, pp. 544ff, or Buck and Buck, pp. 478ff); or appeal to the perhaps more familiar integral formula for the area inside a positively oriented curve,

$$A = \frac{1}{2}\int_{\partial D} (x\,dy - y\,dx),$$

(see, for example, Buck and Buck, p. 496, Exercise 6) and reduce the resulting formula to our version (cf. Littlewood, p. 209). However, we can argue, less elegantly but more concisely, that we necessarily got either the area or its negative, with the same sign for all functions in U, since no properties of the functions were used beyond those implied by the definition. If we had got the area with the wrong sign, we would have had $r^2 \leq \sum_{n=1}^{\infty} n\,|\alpha_n|^2 r^{-2n}$ for *all* elements F of U. However, the function $f(z) = z$ is in U and has all $\alpha_n = 0$, so the preceding inequality would assert that $r^2 \leq 0$. Hence, whether by luck or good management, we must have computed the area correctly.

3. L. de Branges, "Proof of the Bieberbach Conjecture," *Acta Math. 154* (1985): 137–152.

4. This was noticed by G. Szegő.

5. There are many proofs of the isoperimetric inequality. This one was given by M. Mateljević and M. Pavlović, "New Proofs of the Isoperimetric Inequality and Some Generalizations," *J. Math. Analysis and Appl. 98* (1984): 25–30.

Solutions of Exercises

Section 1

1.1. Apply the definition of multiplication of two ordered pairs.
1.2. $c(x, y) = (cx, cy)$; $(c, 0)(x, y) = (cx + 0y, 0x + cy)$.
1.3. The first part is a matter of direct verification. For quotients, multiply numerator and denominator by the conjugate of the denominator before trying to take the conjugate.
1.4. The answers are
$(i+2)/(i-2) = -\frac{3}{5} - \frac{4}{5}i$;
$(1+i\sqrt{3})^2 = -2 + 2\sqrt{3}\,i$;
$1/(1-i) = \frac{1}{2} + \frac{1}{2}i$;
$(2-3i)/(3i+2) = -\frac{5}{13} - \frac{12}{13}i$.
1.5. $|zw|^2 = zw\overline{zw} = z\bar{z}w\bar{w} = |z|^2\,|w|^2$;
$\bar{z} = x - iy$;
$|\bar{z}| = \sqrt{x^2+y^2} = |z|$;
$|z/w|^2 = (z/w)(\bar{z}/\bar{w}) = |z|^2/|w|^2$;
$|\text{Re } z| = |x| < (x^2+y^2)^{1/2}$ unless $y = 0$.
1.6. Use $|z|^2 = z\bar{z}$. Then $|(i+2)/(i-2)| = 1$ because $|z/w| = |z|/|w|$ and $|\bar{z}| = |z|$; and $|1+i\sqrt{3}|^2 = |2[\cos(\pi/3) + i\sin(\pi/3)]|^2 = 4$ because $|\cos\theta + i\sin\theta| = 1$.
$|1/(1-i)| = |1/[2(\cos(\pi/4) - i\sin(\pi/4)]| = 2^{-1/2}$.
$|(2-3i)/(3i+2)| = (\frac{5}{13})^2 + (\frac{12}{13})^2 = 1$. Or, since $\overline{3i+2} = 2-3i$, this is another example of $z/\bar{z}$.
1.7. (*a*) Open disk, center 0, radius 2.
(*b*) Circle, center 0, radius 3, together with its exterior.
(*c*) Open disk of radius 5, center at $2-3i$.
(*d*) Half-plane to the right of the imaginary axis.
(*e*) Half-plane to the right of the line $y = \frac{2}{3}x$.
(*f*) Closed half-plane to the left of the line $y = x$.
(*g*) Ellipse with foci at i and 1, major axis 2.
(*h*) Perpendicular bisector of the line segment connecting $z = i$ and $z = -1$.
(*i*) Parabola, focus at $z = 2$, directrix the imaginary axis.
1.8. The parallelogram law, expressed algebraically, is $(x, y) + (u, v) = (x+u, y+v)$; identify the vectors (x, y) and (u, v) with $x + iy$ and $u + iv$.
1.9. $3 - 4i = 5(\cos\theta + i\sin\theta)$, $\theta = -\tan^{-1}(\frac{4}{3}) = -0.927 = -53.13°$.
$-4 - 11i = 11.7(\cos\theta + i\sin\theta)$, $\theta = \tan^{-1} 2.75 = -1.92$ or $-110°$
$2.236 + 3.142i = 3.856(\cos\theta + i\sin\theta)$, $\theta = \tan^{-1} 1.405 = 0.952$ or $54.6°$.
$(1+i)/(1-i) = i = \cos(\pi/2) + i\sin(\pi/2)$.
$(2+i)/(3-i) = (1+i)/2 = 2^{-1/2}[\cos(\pi/4) + i\sin(\pi/4)]$.
$-1 + i\sqrt{3} = 2[(\cos(2\pi/3) + i\sin(2\pi/3)]$.
$\cos(3\pi/2) + i\sin(3\pi/2) = \cos(-\pi/2) + i\sin(-\pi/2)$.
$2\{(\cos(7\pi/6) + i\sin(7\pi/6)\} = 2\{\cos(-5\pi/6) + i\sin(-5\pi/6)\}$.

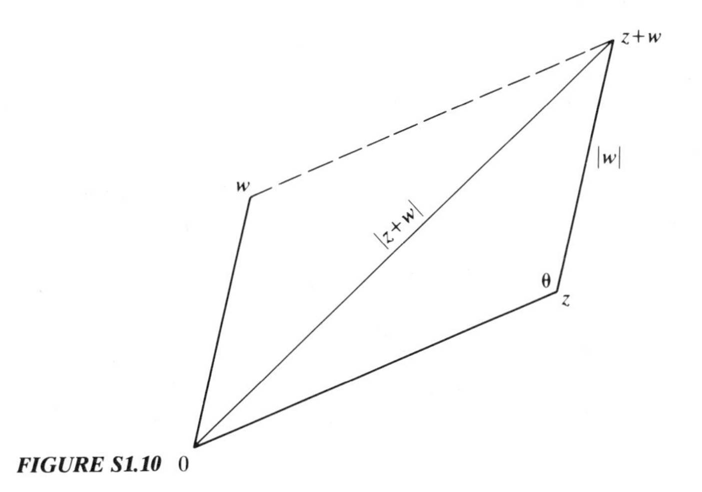

FIGURE S1.10

1.10. A straightforward algebraic proof is possible but rather complicated. It is simpler to argue from plane geometry that a side of a triangle is less than the sum of the other two sides unless the triangle degenerates. The proof can be put in analytic form by using the law of cosines from trigonometry (see Figure S1.10):

$$\begin{aligned}|z+w|^2 &= |z|^2+|w|^2-2\,|z|\,|w|\cos\theta \\ &= (|z|+|w|)^2-2\,|z|\,|w|\,(1+\cos\theta).\end{aligned}$$

Since $1+\cos\theta \geq 0$, it follows that $|z+w|^2 \leq (|z|+|w|)^2$, with equality if and only if $\theta \equiv \pi \pmod{2\pi}$.

See the comment following the statement of the exercise. For $|z-w|$, apply the triangle inequality to $(z-w)$ and w to get $|z| \leq |z-w|+|w|$, hence $|z|-|w| \leq |z-w|$. If $|w|>|z|$, interchange w and z.

1.11. $$\begin{aligned}zw &= r(\cos\theta+i\sin\theta)s(\cos\phi+i\sin\phi) \\ &= rs[(\cos\theta\cos\phi-\sin\theta\sin\phi)+i(\cos\theta\sin\phi+\sin\theta\cos\phi)] \\ &= rs[\cos(\theta+\phi)+i\sin(\theta+\phi)].\end{aligned}$$

1.12. $1+i\sqrt{3}=2\left[\cos\left(\frac{\pi}{3}\right)+i\sin\left(\frac{\pi}{3}\right)\right].$

$$(a)\quad (1+i\sqrt{3})^2=4\left[\cos\left(\frac{2\pi}{3}\right)+i\sin\left(\frac{2\pi}{3}\right)\right]=4\left(-\frac{1}{2}+\frac{i\sqrt{3}}{2}\right).$$

$$\begin{aligned}(b)\quad (1+i\sqrt{3})(\sqrt{3}-i) &= 2\left[\cos\left(\frac{\pi}{3}\right)+i\sin\left(\frac{\pi}{3}\right)\right]2\left[\cos\left(\frac{-\pi}{6}\right)+i\sin\left(\frac{-\pi}{6}\right)\right] \\ &= 4\left[\cos\left(\frac{\pi}{6}\right)+i\sin\left(\frac{\pi}{6}\right)\right] \\ &= 2(\sqrt{3}+i).\end{aligned}$$

(c) $2^{99}\left(\cos\frac{99\pi}{3} + i\sin\frac{99\pi}{3}\right) = 2^{99}(\cos\pi + i\sin\pi) = -2^{99}$.

1.13. Let $z = r(\cos\theta + i\sin\theta)$ and $1/z = \rho(\cos\phi + i\sin\phi)$. Then

$$1 = z\cdot z^{-1} = r\rho[\cos(\theta+\phi) + i\sin(\theta+\phi)].$$

Hence $r\rho = 1$ and $\theta + \phi$ is an integral multiple of 2π; in other words, $\rho = 1/r$ and $\phi = -\theta$ (plus an integral multiple of 2π), or

$$\frac{1}{z} = r^{-1}(\cos\theta - i\sin\theta) = r^{-1}[\cos(-\theta) + i\sin(-\theta)]$$

Now proceed by induction, since $z^{-n} = (1/z)^n$.

1.14. $$\left(\frac{i+\sqrt{3}}{-1-i}\right)^{123} = \left\{\frac{2[\cos(\pi/6) + i\sin(\pi/6)]}{-\sqrt{2}\,[\cos(\pi/4) + i\sin(\pi/4)]}\right\}^{123}$$
$$= -2^{123/2}\left[\cos\left(\frac{-\pi}{12}\right) + i\sin\left(\frac{-\pi}{12}\right)\right]^{123}$$
$$= -2^{61.5}\left[\cos\left(\frac{\pi}{4}\right) - i\sin\left(\frac{\pi}{4}\right)\right]$$
$$= -2^{61}(1-i).$$

1.15. $(u+iv)^2 = x + iy$ implies that $u^2 - v^2 + 2iuv = x + iy$, hence $u^2 - v^2 = x$ and $2uv = y$. Solve the second equation for v and substitute into the first equation to get

$$4u^4 - 4u^2x - y^2 = 0,$$
$$u^2 = \frac{x \pm (x^2+y^2)^{1/2}}{2}.$$

We take the + sign so that u^2 will be positive, and then we get two real values for u; and $v = y/(2u)$, so we get two values (of opposite sign) for $x + iy$.

If $z = 4 + 10i$, then $u^2 = 2 + \sqrt{29}$, $u = \pm 2.718$, $v = 10/(2u) = \pm 1.840$, where both upper or both lower signs are to be used.

1.16. $$\sqrt{i} = \left[\cos\left(\frac{\pi}{2}\right) + i\sin\left(\frac{\pi}{2}\right)\right]^{1/2} = \pm\left[\cos\left(\frac{\pi}{4}\right) + i\sin\left(\frac{\pi}{4}\right)\right]$$
$$= \pm 2^{-1/2}(1+i).$$
$$\sqrt{1+i} = \left\{\sqrt{2}\left[\cos\left(\frac{\pi}{4}\right) + i\sin\left(\frac{\pi}{4}\right)\right]\right\}^{1/2}$$
$$= \pm 2^{1/4}\left[\cos\left(\frac{\pi}{8}\right) + i\sin\left(\frac{\pi}{8}\right)\right]$$
$$= 1.099 + 0.455i.$$
$$4 + 10i = 2\sqrt{29}\,(\cos\theta + i\sin\theta),\quad \theta = \tan^{-1}(\tfrac{5}{2}),$$
$$\sqrt{4+10i} = \pm 2^{1/2}29^{1/4}(\cos\tfrac{1}{2}\theta + i\sin\tfrac{1}{2}\theta),$$

or (if you prefer a more explicit answer),

$$4 + 10i = 10.77(\cos 68.20^\circ + i\sin 68.20^\circ)$$
$$\sqrt{4+10i} = \pm 3.28(\cos 34.10^\circ + i\sin 34.10^\circ) = \pm(2.72 + 1.84i).$$

$$\sqrt{1+i\sqrt{3}} = 2^{1/2}\left[\cos\left(\frac{\pi}{3}\right) + i\sin\left(\frac{\pi}{3}\right)\right]^{1/2}$$
$$= \pm 2^{1/2}\left[\cos\left(\frac{\pi}{6}\right) + i\sin\left(\frac{\pi}{6}\right)\right]$$
$$= \pm(2)^{1/2}(\tfrac{1}{2}\sqrt{3} + \tfrac{1}{2}i)$$
$$= \pm(1.225 + 0.707i).$$
$$(-1)^{1/4} = (\cos\pi + i\sin\pi)^{1/4}$$
$$= \cos\left[\left(\frac{\pi}{4}\right) + k\left(\frac{\pi}{2}\right)\right] + i\sin\left[\left(\frac{\pi}{4}\right) + k\left(\frac{\pi}{2}\right)\right],\ k = 0, 1, 2, 3,$$
$$= 2^{-1/2}(\pm 1 \pm i).$$
$$(8i)^{1/3} = 2\left[\cos\left(\frac{\pi}{2}\right) + i\sin\left(\frac{\pi}{2}\right)\right]^{1/3}$$
$$= 2\left[\cos\left(\frac{\pi}{6} + \frac{2k\pi}{3}\right) + i\sin\left(\frac{\pi}{6} + \frac{2k\pi}{3}\right)\right],\ k = 0, 1, 2,$$
$$= (\pm\sqrt{3} + i), \quad -2i.$$
$$[64(\cos 60° + i\sin 60°]^{1/6} = 2[\cos(10° + k\cdot 60°) + i\sin(10° + k\cdot 60°)]$$
$$= 2[\cos(10°, 70°, 130°, 190°, 250°, 310°) + i\sin(10°, 70°, 130°, 190°, 250°, 310°)$$
$$= 1.970 + 0.347i,\quad 0.684 + 1.879i,\quad -1.286 + 1.532i,\ -1.970 - 0.347i,\quad -0.684 - 1.879i,\quad 1.286 - 1.532i.$$

1.17. The sum of the roots of a polynomial equation $z^n + a_{n-1}z^{n-1} + \cdots + a_0$ is $-a_{n-1}$. [If you are not familiar with this fact, observe that if the roots are r_1, r_2, $r_3, \ldots, r_n$, then the equation is equivalent to $(z - r_1)(z - r_2)\cdots(z - r_n) = 0$; pick out the coefficient of z^{n-1} in the product.] The sum of the nth roots of w is the sum of the roots of the equation $z^n - w = 0$, hence 0.

It is also possible, but tedious, to write the nth roots in trigonometric form and add them up by using algebraic and trigonometric identities.

1.18. Here are two proofs, one algebraic and one geometric.

1. If z_1, z_2, z_3 are the vertices of an equilateral triangle, and λ is the centroid of the triangle, then the three numbers $z_k - \lambda$, $k = 1, 2, 3$, are the three cube roots of some number w, that is, the three roots of the equation $z^3 - w = 0$. Then their sum is zero (compare the solution of Exercise 1.17) and so is the sum of their products in pairs. In other words,

$$(z_1 - \lambda) + (z_2 - \lambda) + (z_3 - \lambda) = 0,$$

that is,

$$z_1 + z_2 + z_3 - 3\lambda = 0;$$

and

$$(z_1 - \lambda)(z_2 - \lambda) + (z_1 - \lambda)(z_3 - \lambda) + (z_2 - \lambda)(z_3 - \lambda) = 0,$$

that is, by the preceding equation,

$$z_1z_2 + z_1z_3 + z_2z_3 = 3\lambda^2.$$

If we square the equation $z_1 + z_2 + z_3 = 3\lambda$, we get

$$z_1^2 + z_2^2 + z_3^2 + 2(z_1z_2 + z_1z_3 + z_2z_3) = 9\lambda^2$$

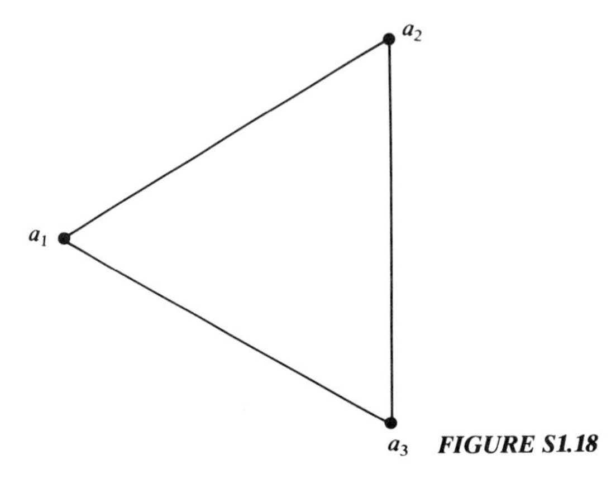

FIGURE S1.18

and using the preceding equation we then get

$$z_1^2 + z_2^2 + z_3^2 = 3\lambda^2 = z_1 z_2 + z_1 z_3 + z_2 z_3.$$

For the converse, determine λ from $z_1 + z_2 + z_3 = 3\lambda$ and work backwards.

2. Let a_1, a_2, a_3 be the vertices of a nondegenerate equilateral triangle. Number the vertices clockwise, as shown in Fig. S1.18. Then the arguments of three equal angles can be chosen so that

$$\arg\frac{a_2 - a_1}{a_3 - a_1} = \arg\frac{a_3 - a_2}{a_1 - a_2}.$$

Since the sides are equal,

$$\left|\frac{a_2 - a_1}{a_3 - a_1}\right| = \left|\frac{a_3 - a_2}{a_1 - a_2}\right|.$$

Hence

$$\frac{a_2 - a_1}{a_3 - a_1} = \frac{a_3 - a_2}{a_1 - a_2},$$

which simplifies to

$$a_1^2 + a_2^2 + a_3^2 = a_1 a_2 + a_1 a_3 + a_2 a_3.$$

Since this equation is symmetric in the indices $1, 2, 3$, it doesn't matter how we numbered the vertices.

Conversely, we can start from the last equation and work back to the equations we started from.

1.19. $\lim z_n = z_0$ means $|z_n - z_0| < \varepsilon$ if $n > N$. Now $|z_n - z_0| \le |x_n - x_0| + |y_n - y_0|$, so if $|x_n - x_0| \le \frac{1}{2}\varepsilon$ and $|y_n - y_0| < \frac{1}{2}\varepsilon$ (which are true for sufficiently large N), we have $|z_n - z_0| < \varepsilon$.

Conversely, $|x_n - x_0| \le |z_n - z_0|$ because $|\operatorname{Re} z| \le |z|$ for every z.

1.20. First, $z_n - z_0 = |z_n - z_0|\,(\cos\theta_n + i\sin\theta_n)$, where θ_n is assumed to approach a limit. Since $|\cos\theta_n + i\sin\theta_n| = 1$, it follows that $z_n - z_0 \to 0$.

The converse is clearer geometrically than algebraically. Suppose that $z_n \to z_0 \neq 0$; then for sufficiently large n, z_n is in the disk in Fig. S1.20, and $\arg z$ and

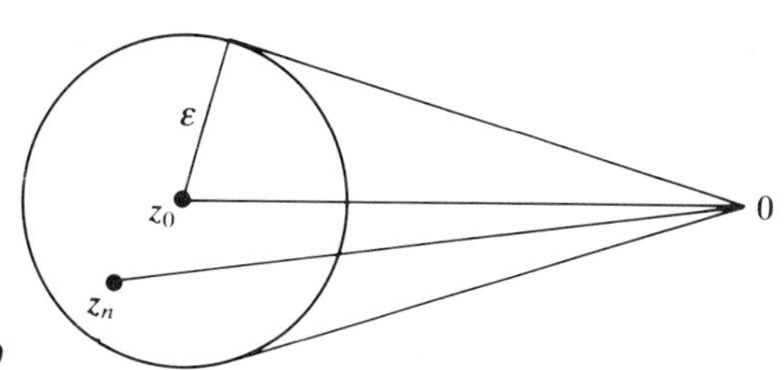

FIGURE S1.20

$\arg z_n$ can be assigned values that differ by less than the angle subtended at 0 by the disk. Hence $|\arg z_n - \arg z_0|$ is at most $2\sin^{-1}(\varepsilon/|z_0|)$. Because $z_0 \neq 0$, this approaches 0 as $\varepsilon \to 0$.

1.21. (*a*) $|z - i|$ is real and nonnegative, so $2z + i$ is real and nonnegative. This means that $2z = r - i$, where r is real and $r \geq 0$. Hence $|\frac{1}{2}r - \frac{1}{2}i - i| = r$, that is, $\frac{1}{4}r^2 + \frac{9}{4} = r^2$. This implies that $r^2 = 3$ and (since $r \geq 0$) $r = \sqrt{3}$. Thus $z = \frac{1}{2}(\sqrt{3} - i)$.

(*b*) $1 + x + iy = 2x + 3y + i(3x - 2y)$. Equate real parts and imaginary parts on both sides to get $x + 1 = 2x + 3y$ and $y = 3x - 2y$. These equations lead to $y = x$, and then $x = \frac{1}{4}$.

(*c*) $y = 3, \quad x = -\frac{1}{2}$.

1.22. The square of the absolute value of $(z - w)/(\bar{w}z - 1)$ is

$$\frac{z - w}{\bar{w}z - 1} \cdot \frac{\bar{z} - \bar{w}}{w\bar{z} - 1} = \frac{|w|^2 - w\bar{z} - \bar{w}z + |z|^2}{|w|^2|z|^2 - w\bar{z} - \bar{w}z + 1}$$

$$= \frac{|w|^2 - 2\,\mathrm{Re}\, w\bar{z} + |z|^2}{|w|^2|z|^2 - 2\,\mathrm{Re}\, w\bar{z} + 1} \leq 1,$$

if

$$|w|^2 + |z|^2 \leq |w|^2|z|^2 + 1,$$

$$|w|^2(1 - |z|^2) \leq 1 - |z|^2.$$

The last inequality holds with $|z| \leq 1$ because $|w| < 1$, and is strict unless $|z| = 1$.

1.23. From Exercise 1.10, we see that

$$|P_n(z)| \geq ||a_n z^n| - (|a_{n-1}z^{n-1}| + \cdots + |a_0|)|$$

$$\frac{|P_n(z)|}{R^n} > |a_n| - (|a_{n-1}|R^{-1} + \cdots + |a_0|R^{-n}).$$

The expression in parentheses can be made arbitrarily small by taking R large enough, and in particular less than $|a_n|/2$.

1.24.
$$\begin{pmatrix} x & y \\ -y & x \end{pmatrix}\begin{pmatrix} u & v \\ -v & u \end{pmatrix} = \begin{pmatrix} xu - yv & xv + yu \\ -xv - yu & xu - yv \end{pmatrix}$$

corresponds to $(xu - yv, xv + yu)$.

1.25. (If necessary, consult a text on analytic geometry.) A rotation about 0 through angle θ has the matrix

$$\begin{pmatrix} \cos\theta & \sin\theta \\ -\sin\theta & \cos\theta \end{pmatrix}.$$

The matrix of a magnification by r is

$$\begin{pmatrix} r & 0 \\ 0 & r \end{pmatrix}$$

and the matrix of the composition is

$$\begin{pmatrix} r\cos\theta & r\sin\theta \\ -r\sin\theta & r\cos\theta \end{pmatrix} = \begin{pmatrix} x & y \\ -y & x \end{pmatrix}.$$

The product

$$\begin{aligned}
&\begin{pmatrix} r\cos\theta & r\sin\theta \\ -r\sin\theta & r\cos\theta \end{pmatrix}\begin{pmatrix} s\cos\phi & s\sin\phi \\ -s\sin\phi & s\cos\phi \end{pmatrix} \\
&= \begin{pmatrix} rs(\cos\theta\cos\phi - \sin\theta\sin\phi) & rs(\cos\theta\sin\phi + \sin\theta\cos\phi) \\ -rs(\sin\theta\cos\phi + \cos\theta\sin\phi) & rs(-\sin\theta\sin\phi + \cos\theta\cos\phi) \end{pmatrix} \\
&= rs\begin{pmatrix} \cos(\theta+\phi) & \sin(\theta+\phi) \\ -\sin(\theta+\phi) & \cos(\theta+\phi) \end{pmatrix}.
\end{aligned}$$

1.26. Obvious except perhaps for $\mathbf{z} \times \mathbf{w}$. Here we have

$$\mathbf{z} \times \mathbf{w} = \begin{vmatrix} \mathbf{i} & \mathbf{j} & \mathbf{k} \\ x & y & 0 \\ u & v & 0 \end{vmatrix} = \mathbf{k}(xv - yu).$$

Section 2

2.1. If

$$\lim_{z \to z_0} \frac{f(z) - f(z_0)}{z - z_0}$$

exists, then the limit exists in particular if $z \to z_0$ through either real or pure imaginary values. In the real case,

$$\lim_{h \to 0} \frac{u(x_0 + h, y_0) - u(x_0, y_0) + iv(x_0 + h, y_0) - iv(x_0, y_0)}{h}$$

exists; by Exercise 1.19, the separate limits

$$\lim \frac{u(x_0 + h, y_0) - u(x_0, y_0)}{h}, \qquad \lim \frac{v(x_0 + h, y_0) - v(x_0, y_0)}{h}$$

exist, which says that

$$f'(z_0) = \left(\frac{\partial u}{\partial x} + i\frac{\partial v}{\partial x}\right)_{z = z_0}.$$

Similarly,

$$\lim_{h \to 0} \frac{u(x_0, y_0 + h) - u(x_0, y_0) + iv(x_0, y_0 + h) - iv(x_0, y_0)}{ih}$$

exists, and so

$$f'(z_0) = \left(-i\frac{\partial u}{\partial y} + \frac{\partial v}{\partial y}\right)_{z=z_0}.$$

Equating real and imaginary parts in the two expressions for $f'(z_0)$, we have the Cauchy–Riemann equations.

If we apply the Cauchy–Riemann equations to the two expressions for $f'(z_0)$, we then have

$$\frac{\partial f}{\partial x} = \frac{\partial v}{\partial y} - i\frac{\partial u}{\partial y} = -i\left(\frac{\partial u}{\partial y} + i\frac{\partial v}{\partial y}\right) = -i\frac{\partial f}{\partial y}.$$

2.2. (*a*) In polar coordinates, $f(z) = r(\cos 4\theta + i \sin 4\theta)$. Clearly, $f(z) \equiv r$ on the real axis and on the imaginary axis, so $\partial f/\partial x = \partial f/\partial y = 0$ at $x = y = 0$. Thus the Cauchy–Riemann equations are satisfied at 0. However, if $\arg z = \pi/4$, we have $f(z) = -r$ and

$$\frac{f(z) - f(0)}{z} = \frac{-r}{r[\cos(\pi/4) + i \sin(\pi/4)]} = \frac{-\sqrt{2}}{1+i},$$

so f does not have a derivative at 0.

(*b*) $f(z) = z\bar{z}/|z|^{1/2} \to 0$ as $z \to 0$, so $f'(0) = 0$.

If $f = u + iv$, then $v \equiv 0$ and the Cauchy–Riemann equations could be satisfied only if u were constant.

2.3. With $f(z) = u(x, y) + iv(x, y)$, we have

$$u(x, y) = u(x_0, y_0) + u_1(x_0, y_0)[1 + o(x - x_0)] + u_2(x_0, y_0)[1 + o(y - y_0)].$$

Since $|x - x_0| \le |z - z_0|$ and $|y - y_0| \le |z - z_0|$, we can combine the o terms into $o(z - z_0)$; and since there is a similar equation for v, we have

$$\begin{aligned} f(z) - f(z_0) = {} & (x - x_0)u_1(x_0, y_0) + (y - y_0)u_2(x_0, y_0) + \\ & i(x - x_0)v_1(x_0, y_0) + i(y - y_0)v_2(x_0, y_0) + o(z - z_0). \end{aligned}$$

In our current notation, the Cauchy–Riemann equations read $u_1 = v_2$, $u_2 = -v_1$. Hence we have

$$\begin{aligned} f(z) - f(z_0) &= (x - x_0)(u_1 - iu_2) + (y - y_0)(u_2 + iu_1) + o(z - z_0) \\ &\qquad \text{(derivatives evaluated at } z_0) \\ &= (x - x_0)(u_1 - iu_2) + i(y - y_0)(u_1 - iu_2) + o(z - z_0) \\ &= (z - z_0)(u_1 - iu_2) + o(z - z_0). \end{aligned}$$

If we now divide by $z - z_0$, we see that $f'(z_0)$ exists.

2.4. With $f(z) = U(r, \theta) + iV(r, \theta)$,

$$\frac{\partial f}{\partial r} = U_r + iV_r = f'(z)\frac{\partial z}{\partial r} = e^{i\theta}f'(z),$$

$$\frac{\partial f}{\partial \theta} = U_r + iV_\theta = f'(z)\frac{\partial z}{\partial \theta} = ire^{i\theta}f'(z).$$

Equate the two expressions for $e^{i\theta}f'(z)$ to get

$$U_r = r^{-1}V_\theta, \qquad V_r = -r^{-1}U_\theta.$$

Equating real parts and then equating imaginary parts in the two expressions for $f'(z)$, we get the required equations. The Cauchy–Riemann equations in polar coordinates, together with the continuity of the partial derivatives, are sufficient for the differentiability of f (if $z \neq 0$), the proof being much the same as in Exercise 2.3.

2.5. If $f(z)=u(x, y)+iv(x, y)$, then $g(z)=u(x, -y)-iv(x, -y)=U(x, y)+iV(x, y)$ (say). Then

$$\frac{\partial U}{\partial x}=u_1(x, -y), \qquad \frac{\partial U}{\partial y}=-u_2(x, -y), \qquad \frac{\partial V}{\partial x}=-v_1(x, -y), \qquad \frac{\partial V}{\partial y}=v_2(x, -y),$$

and since the Cauchy–Riemann equations give us $u_1(x, -y)=v_2(x, -y)$ and $v_2(x, -y)=-u_1(x, -y)$, it follows that

$$\frac{\partial U}{\partial x}=\frac{\partial V}{\partial y} \quad \text{and} \quad \frac{\partial U}{\partial y}=-\frac{\partial V}{\partial x}.$$

2.6. This is like Exercise 2.5 except that the partial differentiation is trickier. Let $f(z)=U(r, \theta)+iV(r, \theta)$ as in Exercise 2.4. Then, since $\arg(1/\bar{z})=\arg z$, $g(z)=U(R^2/r, \theta)-iV(R^2/r, \theta)=P(r, \theta)+iQ(r, \theta)$, say. Now

$$\frac{\partial P}{\partial r}=-\frac{R^2}{r^2}U_1\left(\frac{R^2}{r}, \theta\right), \qquad \frac{\partial Q}{\partial r}=\frac{R^2}{r^2}V_1\left(\frac{R^2}{r}, \theta\right),$$

$$\frac{\partial P}{\partial \theta}=U_2\left(\frac{R^2}{r}, \theta\right), \qquad \frac{\partial Q}{\partial \theta}=-V_2\left(\frac{R^2}{r}, \theta\right).$$

The function f satisfies the Cauchy–Riemann equations of Exercise 2.4, and hence $U_1(r, \theta)=r^{-1}V_2(r, \theta)$. If we replace r by R^2/r and remember that $Q=-V$, we get

$$\frac{\partial P}{\partial r}=-\frac{R^2}{r^2}U_1\left(\frac{R^2}{r}, \theta\right)=-\frac{R^2}{r^2}\frac{r}{R^2}V_2\left(\frac{R^2}{r}, \theta\right)=\frac{1}{r}\frac{\partial Q}{\partial \theta}.$$

Similarly for $\partial Q/\partial r$.

2.7. $f(w)-f(z)=0$, but $f'(t)=4t^3=0$ only for $t=0$, which is not between the points z and w.

2.8, 2.9. Since $|f|^2=u^2+v^2=\text{constant}$, we have

$$u\frac{\partial u}{\partial x}+v\frac{\partial v}{\partial x}=0,$$

$$u\frac{\partial u}{\partial y}+v\frac{\partial v}{\partial y}=0.$$

But by the Cauchy–Riemann equations, the second equation says that

$$-u\frac{\partial v}{\partial x}+v\frac{\partial u}{\partial x}=0.$$

We now have the pair of equations

$$-u\frac{\partial v}{\partial x}+v\frac{\partial u}{\partial x}=0,$$

$$v\frac{\partial v}{\partial x}+u\frac{\partial u}{\partial x}=0,$$

which have a nonzero solution for the two derivatives only if $u^2 + v^2 = 0$. Hence (by the Cauchy–Riemann equations again) all the first derivatives of u and v are zero in D. That is, $f'(z) \equiv 0$ in D. We now establish that *an analytic function whose derivative is identically zero in a region D is a constant in D*, a proposition that will be important later.

In the first place, $f'(z) \equiv 0$ implies that f is constant in a neighborhood of any convenient starting point $z_0 \in D$. To see this, we cannot (as we do for functions on the real line) apply the mean value theorem for f, because we saw in Exercise 2.7 that it is not true. We can, however, apply the ordinary mean value theorem to u and v, as functions of x and of y separately, to see that they are constant in a neighborhood of z_0, and therefore f is constant there. Let $f(z) \equiv c$ in a neighborhood of z_0. Now consider the largest connected subset E of D that contains z_0 and is such that $f(z) = c$ in E. [We can simply define E as the union of all connected subsets of D that contain z_0 and on which $f(z) = c$.] If E is not all of D, then E must have a boundary point in D, that is, a point z_1 of E such that every neighborhood of z_1 contains points of E and points of D not in E. However, reasoning with z_1 as with z_0, we obtain $f(z) = c$ in a neighborhood of z_1, and hence at points of D not in E, contradicting the definition of E.

For Exercise 2.9, notice that the hypothesis means that $v \equiv 0$; then the Cauchy–Riemann equations show that $f' \equiv 0$.

Section 3

3.1. The series $\sum_{k=0}^{\infty} (z + \frac{1}{2})^k$ has radius of convergence 1, so in particular it converges at $z = -1$.

If the powers of $(z + \frac{1}{2})$ are expanded by the binomial theorem, we have

$$\sum_{k=0}^{\infty} \sum_{j=0}^{k} \binom{k}{j} z^j \left(\frac{1}{2}\right)^{k-j},$$

which, if rearranged as a series of powers of z, becomes

$$\sum_{j=0}^{\infty} z^j \sum_{k=j}^{\infty} \binom{k}{j} \left(\frac{1}{2}\right)^{k-j}.$$

With $k = j + l$, this becomes

$$\sum_{j=0}^{\infty} z^j \sum_{l=0}^{\infty} \binom{j+l}{j} \frac{1}{2^l}.$$

The inner sum is the binomial expansion of $(1 - \frac{1}{2})^{-j-1}$, so we obtain

$$\sum_{j=0}^{\infty} 2^{j+1} z^j,$$

and this series diverges when $z = -1$.

3.2. When $|z| \geq 1$, the terms evidently do not approach 0. When $0 < |z| < 1$, we have the same conclusion as soon as n is so large that $n\,|z| > 1$.

Alternatively, the ratio test applies:

$$\frac{n^{n+1}}{n^n} \frac{z^{n+1}}{z^n} = n\,|z| \to \infty.$$

3.3. $\left|\sum_{k=m}^{n} z_k\right| \leq \sum_{k=m}^{n} |z_k| \leq \sum_{k=m}^{n} c_k \to 0.$

3.4. By Exercise 3.3, we have only to show that $\sum |z|^n/n^n$ converges. As soon as $n > N > |z|$, the series is dominated by the geometric series $\sum |z|^n/N^n$. Alternatively, use the ratio test.

3.5. Let $\sum a_n z^n$ diverge for some z. It must diverge for every Z with $|Z| > |z|$, since if $\sum a_n Z^n$ converged, $\sum a_n z^n$ would converge, by what we have just proved.

3.6. Define r as in the statement of the exercise. If r is neither 0 nor ∞, then, by the definition of r, the power series converges in every disk $|z| < r_1$ with $r_1 < r$, and hence whenever $|z| < r$; and diverges outside every disk $|z| \leq r_2$ with $r_2 > r$, hence whenever $|z| > r$. The cases $r = 0$, $r = \infty$ are treated similarly.

3.7. For example, (a) $\sum z^n/n^2$; (b) $\sum nz^n$.

3.8. $\limsup x_n = \infty$ if, for every positive number N, $x_n > N$ for infinitely many values of n.

3.9. First suppose that $0 < R < \infty$. Given $\varepsilon > 0$, we have $|a_n|^{1/n} \leq 1/(R - \varepsilon)$ for all sufficiently large n, so $|a_n| \leq (R - \varepsilon)^{-n}$ and $|a_n z^n| \leq [|z|/(R - \varepsilon)]^n$. Thus, if $|z| < R$, we can choose ε so that $|z| < R - \varepsilon$ and then the power series is dominated by $\sum [|z|/(R - \varepsilon)]^n$.

Similarly, there are infinitely many values of n such that $|a_n| \geq (R + \varepsilon)^{-n}$. If $|z| > R$, then we can choose ε so that $R + \varepsilon < |z|$ and then $|a_n z^n| > (R + \varepsilon)^{-n} |z|^n > 1$. Consequently, $\sum a_n z^n$ diverges because its terms do not approach 0.

If $R = 0$, there is an infinite sequence of integers n for which $|a_n| > K^n$, no matter how large a number K we choose. For any z we choose $K > 1/|z|$; then $|a_n z^n| > 1$ infinitely often, so $\sum a_n z^n$ cannot converge for this value of z (however small $|z|$ may be, as long as it is positive). Thus, if $R = 0$, the power series converges only for $z = 0$.

If $R = \infty$, $\limsup |a_n|^{1/n} = 0$, which is the same as saying that $|a_n|^{1/n} \to 0$. For an arbitrary $\varepsilon > 0$, we then have $|a_n| < \varepsilon^n$ for sufficiently large n, therefore $|a_n z^n| < |\varepsilon z|^n$, and the power series is dominated by a convergent geometric series as long as $|z| < 1/\varepsilon$. Hence the power series converges for every finite z.

3.10. Direct comparison with geometric series shows that $R = 1$ for (a) and $R = 2^{-1/2}$ for (b). In (c), the formula for $1/R$ yields $R = 3^{-1/2}$ (the terms of even index don't have any influence). For (d), $R = 1$ by comparison with $\sum |z|^n$. For (e), the easiest method is to apply the ratio test:

$$\frac{2|z|(2p)!}{(2p+2)!} = \frac{2|z|}{(2p+2)(2p+1)} \to 0$$

for every $z (\neq 0)$, so $R = \infty$.

For (f), the test ratio is

$$\frac{|z|^2 (s+1)^2}{(2s+1)(2s+2)} \to \frac{|z|^2}{4},$$

so $R = 2$.

For (g), the test ratio is

$$\frac{(k+1)|z+2|k^k}{(k+1)^{k+1}} = |z+2|\left(\frac{k}{k+1}\right)^k = |z+2|\left(1+\frac{1}{k}\right)^{-k} \to \frac{|z+2|}{e},$$

so $R = e$.

For (h), if $k=2^n$, we have $a_k=k$; otherwise $a_k=0$. Hence we have $1/R=\limsup k^{1/k}=1$.

For (i), since $2^{\log n}=n^{\log 2}$, we have $R=1$.

For (j), by the ratio test,

$$\frac{|z|^2\,[(2n+1)(2n+2)]^{1/2}}{n+1}\to 2\,|z|^2,$$

so $R=2^{-1/2}$.

For (k), direct application of the formula for $1/R$ yields $1/R=1$.

3.11. (a) $[k^m\,|a_k|]^{1/k}=k^{m/k}\,|a_k|^{1/k}$. Since $k^{m/k}\to 1$, we have $\limsup[k^m\,|a_k|]^{1/k}=\limsup |a_k|^{1/k}$.

(b) $1/R=\limsup[k(k-1)\cdots(k-m)]^{1/k}\,|a_k|^{1/k}$. We have $(k-m)^m\le[k(k-1)\cdots(k-m)]\le k^m$ and $(k-m)^{m/k}\le[k(k-1)\cdots(k-m)]^{1/k}\le k^{m/k}$. The left-hand and right-hand sides both have limit 1.

3.12. If $\sum_0^\infty a_k z^k$ is formally differentiated m times, we get

$$\sum_m^\infty a_k k(k-1)\cdots(k-m+1)z^{k-m},$$

which has the same radius of convergence as

$$\sum_m^\infty a_k k(k-1)\cdots(k-m+1)z^k.$$

Use Exercise 3.11 with m replaced by $m-1$.

3.13. Any nonabsolutely convergent series of numbers, for example, $\sum(-1)^n/n$.

3.14. If $|z|<1-\delta$, then $\sum z^n$ is dominated by $\sum(1-\delta)^n$, a convergent series of numbers.

If we suppose only that $|z|<1$, the remainder of the series after z^{M-1} is $z^M/(1-z)$. If $z=1-(1/M)$, this is arbitrarily close to $1/e$ for large M, so the remainders of the series cannot approach 0 uniformly.

Section 4

4.1. $|\sin i|=\left|i\left(1+\frac{1}{3!}+\frac{1}{5!}+\cdots\right)\right|>1.$

4.2. The nth term of the product series is

$$\sum_{k=1}^{n-1}\frac{(-1)^k}{\sqrt{k}}\cdot\frac{(-1)^{n-k}}{\sqrt{n-k}}=(-1)^n\sum_{k=1}^{n-1}\frac{1}{\sqrt{k}}\cdot\frac{1}{\sqrt{n-k}}.$$

Each square root is less than $\sqrt{n}$, so

$$\sum_{k=1}^{n-1}\frac{1}{\sqrt{k}}\cdot\frac{1}{\sqrt{n-k}}>\frac{n-1}{n}.$$

Hence the nth term of the product series does not approach zero.

4.3. $$\cosh z=\frac{1}{2}\left[\left(1+z+\frac{z^2}{2!}+\cdots\right)+\left(1-z+\frac{z^2}{2!}-\frac{z^3}{3!}+\cdots\right)\right]$$

$$=1+\frac{z^2}{2!}+\frac{z^4}{4!}+\cdots=\cos(iz).$$

4.4. $$e^{z+w} = \sum_{n=0}^{\infty} \frac{(z+w)^n}{n!} = \sum_{n=0}^{\infty} \frac{1}{n!} \sum_{k=0}^{n} \frac{z^k w^{n-k} n!}{k!\,(n-k)!}$$
$$= \sum_{k=0}^{\infty} \frac{z^k}{k!} \sum_{n=k}^{\infty} \frac{w^{n-k}}{(n-k)!}$$
$$= \sum_{k=0}^{\infty} \frac{z^k}{k!} \sum_{j=0}^{\infty} \frac{w^j}{j!}.$$

4.5. Since $e^z e^{-z} = 1$, neither factor can be zero (since neither is infinite).

4.6. $$\frac{d}{dz} \sin z = \frac{d}{dz} \frac{e^{iz} - e^{-iz}}{2i} = \frac{e^{iz} + e^{-iz}}{2} = \cos z.$$
$$\frac{d}{dz} e^z = \frac{d}{dz} \sum_{n=0}^{\infty} \frac{z^n}{n!} = \sum_{n=1}^{\infty} \frac{z^{n-1}}{(n-1)!} = \sum_{n=0}^{\infty} \frac{z^n}{n!}.$$

4.7. $1 - z^2 + \dfrac{z^4}{3} - \dfrac{2z^6}{45} - \dots$.

4.8. $$\frac{e^z}{1-z} = \left(1 + z + \frac{z^2}{2!} + \dots\right)(1 + z + z^2 + \dots)$$
$$= 1 + 2z + z^2\left(1 + 1 + \frac{1}{2!}\right) + z^3\left(1 + 1 + \frac{1}{2!} + \frac{1}{3!}\right) + \dots.$$

The coefficient of z^n is the nth partial sum of the series for e. Similarly, the coefficient of z^n in the power series of $f(z)/(1-z)$ is the sum of the first $n+1$ coefficients of the power series of f.

4.9. The function e^f is differentiable with a continuous derivative. We have $|e^f| = e^{\text{Re} f} =$ constant, so by Exercise 2.8, e^f is constant, whence f is constant.

4.10. $$e^{iz} = 1 + iz - \frac{z^2}{2!} - i\frac{z^3}{3!} + \frac{z^4}{4!} + \dots$$
$$= 1 - \frac{z^2}{2!} + \frac{z^4}{4!} + \dots + i\left(z - \frac{z^3}{3!} + \dots\right)$$
$$= \cos z + i \sin z.$$
$$e^{-iz} = \cos(-z) + i \sin(-z)$$
$$= \cos z - i \sin z.$$

Hence
$$\frac{e^{iz} + e^{-iz}}{2} = \cos z, \qquad \frac{e^{iz} - e^{-iz}}{2i} = \sin z.$$

4.11. $e^{2i\theta} = e^{i\theta} \cdot e^{i\theta}$; therefore we have $\cos 2\theta + i \sin 2\theta = (\cos\theta + i \sin\theta)^2 = \cos^2\theta - \sin^2\theta + 2i \sin\theta \cos\theta$. Equate real and imaginary parts.

4.12. $e^{3i\theta} = (e^{i\theta})^3$. Then
$$\cos 3\theta + i \sin 3\theta = (\cos\theta + i \sin\theta)^3$$
$$= \cos^3\theta + 3i \cos^2\theta \sin\theta - 3 \cos\theta \sin^2\theta - i \sin^3\theta,$$
so
$$\cos 3\theta = \cos^3\theta - 3 \cos\theta \sin^2\theta = 4 \cos^3\theta - 3 \cos\theta,$$
$$\sin 3\theta = 3 \cos^2\theta \sin\theta - \sin^3\theta = 3 \sin\theta - 4 \sin^3\theta.$$

4.13. If $e^{x+iy}=1$, then $|e^x e^{iy}|=1$. But $|e^{iy}|=1$, so $e^x=1$ because $e^x>0$ for real x. Since e^x is strictly increasing, $x=0$, and $z=e^{iy}$. This means that $\cos y+i\sin y=1$, which implies that $\sin y=0$ and $\cos y=1$. If $\sin y=0$, y is an integral multiple of π, and the integer is even since $\cos y=1$ only for even multiples of π. Thus $z=2k\pi i$, k integral.

4.14.
$$\begin{aligned}\sin z=\sin(x+iy)&=\sin x\cos iy+\cos x\sin iy\\&=\sin x\cosh y+i\sinh y\cos x.\\|\sin z|^2&=\sin^2 x\cosh^2 y+\sinh^2 y\cos^2 x\\&=\cosh^2 y(1-\cos^2 x)+\sinh^2 y\cos^2 x\\&=\cosh^2 y+\cos^2 x(\sinh^2 y-\cosh^2 y)\\&=\cosh^2 y-\cos^2 x=\sinh^2 y+\sin^2 x.\end{aligned}$$

$$|\cos z|^2=\left|\sin\left(\frac{\pi}{2}-z\right)\right|^2=\cosh^2 y-\sin^2 x=\cos^2 x-\sinh^2 y.$$

Thus

$$|\sin(x+iy)|^2=\tfrac{1}{4}(e^{2y}+e^{-2y}+2)-\cos^2 x$$

and

$$\lim_{|y|\to\infty}\frac{|\sin(x+iy)|^2}{e^{2|y|}}=\frac{1}{4}.$$

4.15. If $f(z)=\sum_{k=0}^{\infty}a_n(z-z_0)^n$, we have to show that $a_n=f^{(n)}(z_0)/n!$. Since we know that the series can be correctly differentiated term by term, we just hâve to write the series for $f^{(n)}(z)$ and then set $z=z_0$.

4.16. For $z\neq 0$, the function is evidently differentiable. We also have $f(x)=f(ix)$, and we can appeal to the definition of f' to show that $f'(0)=0$. However, we can show by induction that all the derivatives of f are 0 at 0, so that f cannot be represented by its Maclaurin series.

4.17. Let

$$S_n=1+e^{i\theta}+\cdots+e^{in\theta}=\frac{1-e^{i(n+1)\theta}}{1-e^{i\theta}},$$

$$\begin{aligned}\operatorname{Re}S_n&=\frac{1}{2}\left[\frac{1-e^{i(n+1)\theta}}{1-e^{i\theta}}+\frac{1-e^{-i(n+1)\theta}}{1-e^{-i\theta}}\right]\\&=\frac{1}{2}\frac{1-\cos\theta+\cos n\theta-\cos(n+1)\theta}{1-\cos\theta}\end{aligned}$$

$$\operatorname{Re}S_n-\frac{1}{2}=\frac{1}{2}\frac{\cos n\theta-\cos(n+1)\theta}{1-\cos\theta}.$$

Now we use the formulas

$$\cos A-\cos B=-2\sin\frac{A+B}{2}\sin\frac{A-B}{2},$$

$$1-\cos\theta=2\sin^2\left(\frac{\theta}{2}\right)$$

to obtain

$$\operatorname{Re}S_n-\frac{1}{2}=\frac{\sin[(2n+1)/2]\theta}{2\sin\theta/2}.$$

Similarly,

$$\begin{aligned}\sin\theta+\sin 2\theta+\cdots+\sin n\theta &= \operatorname{Im}(S_n)\\ &= \frac{1}{2i}\left[\frac{1-e^{i(n+1)\theta}}{1-e^{i\theta}}-\frac{1-e^{-i(n+1)\theta}}{1-e^{-i\theta}}\right]\\ &= \frac{1}{2}\frac{\sin\theta+\sin n\theta-\sin(n+1)\theta}{1-\cos\theta}\\ &= \frac{1}{2}\frac{\sin\theta/2\cos\theta/2-\cos[(2n+1)/2]\theta\sin\theta/2}{\sin^2\theta/2}\\ &= \frac{1}{2}\frac{\cos\theta/2-\cos[(2n+1)/2]\theta}{\sin\theta/2}\\ &= \frac{\sin[n/2]\theta\sin[(n+1)/2]\theta}{\sin\theta/2}.\end{aligned}$$

4.18. The sums $\sum_{k=1}^{n}\cos k\theta$ and $\sum_{k=1}^{n}\sin k\theta$ are bounded (uniformly in n) on any interval $(\delta, \pi-\delta)$ with $0<\delta<\pi$, by the formulas in Exercise 4.17. Consequently, Abel's test (Exercise 18.1) shows that the real and imaginary parts of $\sum_{n=1}^{\infty} e^{in\theta}/n$ are convergent except perhaps for $\theta=0$ or π. For $\theta=0$, the real part diverges; for $\theta=\pi$, the imaginary part is zero and the real part is the convergent series $\sum(-1)^n/n$.

Thus we have an example of a power series that converges except for one point on its circle of convergence. Clearly, the coefficients $1/n$ can be replaced by any numbers $\phi(n)$ as long as $\phi(n)$ decreases monotonically to 0.

4.19. The distance from 1 to a root $\exp(2\pi ik/n)$ is $2\sin(k\pi/n)$ (see Fig. S4.19) and the required sum S is $2\sum_{k=1}^{n}\sin(k\pi/n)$ (the $k=n$ term is zero).

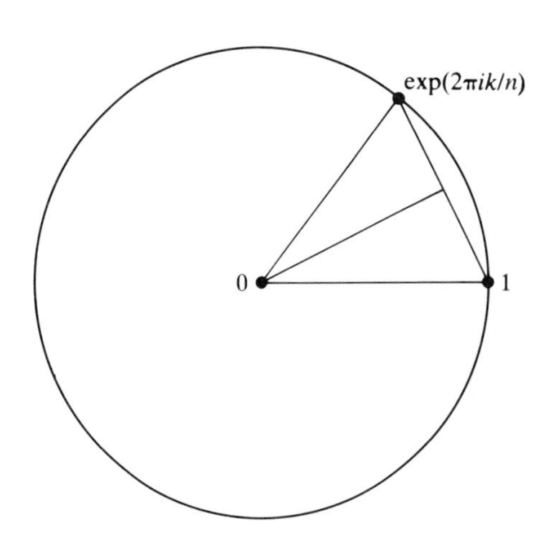

FIGURE S4.19

By Exercise 4.17, with $\theta = \pi/n$,

$$2\sum_{k=1}^{n} \sin\frac{k\pi}{n}\Big)$$

$$= 2\sin\left(\frac{\pi}{2}\right)\sin\left(\frac{n+1}{n}\frac{\pi}{2}\right)$$

$$= 2\cos\left(\frac{\pi}{2n}\right)\Big/\sin\left(\frac{\pi}{2n}\right)$$

$$= 2\cot\left(\frac{\pi}{2n}\right).$$

There are $n-1$ roots; their average is their sum divided by $n-1$, that is,

$$\frac{2}{n-1}\cot\frac{\pi}{2n} \to \frac{4}{\pi}.$$

As $n \to \infty$, the roots are more and more densely distributed on the circle, so it is intuitively plausible that $4/\pi$ is the average distance from 1 to a point on the circle. We can check this by integration: The distance from 1 to $e^{i\theta}$ is $|2\sin\theta/2|$, and if we average this over $(0, 2\pi)$, we do get $4/\pi$.

4.20. The sum of the squared distances is

$$4\sum_{k=1}^{n} \sin^2\left(\frac{k\pi}{n}\right) = 2\sum_{k=1}^{n}\left[1 - \cos\left(\frac{2k\pi}{n}\right)\right]$$

$$= 2n - 2\sum_{k=1}^{n} \operatorname{Re} e^{2k\pi i/n}$$

$$= 2n - 2\operatorname{Re}\left\{\sum_{k=1}^{n} e^{2k\pi i/n}\right\}$$

$$= 2n,$$

since the sum of all the nth roots of 1 is zero (Exercise 1.17).

4.21. $\cos(z+3i) = \cos[x+(y+3)i]$

$= \cos x \cos[i(y+3)] - \sin x \sin[i(y+3)]$.

From Euler's formulas, if R stands for any real number, then $\cos(iR)$ is real $(= \cosh R)$ and $\sin(iR)$ is pure imaginary. Hence

$$\operatorname{Re}[\cos(z+3i)] = \cos x \cosh(y+3).$$

4.22. Clearly, $g(z)=g(\omega z)=g(\omega^2 z)$. So

$$3g(z)=g(z)+g(\omega z)+g(\omega^2 z).$$

Let $g(z)=\sum a_n z^n$. Then

$$3g(z)=\sum a_n(1+\omega^n+\omega^{2n})z^n.$$

If n is a multiple of 3, then ω^n and ω^{2n} are 1. If n is not a multiple of 3, then ω^n and ω^{2n} are equal to ω and ω^2 in one order or the other, and so $1+\omega^n+\omega^{2n}=0$ (Exercise 1.17).

Section 5

5.1. The trace of a closed curve is a bounded set, since continuous functions on compact sets are bounded. Let z_n be a sequence of different points of the trace with limit point z_0. Since the curve is simple, $z_n=z(t_n)$, where t_n is uniquely determined by z_n. The set $\{t_n\}$ (being bounded) has a limit point t_0 and a subsequence $\{t_{n_k}\}$ of $\{t_n\}$ approaches t_0. Since $z(t)$ is continuous, $z_{n_k}=z(t_{n_k})\to z(t_0)$. Since $z_n\to z_0$ and $z_{n_k}\to z(t_0)$, it follows that $z_0=z(t_0)$ and z_0 is a point of the trace. Hence the trace is a closed set.

5.2. (*a*) C_1: $z(t)=x$ from $x=1$ to $x=0$; $z(t)=iy$ from $y=0$ to $y=1$.

$$\int_{C_1}\bar{z}\,dz=\int_1^0 x\,dx+\int_0^1 y\,dy=0.$$

(*b*) C_2: $z=(1+i)t$, $0\le t\le 1$.

$$\int_{C_2}\bar{z}\,dz=\int_0^1(1-i)t(1+i)\,dt=2\int_0^1 t\,dt=1.$$

(*c*) $\int_0^1 x\,dx+i\int_0^1(1+iy)\,dy+\int_1^0(x+i)\,dx+i\int_1^0 iy\,dy=0.$

(*d*) $z(t)=e^{it}$, $0\le t\le 2\pi$.

$$\int_C z^{-1}\,dz=\int_0^{2\pi}e^{-it}ie^{it}\,dt=i\int_0^{2\pi}dt=2\pi i.$$

(*e*) $z=t$, $0\le t\le 2$; $z=2e^{i\theta}$, $0\le\theta\le\pi/2$; $z=it$, $2\ge t\ge 0$.

$$\int_C\bar{z}\,dz=\int_0^2 t\,dt+2\int_0^{\pi/2}e^{-i\theta}2ie^{i\theta}\,d\theta-\int_0^2 t\,dt$$
$$=2\pi i.$$

5.3. (*a*) $z=\pm\frac{1}{2}(a+ae^{i\theta})$, $0\le\theta\le 2\pi$; $\pm\frac{1}{2}(-a+ae^{i\phi})$, $0\le\phi\le 2\pi$; any choice of signs.

(*b*) $z=x$, $0\le x\le 1$; $z=e^{i\theta}$, $0\le\theta\le 2\pi$; $z=u$, $1\ge u\ge 0$.

(*c*) $z=x$ for $-R\le x\le -\varepsilon$ and $\varepsilon\le x\le R$; $z=\varepsilon e^{i\phi}$, $\pi\ge\phi\ge 0$; $z=Re^{i\phi}$, $0\le\phi\le\pi$.

5.4. $f(z)=1/z$, for example.

5.5. Write the integrand as $2(1-z)^{-1}-1$, and parametrize C by $z=1+e^{i\theta}$. Then the integral equals $-4\pi i$.

5.6. We have to show that

$$\int_a^b f[z(t)]z'(t)\,dt = \int_p^q f\{z[t(u)]\}z'[t(u)]t'(u)\,du,$$

which follows by elementary integration theory.
5.7. The mean value theorem would say that $(1+i)z = i$ for some value of z on the contour. However, this would mean that $z = i/(1+i) = (1+i)/2$, which is not on the contour.

Section 6

6.1. We can write

$$F(z) = \int_{z_0}^{x_1+iy} f(w)\,dw + \int_{x_1+iy}^{x+iy} f(w)\,dw,$$

where x_1 is chosen (and then fixed) so that the second integral is along a line segment parallel to the real axis. Then if $f(z) = u(x, y) + iv(x, y)$, we can write

$$F(z) = \int_{z_0}^{x_1+iy} f(w)\,dw + \int_0^{x-x_1} u(x_1+t, y)\,dt + i\int_0^{x-x_1} v(x_1+t, y)\,dt.$$

We can then compute $\partial F/\partial x$ by applying the fundamental theorem of calculus to the last two integrals separately:

$$\frac{\partial F}{\partial x} = u(x, y) + iv(x, y) = f(x+iy).$$

Similarly, by taking the second integral along a line segment parallel to the imaginary axis, we get

$$\frac{\partial F}{\partial y} = if(x+iy)$$

so that

$$-i\frac{\partial F}{\partial y} = \frac{\partial F}{\partial x}.$$

Consequently, F satisfies the complex form of the Cauchy–Riemann equations (p. 16) and hence is differentiable. Its derivative is $\partial F/\partial x$, which we showed is $f(z)$.
6.2. $\int_{|z|=1} z^{-2}\,dz = i\int_0^{2\pi} e^{-2i\theta}e^{i\theta}\,d\theta = i\int_0^{2\pi} e^{-i\theta}\,d\theta = 0$. When $|z| = 1$, $z\bar{z} = 1$, so $\bar{z} = 1/z$, and the first integral reduces to the second.
6.3. $P(z) = \sum_{k=0}^{n} a_k z^k$, $\bar{P}(z) = \sum_{k=0}^{n} \bar{a}_k z^k$. If z is real, $\bar{P}(z) = \overline{P(z)}$, so $P(z)\bar{P}(z) = |P(z)|^2$ for real z, that is, $P\bar{P}$ is real (and positive) for real z. For all z, $\bar{P}(z) = \overline{P(\bar{z})}$, so if $\bar{P}$ had a zero at z_0, P would have a zero at $\bar{z}_0$.
6.4. If ϕ is any continuous function such that $\iint_D \phi(x, y)\,dx\,dy = 0$ for every disk D (or every rectangle) then $\iint_D \operatorname{Re}\phi = 0$ and $\iint_D \operatorname{Im}\phi = 0$. But if $\operatorname{Re}\phi \neq 0$ at some point z_0, we have $\operatorname{Re}\phi$ either positive or negative at z_0, hence (because it is

continuous) positive (or negative) in a neighborhood of z_0. This makes it impossible to have $\iint_D \operatorname{Re} \phi = 0$ for *every* D. The same argument applies to $\operatorname{Im} \phi$.

Section 7

7.1. If $f_n(w) \to f(w)$ uniformly on C, then, given $\delta > 0$, we have $|f_n(w) - f(w)| < \delta$ for sufficiently large n (independent of w), and so

$$\left| \int_C f_n(w)\, dw - \int_C f(w)\, dw \right| \le \left| \int_C [f_n(w) - f(w)]\, dw \right| < \delta \cdot (\text{length of } C).$$

7.2. If z is outside C, then $f(w)/(z - w)$ is analytic (as a function of w) for w inside and on C, so the integral is 0 by Cauchy's theorem.

7.3. Since

$$g'(z) = \int_C \frac{g(w)}{(w-z)^2}\, dw,$$

we have

$$\begin{aligned} g'(z_1) - g'(z) &= \int_C g(w) \left\{ \frac{1}{(w-z_1)^2} - \frac{1}{(w-z)^2} \right\} dw \\ &= \int_C g(w) \frac{z^2 - z_1^2 + 2wz_1 - 2wz}{(w-z_1)^2 (w-z)^2}\, dw. \end{aligned}$$

Consequently,

$$\frac{g'(z_1) - g'(z)}{z_1 - z} = \int_C g(w) \frac{2w - (z + z_1)}{(w-z_1)^2 (w-z)^2}\, dw.$$

Since the denominator is bounded away from 0, we can let $z_1 \to z$ under the integral sign. Hence

$$g''(z) = 2\int_C g(w) \frac{w - z}{(w-z)^4}\, dw = 2\int_C g(w) \frac{dw}{(w-z)^3}.$$

7.4. The argument does not show that $\sum_{k=m}^{n} |u_k(z)| \to 0$ uniformly in z. For a counterexample, we may (for instance) take $u_n(z) = x^n - x^{n+1}$, where $0 \le x \le 1$. The partial sums of $\sum_{n=0}^{\infty} (x^n - x^{n+1})$ are

$$\begin{aligned} S_n &= (1-x) + x(1-x) + \cdots + x^n(1-x) \\ &= (1-x)\frac{1 - x^{n+1}}{1-x} = 1 - x^{n+1}. \end{aligned}$$

The sums S_n do not converge uniformly on $0 \le x \le 1$, because $S_n \to 1$ for $0 \le x < 1$ but $S_n \to 0$ for $x = 1$. Thus the limit of the sequence $\{S_n\}$ of continuous functions is discontinuous, which would be impossible if the convergence were uniform. [Alternatively, compute $\sup_x |S_n(x) - 1|$ on $(0, 1)$ and show that it does not tend to 0.] However, $\{u_n\}$ does satisfy the condition in the exercise, since

$$\sum_{n=0}^{\infty} |u_n(x)| = 1 \le \sum_{n=0}^{\infty} 2^{-n-1} = \sum M_n.$$

7.5. We do not know that $[f(w)-f(z)]/(w-z)$ is analytic at z (but you may look ahead to Sec. **8C**). If you write the integral as

$$\frac{1}{2\pi i}\int_C \frac{f(w)\,dw}{(w-z)^2}-\frac{1}{2\pi i}f(z)\int_C \frac{dw}{(w-z)^2},$$

you should see at once that its value is $f'(z)$.
7.6. To simplify the formulas, take $z_0=0$. Then

$$f'(0)=\frac{1}{2\pi i}\int_C \frac{f(w)\,dw}{w^2},$$

where C is the boundary of the square. The length of C is $4L$, and $|w|\geq L/2$ on C, so

$$|f'(0)|\leq \frac{1}{2\pi}\cdot 4LM\cdot\frac{4}{L^2}=\frac{8M}{\pi L}.$$

7.7. Let $P(z)=a_nz^n+a_{n-1}z^{n-1}+\cdots+a_0$ with $a_n\neq 0$. If P has no zeros, $1/P(z)$ is analytic in the whole finite plane. By Exercise 1.23, $|P_n(z)|\geq \frac{1}{2}|a_n|$, uniformly for sufficiently large $|z|$, and therefore $1/|P_n(z)|$ is bounded for large $|z|$; but $1/|P(z)|$ is continuous and hence bounded for z in any disk. By Liouville's theorem, $1/P(z)=c$, a constant. This is impossible if $c=0$; and if $c\neq 0$, it implies that $P(z)=1/c$.
7.8. If we subtract from $f(z)$ the first m terms, $m\geq c$, of its Maclaurin series, the difference is $z^m\phi(z)$, where $\phi(z)$ is a power series that converges everywhere and so is analytic everywhere. Since we have

$$\phi(z)=\frac{f(z)-(a_0+a_1z+\cdots+a_{m-1}z^{m-1})}{z^m},$$

we obtain

$$|\phi(z)|\leq 1+\left|\frac{a_0}{z^m}+\cdots+\frac{a_m}{z}\right|\leq B$$

for some constant B when $|z|\geq A$; and when $|z|\leq A$, $|\phi(z)|$ is bounded. Liouville's theorem tells us that $\phi(z)$ is a constant. This means that f is a polynomial of degree at most m.
7.9. The proof is the same as for Liouville's theorem except that we estimate $|f(z)-f(0)|$ by

$$\frac{|z|}{2\pi(R-|z|)}\int_0^{2\pi}|f(Re^{i\theta})|\,d\theta.$$

7.10. If $r<1$, we have $|f(re^{i\theta})|\leq 1-r$. For any given z with $|z|<r$, Cauchy's formula gives us

$$|f(z)|\leq\frac{1}{2\pi}\int_0^{2\pi}\frac{(1-r)r}{r-|z|}\,d\theta.$$

Fixing z and letting $r\to 1$ leads to $|f(z)|=0$; hence $f(z)\equiv 0$.
7.11. Suppose that there is such an f. If $|f(z)|>(1-|z|)^{-1}$, then $f(z)\neq 0$, so

$1/f(z)$ satisfies the hypotheses of Exercise 7.10 and hence is identically 0, which is impossible.

7.12. $f^{(n)}(z)\to g(z)$ uniformly for $|z|\le R$, so $f^{(n+1)}(z)\to g(z)$ uniformly. On the other hand, the differentiation theorem (Sec. **7C**) says that $f^{(n+1)}(z)=[f^{(n)}(z)]'\to g'(z)$. Therefore $g(z)=g'(z)$, that is, $g(z)-g'(z)=0$. Then

$$\frac{d}{dz}[e^{-z}g(z)]=e^{-z}g'(z)-e^{-z}g(z)=0.$$

This implies that $e^{-z}g(z)=c$, a constant, and $g(z)=ce^{z}$.

7.13. Confine z and w to a compact set. Then all points zt and wt are in some compact set S. Since ϕ is uniformly continuous on S, we have $g(t)\phi(wt)\to g(t)\phi(zt)$ uniformly as $w\to z$. By the uniform convergence theorem,

$$\int_a^b g(t)\phi(wt)\,dt\to\int_a^b g(t)\phi(zt)\,dt.$$

7.14. Since $\pi/4<1$, the integral is $\frac{1}{2}\cdot 2\pi i$ times the second derivative of $\sin w$ at $w=\pi/4$, hence $-\pi i/\sqrt{2}$.

7.15. The integral equals $(1/3!)$ times $2\pi i$ times the third derivative of e^{3z} at $z=1$, hence $9\pi i e^{3}$.

7.16. Fix w and show that

$$\sin(z+w)-[\sin z\cos w+\cos z\sin w)]=0$$

for every z. Now fix z and argue in the same way with w.

7.17. Since $\Gamma(1+z)$ is analytic and not 0 for $\operatorname{Re} z=0$, so is $\Gamma(-z)=-\pi/[\Gamma(1+z)\sin\pi z]$ except at $z=0$.

Section 8

8.1. For $z=x>0$, $e^{1/x}\to\infty$; for $z=x<0$, $e^{1/x}\to 0$; for $z=iy$ (real y), $e^{1/z}=e^{-i/y}$ has no limit.

8.2. $\tan(1/z)$ is undefined at points where $1/z$ is an odd integral multiple of $\pi/2$, so all the points $z=2/[(2k+1)\pi]$ are singular.

8.3. (*a*) Near $z=\pi$, $\sin z=-\sin(z-\pi)=-(z-\pi)+[(z-\pi)^3/6]-\cdots$, so $(\sin z)/(z-\pi)\to -1$; the singular point is removable.

(*b*) Here

$$\frac{\sin z}{(z-\pi)^2}=-\frac{1}{z-\pi}+\cdots\to\infty;$$

the singular point is not removable.

(*c*) Near $z=0$,

$$1-\cos z=\frac{z^2}{2!}-\frac{z^4}{4!}+\cdots,\qquad \frac{1-\cos z}{z}\to 0;$$

the singular point is removable.

(*d*)
$$z\cot z=z\frac{\cos z}{\sin z}\to 1\qquad\text{as } z\to 0;$$

the singular point is removable.

(e)
$$\frac{6 - z - z^2}{2 - z} = \frac{(3 + z)(2 - z)}{2 - z} \to 5;$$

the singular point is removable.

(f)
$$\frac{\sin z}{z^2 + z} = \frac{\sin z}{z} \cdot \frac{1}{z + 1} \to 1;$$

the singular point is removable.

(g) Near $z = \pi/2$,

$$\cos z = -\sin(z - \tfrac{1}{2}\pi), \qquad \sin z = \cos(z - \tfrac{1}{2}\pi);$$

$$\frac{\cos z}{1 - \sin z} = -\frac{\sin(z - \frac{1}{2}\pi)}{1 - \cos(z - \frac{1}{2}\pi)} = \frac{-(z - \frac{1}{2}\pi) + \frac{1}{6}(z - \frac{1}{2}\pi)^3 - \ldots}{\frac{1}{2}(z - \frac{1}{2}\pi)^2 - \ldots};$$

the singular point is not removable.

8.4. (a) 1; (b) 0; (c) not a pole; (d) 2; (e) The series represent $-z/(1 + z)^2$, so order 2; (f) 6.

8.5. (a)

$$f(z) = \frac{1}{z - z_0}(R + (z - z_0)\varphi(z))$$

with φ analytic at z_0. Hence

$$f(z) - R/(z - z_0) \text{ is bounded near } z_0.$$

(b) The integral that defines the residue is the same whether or not the singularity has been removed; but after the singularity has been removed, the integral is zero by Cauchy's theorem.

8.6. The pole is simple, and the residue is

$$\lim_{z \to z_0} g(z) \frac{z - z_0}{h(z)}.$$

8.7. Let f have a pole of order n. The residue is

$$\frac{1}{2\pi i} \int_C \frac{\phi(z)}{(z - z_0)^n}\, dz = \frac{1}{2\pi i} \int_C \frac{\phi(z)(z - z_0)^{m-n}}{(z - z_0)^m}\, dz$$
$$= \frac{1}{(m-1)!} \left(\frac{d}{dz}\right)^{m-1} [\phi(z)(z - z_0)^{m-n}]|_{z=z_0}$$
$$= \frac{1}{(m-1)!} \sum_{k=0}^{m-1} \binom{m-1}{k} \left(\frac{d}{dz}\right)^k (z - z_0)^{m-n} \times$$
$$\left(\frac{d}{dz}\right)^{m-k-1} \phi(z)|_{z_0}.$$

The indicated derivatives of $(z - z_0)^{m-n}$ are all 0 at z_0 except when $k = m - n$, so only the term with $k = m - n$ survives, and the result is

$$\frac{1}{(m-1)!}(m-n)! \frac{(m-1)!\, \phi^{(n-1)}(z_0)}{(m-n)!\,(n-1)!} = \frac{\phi^{(n-1)}(z_0)}{(n-1)!},$$

as before.

8.8. If there is a pole of order n,

$$f(z) = (z - z_0)^{-n}\phi(z), \qquad \phi(z_0) \neq 0.$$

If we multiply by $(z - z_0)^m$, $m < n$, then

$$f(z)(z - z_0)^m = (z - z_0)^{m-n}\phi(z),$$

and $m - 1$ differentiations leave a negative power of $z - z_0$ in every term, so every term $\to \infty$ as $z \to \infty$.

8.9. One counterexample will suffice for each part. For the first part, take $g(z) = 1$, $h(z) = z^2$, $z_0 = 0$. Then the residue of g/h is 0, but $g(z_0)/h''(z_0) = \frac{1}{2}$.

For the second part, let $g(z) = 1 + z + z^2$, $h(z) = 1 + z$, $z_0 = 0$. Then $g(z)z^{-2}/h(z) = z^{-2} + (1 + z)^{-1}$ with residue 0, whereas $h(0) = 1$ and the residue of $g(z)/z^2$ is 1.

8.10. If $g(z) = a_0 + a_1(z - z_0) + a_2(z - z_0)^2 + \ldots$, the residue of $g(z)(z - z_0)^{-k}$ is the coefficient of $1/(z - z_0)$ in the series for $g(z)/(z - z_0)^k$; that is, it is a_{k-1}.

For $e^{-z^2} = 1 - z^2 + (z^4/2!) - (z^6/3!) + \ldots$, the residue of $z^{-7}e^{-z^2}$ is $-1/3!$.

8.11. $R = \dfrac{1}{(n-1)!}\left(\dfrac{d}{dz}\right)^{n-1}[f(z)(z - z_0)^n]|_{z=z_0}$

and

$$S = \frac{1}{(n-1)!}\left(\frac{d}{dz}\right)^{n-1}[f(z)(z - \bar{z}_0)^n]|_{z=\bar{z}_0}.$$

The case $n = 2$ is representative:

$$\begin{aligned}
S &= \lim_{z\to\bar{z}_0}\frac{d}{dz}[f(z)(z - \bar{z}_0)^2] \\
&= \lim_{w\to\bar{z}_0}\frac{d}{dw}[f(w)(w - \bar{z}_0)^2] \\
&= \lim_{w\to\bar{z}_0}[f'(w)(w - \bar{z}_0)^2 + 2f(w)(w - \bar{z}_0)] \\
&= \lim_{\bar{z}\to\bar{z}_0}[f'(\bar{z})(\bar{z} - \bar{z}_0)^2 + 2f(\bar{z})(\bar{z} - \bar{z}_0)] \\
&= \lim_{z\to z_0}\overline{[f'(z)(z - z_0)^2 + 2f(z)(z - z_0)]} \\
&= \overline{R}.
\end{aligned}$$

8.12. Use the rule in Exercise 8.6 whenever it is applicable.

(*a*) $-\dfrac{2}{51}, -\dfrac{3}{68}$ (*b*) e^{-1}

(*c*) $4.9i$ (*d*) $-\dfrac{1}{32}$

(*e*) $\dfrac{1}{2}$ (*f*) $\dfrac{1}{7!}$

(*g*) 1 (*h*) $\frac{-\pi}{2}$

(*i*) $2^{-5/2}(1-i)$ (*j*) $-\frac{1}{32}i$

8.13. (*a*) Yes, from the definition of the residues as integrals.
(*b*) Not always. Example, $f(z) = g(z) = 1/z$ at $z = 0$.
(*c*) Near $z = a$, we have

$$f(z) = \frac{b}{z-a} + \phi(z),$$

with $\phi(z)$ analytic at $z = a$. The residue of f is b.
Near $a^{1/2}$,

$$zf(z^2) = \frac{bz}{z^2 - a} + z\phi(z^2)$$
$$= \frac{bz}{2a^{1/2}}\left(\frac{1}{z-a^{1/2}} - \frac{1}{z+a^{1/2}}\right) + z\phi(z).$$

The residue of $zf(z^2)$ is $b/2$.

8.14. Let C surround all the singular points of f in the finite plane. Then $(2\pi i)^{-1}\int_C f(z)\,dz$ equals the sum of the residues in the finite plane, and its negative is the residue at ∞.

8.15. Direct computation; or use Exercise 8.14 (z^{-2} has one finite singular point, with residue 0).

8.16. Let $P(z) = \sum_{k=0}^{m} p_k z^k$, $Q(z) = \sum_{k=0}^{n} q_k z^k$, $n \geq m+2$, $q_n \neq 0$. For large $|z| = R$, by the triangle inequality (Exercise 1.10) and Exercise 1.23, we have $|P(z)|/|Q(z)| \leq AR^m/(BR^n)$, where A and B depend only on the coefficients. Hence the absolute value of the integral whose limit is the residue does not exceed $AB^{-1}R^{m-n+1}$. Since $m - n + 1 \leq -1$, the integral approaches 0 as $R \to \infty$.

8.17. If f is analytic at ∞, $f(1/z)$ is analytic at 0 and hence bounded in a disk about 0. Hence f is bounded outside some disk. But f is also bounded inside that disk, hence bounded in the whole finite plane, and hence constant by Liouville's theorem.

8.18. From the definition of the order of a pole, we have, for each finite pole z_k, a polynomial $P_k(z)$ such that $f(z) - (z - z_k)^{-n_k}P_k(z)$ has no pole at z_k. Then $g(z) = f(z) - \sum (z - z_k)^{-n_k}P_k(z)$ has no poles in the finite plane; g minus a polynomial has no poles in the extended plane; apply Exercise 8.17.

Section 9

9.1. (*a*) See the solution of Exercise 9.1e, below.

(*b*)
$$\int_0^{2\pi} \frac{d\theta}{1 - 2r\cos\theta + r^2} = \int_0^{2\pi} \frac{d\theta}{1 - r(e^{i\theta} + e^{-i\theta}) + r^2}$$
$$= \int_0^{2\pi} \frac{-e^{i\theta}\,d\theta}{re^{2i\theta} - (1+r^2)e^{i\theta} + r} = i\int_C \frac{dz}{rz^2 - (1+r^2)z + r}.$$

The zeros of the denominator are at $z = r$ and $z = 1/r$; only the first is inside $|z| = 1$. The residue at $z = r$ is

$$\frac{i}{2rz - (1 + r^2)}\bigg|_{z=r} = \frac{i}{r^2 - 1},$$

so the contour integral equals $2\pi/(1 - r^2)$ and the original integral equals $\pi/(1 - r^2)$.

(*c*) $$\int_0^{2\pi} \frac{\cos^2\theta}{5 + 3\sin\theta}\,d\theta = \int_C \frac{(z + z^{-1})^2\,dz}{4iz[5 + 3(2i)^{-1}(z - z^{-1})]} = \int_C \frac{(z^2 + 1)^2\,dz}{6z^2(z + 3i)(z + (i/3))}.$$

Residue at $-i/3$: $$\frac{[(-1/9) + 1]^2}{6\cdot(-1/9)}\,\frac{1}{(-i/3) + 3i} = \frac{4i}{9}.$$

Residue at 0: $$\frac{1}{2}\frac{d}{dz}\frac{(z^2 + 1)^2}{3z^2 + 10iz - 3}\bigg|_{z=0} = \frac{1}{2}\cdot\frac{-10i}{9} = -\frac{5}{9}i.$$

(Note on technique: When evaluating a residue at a multiple pole, do not write out the algebraic form of the derivative; just substitute the corresponding value of z as you go.)

The contour integral is thus equal to $(\frac{4}{9} - \frac{5}{9})i \cdot 2\pi i = 2\pi/9$.

(*d*) $$\int_0^{\pi} \frac{d\theta}{(3 + 2\cos\theta)^2} = \frac{1}{2}\int_C \frac{dz}{iz}\,\frac{1}{(3 + z + z^{-1})^2} = \frac{1}{2i}\int_C \frac{z\,dz}{(z^2 + 3z + 1)^2}.$$

The zeros of $z^2 + 3z + 1$ are a pair ω, $\omega' = 1/\omega$, where $|\omega| < 1$, $\omega + \omega' = -3$, and $\omega - \omega' = \sqrt{5}$. We want π times the residue of the integrand at ω. This is

$$\frac{d}{dz}\frac{z}{(z - \omega')^2}\bigg|_{z=\omega} = \frac{-(\omega + \omega')}{(\omega - \omega')^3} = \frac{3}{5\sqrt{5}},$$

and the original integral equals $3\pi/5^{3/2}$.

(*e*) We may assume that $a > 0$, since otherwise we can consider the negative of the integral that we are to evaluate, and then change the sign of the answer. The corresponding contour integral is

$$\int_C \frac{dz}{iz}\,\frac{2i}{2ia + bz - (b/z)} = \int_C \frac{2_{dz}}{bz^2 + 2iaz - b}.$$

The zeros of the denominator are ω and ω', given by

$$\frac{-i}{b}(a \pm \sqrt{a^2 - b^2}).$$

Since their product is 1, only one of them is inside the unit circle. Let ω be this zero, which is clearly the one with the minus sign. Then the integrand is

$$\frac{2}{b(z-\omega)(z-\omega')},$$

and the residue at ω is

$$\frac{2}{b(\omega-\omega')}=\frac{1}{i\sqrt{a^2-b^2}}.$$

The integral therefore equals

$$\frac{2\pi}{\sqrt{a^2-b^2}}.$$

9.2. (*a*)

$$\int_{-\infty}^{\infty}\frac{x+4}{(x^2+2x+2)(x^2+9)}\,dx.$$

The relevant poles are at $-1+i$ and $3i$. The residues are
At $3i$:

$$\frac{3i+4}{(-9+6i+2)\cdot 6i}=\frac{3-4i}{6(-7+6i)}=\frac{-9+2i}{102};$$

At $-1+i$:

$$\frac{-1+i+4}{[2(-1+i)+2][(-1+i)^2+9]}=\frac{3+i}{2(2+9i)}=\frac{9-15i}{102}.$$

The sum of the residues is $-13i/102$, and the integral equals $13\pi/51$.

(*b*) The integrand has poles at $i\pm 1$ and $-i$, the latter being irrelevant. However, it is a simplification to notice that the residue at ∞ is 0 (Exercise 8.16) and so the sum of the two residues in the upper half plane is the negative of the residue at $-i$ (Exercise 8.14), so we only have to calculate this one residue, which we see immediately is $-\frac{1}{5}$. Hence the original integral equals $2\pi i\cdot\frac{1}{5}=2\pi i/5$.

(*c*) The integrand is

$$\frac{1}{x^4+6x^2+8}=\frac{1}{(x^2+2)(x^2+4)},$$

with poles (in the upper half plane) at $i\sqrt{2}$ and $2i$. The sum of the residues is

$$\frac{1}{-4i\cdot 2\sqrt{2}+12i\sqrt{2}}+\frac{1}{-32i+24i}=\frac{-i}{4}\frac{\sqrt{2}-1}{2},$$

and the integral equals $\pi(\sqrt{2}-1)/8$.

9.3. For $\int_{-\infty}^{\infty}(1+x^4)^{-1}$ there are poles at $e^{i\pi/4}$ and $e^{3\pi i/4}$. The sum of the residues is

$$\frac{1}{4e^{3\pi i/4}}+\frac{1}{4e^{9\pi i/4}}=-\frac{i}{4}\sqrt{2},$$

and the integral equals $\pi\sqrt{2}/2$.

From integral tables, we find

$$\int \frac{dx}{1+x^4} = \frac{1}{4\sqrt{2}} \left\{ \log \frac{x^2 + x\sqrt{2} + 1}{x^2 - x\sqrt{2} + 1} + 2\tan^{-1} \frac{x\sqrt{2}}{1-x^2} \right\}.$$

At $\pm\infty$ the logarithm is 0, so, apparently,

$$\int_{-\infty}^{\infty} \frac{dx}{1+x^4} = \frac{\sqrt{2}}{4} \tan^{-1} \frac{x\sqrt{2}}{1-x^2} \Big|_{-\infty}^{\infty} = 0.$$

Note, however, that the indefinite integral must be a continuous function. Consequently, the inverse tangent cannot always have its principal value, since $x\sqrt{2}/(1-x^2)$ becomes infinite at $x = \pm 1$. If we start the inverse tangent with the value 0 at $-\infty$, it will be positive as x increases until $x = -1$, when its principal value jumps from $\pi/2$ to $-\pi/2$; hence, in order to be continuous, the inverse tangent must take values greater than $\pi/2$. At $x = +1$ we have the same phenomenon, and the inverse tangent now has to have values greater than $3\pi/2$; as $x \to +\infty$, the inverse tangent approaches 2π. Hence the definite integral equals $\pi/\sqrt{2}$.

9.4.

$$\int_R^{2R} \frac{x\,dx}{1+x^2} = \frac{1}{2}\log(1+x^2)|_{-R}^{2R} = \frac{1}{2}\log\frac{1+4R^2}{1+R^2} \to \log 2,$$

$$\int_{-R}^{R} \frac{x\,dx}{1+x^2} = \frac{1}{2}\log(1+x^2)_{-R}^{R} = 0.$$

9.5. The integral over the large semicircle approaches 0, as before, and the contour integral is equal to $2\pi i$ times the sum of the residues of the integrand in the upper half plane. On the small semicircle, $z = c + \varepsilon e^{i\theta}$ and $Q(z) = A(z)(z-c)$, where $A(z)$ approaches the residue of Q at c when $\varepsilon \to 0$. But $P(c)/A(c)$ is the residue of P/Q at c.

9.6. z^{-2} has a double pole at 0, and

$$\left(\int_{-\infty}^{-\varepsilon} + \int_{\varepsilon}^{\infty}\right) \frac{dx}{x^2} = \frac{2}{\varepsilon} \to \infty \quad \text{as} \quad \varepsilon \to 0.$$

9.7. (*a*) The relevant poles are at 2 and $2i$, and the respective residues are

$$-\frac{1}{8} \quad \text{and} \quad \frac{1}{8(1+i)}.$$

Hence the integral equals

$$-\frac{i\pi}{8} + \frac{\pi i}{4(1+i)} = \frac{\pi}{8}.$$

(*b*)

$$\frac{1}{x^3+4x+5} = \frac{1}{(x+1)(x^2-x+5)}.$$

The relevant poles are at -1 and $\frac{1}{2}(1+i\sqrt{19})$. The residues are $\frac{1}{7}$ and

$$\frac{1}{\{[(1+i\sqrt{19})/2]+1\}(1+i\sqrt{19}-1)} = \frac{2}{(3i-\sqrt{19})\sqrt{19}} = -\frac{3i+\sqrt{19}}{14\sqrt{19}}.$$

Hence the integral equals

$$2\pi i\left(\frac{1}{14}+\frac{-3i-\sqrt{19}}{14\sqrt{19}}\right)=\frac{3\pi}{7\sqrt{19}}.$$

9.8. As in Exercise 9.5, we consider C with the corner cut out by a circular arc of radius ε, and write $f(z)=g(z)/(z-z_0)$. With $z-z_0=\varepsilon e^{i\theta}$, we have

$$\int \frac{g(z)}{\varepsilon e^{i\theta}} i\varepsilon e^{i\theta}\,d\theta = i\int g(z_0+\varepsilon e^{i\theta})\,d\theta,$$

with integration over a θ interval of length α. When $\varepsilon \to 0$, the limit of the last integral is $\alpha i g(z_0)$, and $g(z_0)$ is the residue of f at z_0.
9.9. Here we consider (as in Sec. **9B**)

$$\lim_{R\to\infty}\left\{\int_{-R}^{R}\frac{2x+3}{x^2+2x+2}\,dx+i\int_0^{\pi}\frac{2\,Re^{i\theta}+3}{R^2e^{2i\theta}+2\,Re^{i\theta}+2}Re^{i\theta}\,d\theta\right\},$$

that is, the integral around a large semicircle. This equals $2\pi i$ times the sum of the residues of

$$\frac{2z+3}{z^2+2z+2}$$

in the upper half plane. We can write the second integral as

$$i\int_0^{\pi}\frac{2e^{2i\theta}+3e^{i\theta}/R}{e^{2i\theta}+(2/R)e^{i\theta}+(2/R^2)}\,d\theta,$$

and its limit can be found, because a uniformly convergent sequence can be integrated term by term, to be $i\int_0^{\pi} 2\,d\theta = 2\pi i$. The only pole of

$$\frac{2z+3}{z^2+2z+2}$$

in the upper half plane is at $-1+i$, and the residue there is $-\frac{1}{2}i+1$. Consequently,

$$\mathrm{PV}\int_{-\infty}^{\infty}\frac{2x+3}{x^2+2x+2}\,dx+2\pi i=2\pi i\left(1-\frac{1}{2}i\right),$$

and the integral in the problem is equal to π.
9.10. Here we take the contour as indicated in Fig. S9.10; the integral along the segment $(-R, R)$ of the real axis contributes a principal value as in the text. The

FIGURE S9.10

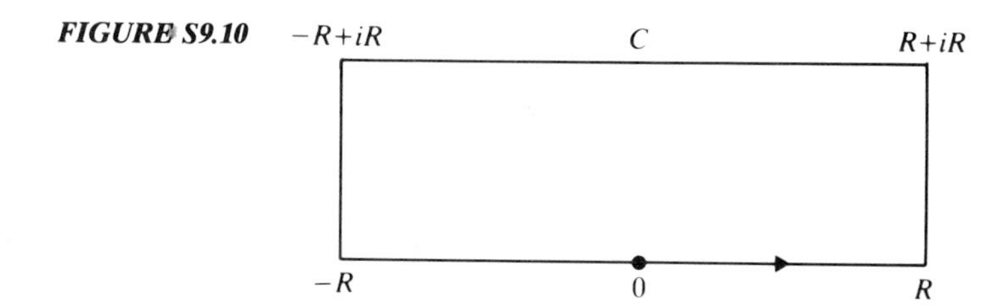

integral along $(R, R+iR)$ can be written as

$$i\int_0^R \frac{e^{i(R+iy)}}{R+iy}\,dy.$$

Its absolute value does not exceed

$$\int_0^R \frac{e^{-y}}{|R+iy|}\,dy = \int_0^R \frac{e^{-y}\,dy}{(R^2+y^2)^{1/2}} \le \frac{1}{R}\int_0^R e^{-y}\,dy = \frac{1}{R}(1-e^{-R}) \to 0.$$

The integral along the other end of the rectangle is estimated in the same way. The remaining integral is

$$\int_{R+iR}^{-R+iR} \frac{e^{iz}}{z}\,dz = -\int_{-R}^{R} \frac{e^{ix-R}}{x+iR}\,dx,$$

and its absolute value does not exceed

$$\int_{-R}^{R} \frac{e^{-R}}{(x^2+R^2)^{1/2}}\,dx \le \frac{e^{-R}}{R}\int_{-R}^{R} dx = 2e^{-R} \to 0.$$

9.11. (*a*) We consider

$$\int_C \frac{ze^{iz}}{16z^2+9}\,dz.$$

The only relevant pole is at $z = \frac{3}{4}i$, and the residue there is $\frac{1}{32}e^{-3/4}$. Hence

$$\int_0^\infty \frac{x\sin x}{16x^2+9}\,dx = \frac{1}{32}\pi e^{-3/4}.$$

(*b*) Consider the contour integral

$$\int_C \frac{e^{iz}\,dz}{(z+1)(z^2+4)}.$$

The relevant poles are at $z=-1$ and $z=2i$. The residues are $e^{-i}/5$ at $z=-1$, and

$$\frac{e^{-2}}{(2i+1)\cdot 4i} = -\frac{i+2}{20e^2}$$

at $z=2i$. Then

$$\mathrm{PV}\int_{-\infty}^{\infty} \frac{e^{ix}\,dx}{(x+1)(x^2+4)} = \frac{i\pi}{5}(\cos 1 - i\sin 1) - \frac{2\pi i}{20e^2}(i+2),$$

and the required integral is its real part,

$$\frac{1}{10}\pi(2\sin 1 + e^{-2}).$$

We also get the value of

$$\int_{-\infty}^{\infty} \frac{\sin x\,dx}{(x+1)(x^2+4)},$$

namely, $\frac{1}{5}\pi(\cos 1 - e^{-2})$.

(*c*) Here the only relevant (double) pole is at $2i$, and the residue there is

$$\frac{d}{dz}\frac{e^{iz}}{(z+2i)^2}\bigg|_{z=2i}=\frac{3}{32e^2 i}.$$

Hence the original integral [over $(0,\infty)$] equals

$$\frac{3\pi}{32e^2}.$$

The integral with $\sin x$ is of course 0, since the integrand is an odd function.

(*d*) The poles of $ze^{iz}/(4-z^2)$ are at ± 2, and the residues are $-\frac{1}{2}e^{2i}$ and $-\frac{1}{2}e^{-2i}$. The sum of the residues is $-\cos 2$. The original integral is therefore $-\pi\cos 2$.

(*e*) The residues of

$$\frac{e^{iz}}{(z-1)(z+2)}$$

are $e^i/3$ and $-e^{-2i}/3$, so the integral is the imaginary part of $\frac{1}{3}\pi i[(\cos 1+i\sin 1)-(\cos 2-i\sin 2)]$, that is, $\frac{1}{3}\pi(\cos 1-\cos 2)=1.00159$.

9.12. We can write the integral as

$$\int_0^\infty \frac{\sin xt}{xt}\,d(xt).$$

If $x>0$, we can take $u=xt$ to get $\int_0^\infty u^{-1}\sin u\,du=\pi/2$. If $x<0$, the integral is $-\pi/2$. Thus the integral represents $\frac{1}{2}\pi\operatorname{sgn} x$. Notice that the integrand is continuous, but the integral is not.

Section 10

10.1. If $u(x,y)=\ln(x^2+y^2)^{1/2}=\frac{1}{2}\ln(x^2+y^2)$ and $v(x,y)=\tan^{-1}(y/x)$, we have, by elementary calculus,

$$\frac{\partial u}{\partial x}=\frac{x}{x^2+y^2}=\frac{\partial v}{\partial y},$$

and

$$\frac{\partial u}{\partial y}=\frac{y}{x^2+y^2}=-\frac{\partial v}{\partial x}.$$

10.2. We have

$$\log z_1=\ln|z_1|+i\arg z_1,$$
$$\log z_2=\ln|z_2|+i\arg z_2,$$
$$\log(z_1z_2)=\ln|z_1|+\ln|z_2|+i\arg(z_1z_2),$$

where the three arg's are determined by the region and by $\arg z_0$ for a specified point in the region. Then some value of the logarithm of the product is the sum of some value of the logarithm of the first factor and some value of the logarithm of

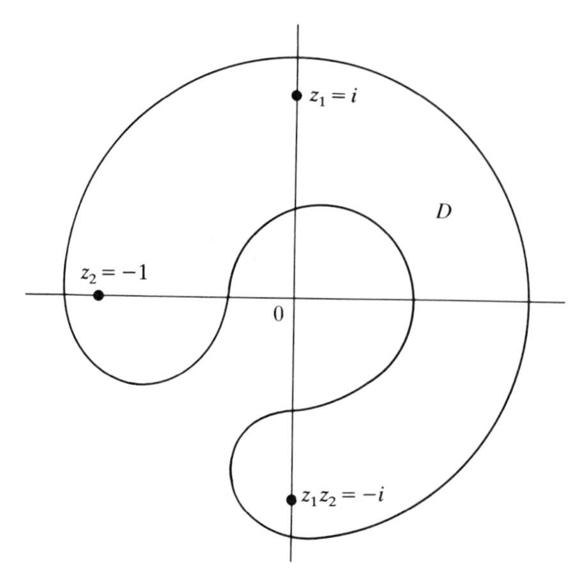

FIGURE S10.2

the second factor. However, if the region looks like Fig. S10.2, we could have $\arg z_1 = \pi/2$, $\arg z_2 = \pi$, but $\arg(z_1 z_2) = -\pi/2$.

10.3. For example, the principal logarithm of -1 is $i\pi$, but the principal logarithm of $(-1)(-1) = +1$ is 0.

10.4. The derivative of $w = e^{\log z}$ is $z^{-1}e^{\log z} = w/z$. Hence $z(dw/dz) - w = 0$, $(d/dz)(w/z) = 0$ (except at $z = 0$), and consequently $w/z = c$, a constant. At $z = 1$, $w = 1$, so $c = 1$, and $w = z$, $z \neq 0$.

10.5.
$$\log(-1) = (2k+1)\pi i, \qquad k = 0, \pm 1, \pm 2, \ldots .$$
$$\log i = i\left(\frac{1}{2}\pi + 2k\pi\right).$$
$$\log\left(\frac{1}{2} + \frac{\sqrt{3}}{2}i\right) = \log e^{i\pi/3} = \frac{i\pi}{3} + 2k\pi i.$$

10.6. $i^{-2i} = e^{-2i(\log i)} = e^{-2i(i\pi/2 + 2k\pi i)} = e^{\pi(4k+1)}$. Notice that all the values are real.

$$i^{(1+i)} = \exp\left[(1+i)i\left(\frac{1}{2}\pi + 2n\pi\right)\right] = \exp\left[i\left(\frac{1}{2}\pi + 2n\pi\right) - \frac{1}{2}\pi - 2n\pi\right]$$
$$= i\exp\left(-\frac{1}{2}\pi - 2n\pi\right)$$
$$(1+i)^i = e^{i\log(1+i)} = \exp[i(\ln\sqrt{2} + \log e^{i\pi/4})]$$
$$= \exp\left[\frac{i}{2}\ln 2 + i\left(\frac{i\pi}{4} + 2k\pi i\right)\right]$$
$$= \left[\cos\left(\frac{1}{2}\ln 2\right) + i\sin\left(\frac{1}{2}\ln 2\right)\right]e^{-\pi/4 - 2k\pi}.$$

$$\begin{aligned}
i^{-\log i} &= e^{-(\log i)(\log i)} \\
&= \exp\left[-\left(\frac{i\pi}{2} + 2k\pi i\right)\left(\frac{i\pi}{2} + 2m\pi i\right)\right] \\
&= \exp\left[-\left(\frac{-\pi^2}{4} - k\pi^2 - m\pi^2 - 4mk\pi^2\right)\right] \\
&= \exp\left[\pi^2\left(\frac{1}{4} + n\right)\right],
\end{aligned}$$

since all integers n can be represented as $k + m + 4mk$.

For i^{i^i}, first find $i^i = e^{i \log i} = e^{-(\pi/2)+2k\pi}$. Then

$$i^{(i^i)} = \exp\left[e^{-(\pi/2)+2k\pi}\left(\frac{\pi}{2} + 2m\pi\right)i\right],$$

where k and m run through all the (positive and negative) integers independently.

Note that $(i^i)^i = -ie^{2n\pi}$.

10.7. $z^{1/n} = \exp(n^{-1} \log z) = r^{1/n} \exp(i\theta/n + 2k\pi i/n)$. There are only n different values for $e^{2\pi ki/n}$, since $e^{2\pi i(k/n)} = e^{2\pi i(k+n)/n}$.

10.8.

$$\begin{aligned}
\sqrt[i]{-1} &= (-1)^{-i} = e^{-i \log(-1)} \\
&= e^{-i(i\pi + 2k\pi i)} = e^{\pi(2k+1)}.
\end{aligned}$$

(I hope you didn't misread the problem as $i^{1/i}$.)

10.9. The quoted calculation makes two unjustified assumptions: that $a^{bc} = (a^b)^c$ and that $1^z = 1$. Neither assumption is always correct.

$$\begin{aligned}
a^{bc} &= e^{bc \log a} = \exp\{bc[\ln |a| + i(\arg a + 2k\pi)]\}, \\
(a^b)^c &= \exp(c \log a^b) \\
&= \exp[c \log(e^{b \log a})] \\
&= \exp\{c \log[e^{b(\ln|a| + i \arg a + 2k\pi i)}]\} \\
&= \exp[c(b \ln |a| + ib \arg a + 2bk\pi i + 2m\pi i)] \\
&= \exp[bc \ln |a| + ibc(\arg a + 2k\pi) + 2cm\pi i].
\end{aligned}$$

Thus $(a^b)^c$ does have more values than a^{bc}.

Also, $1^z = e^{z \log 1} = e^{z \cdot 2k\pi i}$.

Note that $a^b a^c = e^{b \log a} e^{c \log a}$

$$= |a|^{b+c} \exp[i(b \arg a + c \arg a + 2k\pi b + 2m\pi c)],$$

whereas

$$a^{b+c} = |a|^{b+c} \exp[i(b + c)(\arg a + 2k\pi)].$$

10.10. We are to solve

$$\frac{e^{iz} + e^{-iz}}{2} = w,$$

that is,

$$\begin{aligned}
e^{2iz} - 2we^{iz} + 1 &= 0, \\
e^{iz} &= w \pm \sqrt{w^2 - 1}.
\end{aligned}$$

The two roots of the quadratic equation are reciprocals (because their product is the constant term 1), so

$$iz = \pm\log[w + (w^2 - 1)^{1/2}],$$

where we choose either value of the square root and then take all values of the logarithm.

10.11. If $w = \tan z = -i(e^{iz} - e^{-iz})/(e^{iz} + e^{-iz})$, then

$$e^{iz}w + e^{-iz}w = -ie^{iz} + ie^{-iz},$$

$$e^{2iz} = \frac{i - w}{i + w}.$$

Since the exponential function never takes the values 0 and ∞, $\tan z$ cannot take the values $\pm i$. For other values of w we have

$$\tan^{-1} w = -\frac{1}{2} i \log\frac{i - w}{i + w} = -\frac{1}{2} i \log\frac{1 + iw}{1 - iw},$$

where all values of the logarithm are admissible.

Section 11

11.1. (*a*) Take $x^2 = a^2 t$. The integral becomes

$$\frac{1}{2} a^{\lambda - 1} \int_0^\infty \frac{t^{-1+(1+\lambda)/2}}{1 + t}\, dt.$$

By the illustrative example in Sec. **11A**, this is

$$\frac{\pi a^{\lambda-1}}{2 \cos \frac{1}{2}\pi\lambda}$$

if $-1 < \lambda < 1$, so that $0 < \frac{1}{2}(1 + \lambda) < 1$.

(*b*)
$$\int_C \frac{\log z\, dz}{z^\lambda(1 + z)} = \int_0^\infty \frac{\ln x\, dx}{x^\lambda(1 + x)} - \int_0^\infty \frac{(\ln x + 2\pi i)\, dx}{x^\lambda e^{2\pi i\lambda}(1 + x)}.$$

The residue at -1 is $i\pi e^{-i\pi\lambda}$. So

$$-2\pi^2 e^{-i\pi\lambda} = (1 - e^{-2\pi i\lambda})I - 2\pi i e^{-2\pi i\lambda} \int_0^\infty \frac{dx}{x^\lambda(1 + x)}$$

$$-2\pi^2 = 2iI \sin \pi\lambda - 2\pi i e^{-\pi i\lambda} \frac{\pi}{\sin \pi\lambda},$$

where I is the original integral, by the text example with $\alpha = 1 - \lambda$.

The imaginary part of the right-hand side must be 0, and therefore we have

$$2I \sin \pi\lambda - 2\pi \cos \pi\lambda \frac{\pi}{\sin \pi\lambda} = 0,$$

$$I = \pi^2 \cot \pi\lambda \csc \pi\lambda.$$

Notice that $I > 0$ when $\lambda < \frac{1}{2}$, $I < 0$ when $\lambda > \frac{1}{2}$, and $I = 0$ when $\lambda = \frac{1}{2}$. This is not very surprising, since the substitution $y = 1/x$ gives

$$I = -\int_0^\infty \frac{\ln x\, dx}{x^{1-\lambda}(x+1)},$$

and when $\lambda = \frac{1}{2}$ we have $I = -I$.

(*c*)
$$I = \int_0^\infty \frac{\ln x\, dx}{(a+x)(b+x)}.$$

If you start from $\int_C \log z\, dz/(a+z)(b+z)$, you will find that you get no information about the original integral. Instead, consider

$$\int_C \frac{(\log z)^2\, dz}{(a+z)(b+z)} = 2\pi i(\text{sum of residues at } -a \text{ and } -b)$$

$$= \int_0^\infty \frac{(\ln x)^2\, dx}{(a+x)(b+x)} - \int_0^\infty \frac{(\ln x + 2\pi i)^2}{(a+x)(b+x)}\, dx$$

$$= -4\pi i I + 4\pi^2 \int_0^\infty \frac{dx}{(a+x)(b+x)}.$$

The residues are

$$\text{At } -a\text{: } \frac{(\ln a + i\pi)^2}{b-a}; \qquad \text{At } -b\text{: } \frac{(\ln b + i\pi)^2}{a-b}.$$

Hence

$$2\pi i\left[\frac{(\ln a)^2 + 2\pi i \ln a - \pi^2}{b-a} - \frac{(\ln b)^2 + 2\pi i \ln b - \pi^2)}{b-a}\right]$$

$$= -4\pi i I + 4\pi^2 \int_0^\infty \frac{dx}{(a+x)(b+x)}.$$

Equate the imaginary parts of the two sides:

$$\frac{2\pi}{b-a}[(\ln a)^2 - (\ln b)^2] = -4\pi I,$$

$$I = \frac{(\ln b)^2 - (\ln a)^2}{2(b-a)} = \frac{\ln(ab)\ln(b/a)}{2(b-a)}.$$

Since

$$\lim_{b\to a} \frac{\ln b - \ln a}{b-a} = \frac{1}{a},$$

we have

$$\int_0^\infty \frac{\ln x\, dx}{(a+x)^2} = \frac{\ln a}{a},$$

as we could find directly by integration by parts.

11.2.
$$\int_0^\infty \frac{dx}{1+x^4} = \frac{1}{4}\int_0^\infty \frac{t^{-3/4}\, dt}{1+t}.$$

The new integral is the text example with $a = \frac{1}{4}$, so

$$\int_0^\infty \frac{dx}{1+x^4} = \frac{\pi}{4\sin(\pi/4)} = \frac{\pi}{2\sqrt{2}}.$$

11.3. (*a*)

$$I = \int_0^\infty \frac{dx}{x^4 + 2x^2 + 2}.$$

Let $x^2 = t$, so that

$$I = \frac{1}{2}\int_0^\infty \frac{dt}{\sqrt{t}\,(t^2 + 2t + 2)}.$$

Consider

$$\frac{1}{2}\int_C \frac{dz}{z^{1/2}(z^2 + 2z + 2)},$$

where C is the limit of the keyhole contour of Fig. 11.1, and $0 \le \arg z \le 2\pi$. This integral equals

$$\begin{aligned}\frac{1}{2}\cdot 2\pi i \text{ (sum of residues)} &= \frac{1}{2}\int_0^\infty \frac{dx}{x^{1/2}(x^2+2x+2)} \\ &\qquad - \frac{1}{2}\int_0^\infty \frac{dx}{x^{1/2}e^{i\pi}(x^2+2x+2)} \\ &= \int_0^\infty \frac{dx}{x^{1/2}(x^2+2x+2)} = 2I.\end{aligned}$$

The poles are at

$$\begin{aligned} z &= -1 + i = e^{3\pi i/4}\sqrt{2}, \\ z &= -1 - i = e^{5\pi i/4}\sqrt{2}. \end{aligned}$$

Then the residues are the values of $z^{-1/2}(2z+2)$ at each pole, namely,

$$\frac{e^{-3\pi i/8}2^{-1/4}}{2i} \quad \text{and} \quad \frac{e^{-5\pi i/8}2^{-1/4}}{-2i}.$$

The sum of residues is

$$\begin{aligned}\frac{2^{-1/4}}{2i}(e^{-3\pi i/8} - e^{-5\pi i/8}) &= \frac{2^{-1/4}}{2}(e^{-3\pi i/8}e^{-i\pi/2} - e^{-5\pi i/8}e^{-i\pi/2}) \\ &= \frac{2^{-1/4}}{2}(e^{-7\pi i/8} - e^{-9\pi i/8}) \\ &= \frac{-2^{-1/4}}{2}(e^{\pi i/8} - e^{-i\pi/8}) \\ &= -2^{-1/4} i \sin\frac{\pi}{8}.\end{aligned}$$

Then $I = 2^{-5/4}\pi \sin \pi/8$.

(*b*)
$$I = \int_0^\infty \frac{x^3\,dx}{1+x^6}$$

Let $x^6 = t$; then

$$I = \frac{1}{6}\int_0^\infty \frac{t^{-1/3}}{1+t}\,dt = \frac{1}{6}\frac{\pi}{\sin \pi/3} = 3^{-3/2}\pi,$$

by the theorem in Sec. **11A**.

(*c*)
$$I = \int_0^\infty \frac{dx}{(x^3+a^3)^3}, \qquad a > 0.$$

First let $x^3 = t$; then

$$I = \frac{1}{3}\int_0^\infty \frac{t^{-2/3}\,dt}{(t+a^3)^3}.$$

Now consider

$$\frac{1}{3}\int_C \frac{z^{-2/3}\,dz}{(z+a^3)^3} = \frac{1}{3}\cdot 2\pi i \cdot \sum [\text{residues of } z^{-2/3}(z+a^3)^{-3}].$$

There is a pole of order 3 at $z = -a^3$. The residue is

$$\frac{1}{2}\left(\frac{d}{dz}\right)^2 z^{-2/3}\Big|_{z=-a^3} = \frac{1}{2}\left(-\frac{2}{3}\right)\left(-\frac{5}{3}\right) z^{-8/3}\Big|_{z=-a^3}$$
$$= \frac{5}{9}a^{-8}e^{-8i\pi/3} = -\frac{5}{9}a^{-8}e^{\pi i/3},$$

and

$$\frac{-2\pi i}{3}\frac{5}{9}a^{-8}e^{\pi i/3} = \frac{1}{3}\int_0^\infty \frac{x^{-2/3}\,dx}{(x+a^3)^3} - \frac{1}{3}\int_0^\infty \frac{x^{-2/3}e^{-4\pi i/3}\,dx}{(x+a^3)^3},$$
$$\frac{-10\pi i}{9}a^{-8} = (e^{-\pi i/3} - e^{\pi i/3})\int_0^\infty \frac{x^{-2/3}\,dx}{(x+a^3)^3}$$
$$= -2i\sin\frac{\pi}{3}\int_0^\infty \frac{x^{-2/3}\,dx}{(x+a^3)^3}.$$

Consequently,

$$\frac{5\pi}{9}a^{-8} = \frac{\sqrt{3}}{2}\int_0^\infty \frac{x^{-2/3}\,dx}{(x+a^3)^3} = \frac{\sqrt{3}}{2}\cdot 3I,$$
$$I = \frac{10\pi}{27\sqrt{3}\,a^8}.$$

(*d*)
$$\int_0^\infty \frac{\log x\,dx}{1+x^4} = \frac{1}{16}\int_0^\infty \frac{\log t\,dt}{(1+t)t^{3/4}} = \frac{1}{16}I.$$

Using the contour in Fig. 11.1, we have for the contour integral

$$\int_0^\infty \frac{\log x\,dx}{(1+x)x^{3/4}} - i\int_0^\infty \frac{\log x + 2\pi i}{(1+x)x^{3/4}}\,dx = (1-i)I + 2\pi\int_0^\infty \frac{dx}{(1+x)x^{3/4}}$$
$$= (1-i)I + 2^{3/2}\pi^2$$

by the integral in Sec. **11A**. But the contour integral equals $2\pi i$ times the residue at -1, which is

$$2\pi i \frac{\log(-1)}{(-1)^{3/4}} = -2\pi^2 e^{-3\pi i/4} = 2^{1/2}\pi^2(1+i).$$

Therefore

$$(1-i)I + 2^{3/2}\pi^2 = 2^{1/2}\pi^2(1+i),$$
$$(1-i)I = 2^{1/2}\pi^2(i-1)$$
$$I = -2^{1/2}\pi^2,$$

and the original integral equals $-2^{1/2}\pi^2/16$.

11.4. Consider

$$\int_C \frac{\log(z-a)}{1+z^2}\,dz = 2\pi i \sum \text{residues}$$
$$= \int_a^\infty \frac{\ln(x-a)\,dx}{1+x^2} - \int_a^\infty \frac{\ln(x-a)+2\pi i}{1+x^2}\,dx$$
$$= -2\pi i \int_a^\infty \frac{dx}{1+x^2}.$$

The residues are

At i: $\dfrac{\log(i-a)}{2i}$;

At $-i$: $\dfrac{\log(-i-a)}{-2i}$.

The sum of the residues is

$$\frac{1}{2i}[i \arg(-a+i) - i \arg(-a-i)] = -\int_a^\infty \frac{dx}{1+x^2}.$$

Then (see Fig. S11.4)

$$\int_a^\infty \frac{dx}{1+x^2} = \frac{1}{2}[\arg(-a-i) - \arg(-a+i)]$$
$$= \cot^{-1} a = \frac{\pi}{2} - \tan^{-1} a.$$

11.5. The original integral is

$$\lim_{R\to\infty} \int_0^R \frac{dx}{x^2-i-1} = \lim I_1.$$

The "transformed" integral is

$$\lim \int \frac{-i\,dy}{y^2+i+1} = \lim I_2.$$

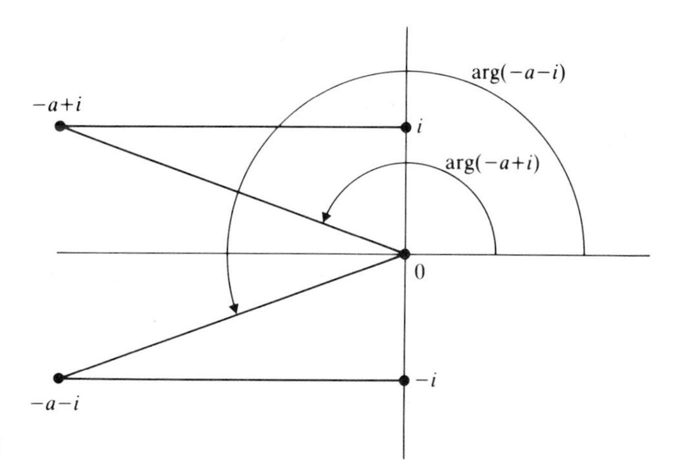

FIGURE S11.4

We can interpret I_2 as

$$-\int_{iR}^{0} \frac{dz}{z^2 - i - 1}.$$

Now consider

$$\int_C \frac{dz}{z^2 - i - 1},$$

where C is indicated in Fig. S11.5. The integral along the quarter circle tends to 0 as $R \to \infty$. However, the contour integral equals $2\pi i$ times the residue of the integrand at $(i+1)^{1/2}$, which is not 0. Consequently $\lim I_1 \neq \lim I_2$.

Section 12

12.1. If z_0 is a zero of f, then, near z_0, we have $f(z) = (z - z_0)^k \phi(z)$, where ϕ is analytic at z_0 and $\phi(z_0) \neq 0$. Then

$$f'(z) = k(z - z_0)^{k-1}\phi(z) + (z - z_0)^k \phi'(z)$$

FIGURE S11.5

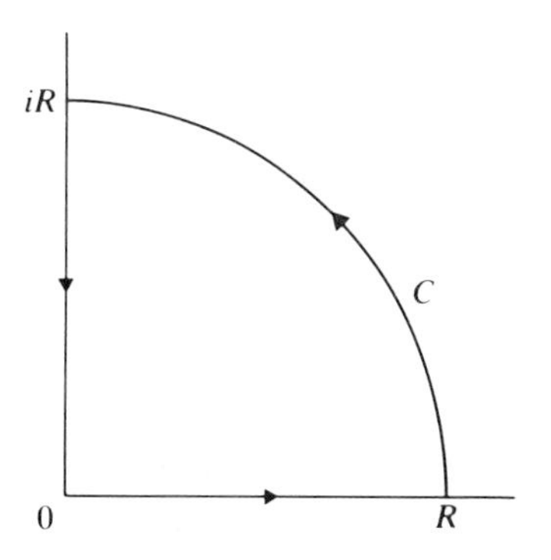

and

$$\frac{f'(z)}{f(z)} = \frac{k}{z - z_0} + \frac{\phi'(z)}{\phi(z)}.$$

Since ϕ'/ϕ is analytic at z_0, the residue of f'/f at z_0 is k, the order of the zero.

For a pole, we have a similar calculation with $k < 0$.

12.2. (*a*) There are evidently no zeros on the negative real axis. On the positive real axis, we can write

$$P(x) = x^4 - 4x^3 + 11x^2 - 14x + 10$$
$$= x^2(x^2 - 4x + 4) + 7(x^2 - 2x + 1) + 3 > 0,$$

so that $P(x) \neq 0$ on the positive real axis. If we start $\arg P(z)$ with the value 0 at the origin, it becomes $4(\pi/2) = 2\pi$ for $z = iy$, y large. On the imaginary axis,

$$P(iy) = y^4 + 4iy^3 - 11y^2 - 14iy + 10,$$

$$\Theta = \arg P(iy) = \tan^{-1}\frac{4y^3 - 14y}{y^4 - 11y^2 + 10} = \tan^{-1}\frac{2y(2y^2 - 7)}{(y^2 - 10)(y^2 - 1)}.$$

Then $\Theta = \arg P(iy)$ changes sign successively at $\sqrt{10}$, $\sqrt{\tfrac{7}{2}}$, and 1. Thus Θ starts near 2π and in the first quadrant for y near $+\infty$, goes into the second quadrant as y crosses $\sqrt{10}$, into the third quadrant at $\sqrt{\tfrac{7}{2}}$, and into the fourth quadrant at 1. Hence $\arg P(z)$ increased by 4π around a large quarter disk in the first quadrant. Therefore there are two zeros in the first quadrant and two in the fourth.

(*b*) We have $P(x) > 0$ for $x \geq 0$. If $z = -x$ $(x > 0)$, $P(z) = x^4 - 2x^3 + 3x^2 - x + 2$.

When $0 < x < 2$, $P(-x) \geq x^2(x^2 - 2x + 3) > x^2(x - 1)^2 > 0$.

When $x > 2$, $P(-x) \geq x^3(x - 2) + x(3x - 1) > 0$.

Hence P has no zeros on the real axis. The table shows how $\Theta = \arg P(iy)$ varies as y goes from $+\infty$ to 0. See Fig. S12.2.

y	$\tan\Theta$	*quadrant of* Θ	Θ
Near $+\infty$	$-$	IV	Near 2π
$\infty > y > 2^{1/2}$	$-$	IV	$2\pi > \Theta > 3\pi/2$
$2^{1/2}$	∞	IV to III	$3\pi/2$
$2^{1/2} > y > 1$	$+$	III	$3\pi/2 > \Theta > 2\pi$
1	∞	III to IV	$3\pi/2$
$1 > y > 2^{-1/2}$	$-$	IV	$3\pi/2 < \Theta < 2\pi$
$2^{-1/2}$	0	IV to I	2π
$2^{-1/2} > y > 0$	$+$	I	$2\pi < \Theta < 5\pi/2$
0	0		2π

Hence there is one zero in the first quadrant and therefore one in each of the other quadrants.

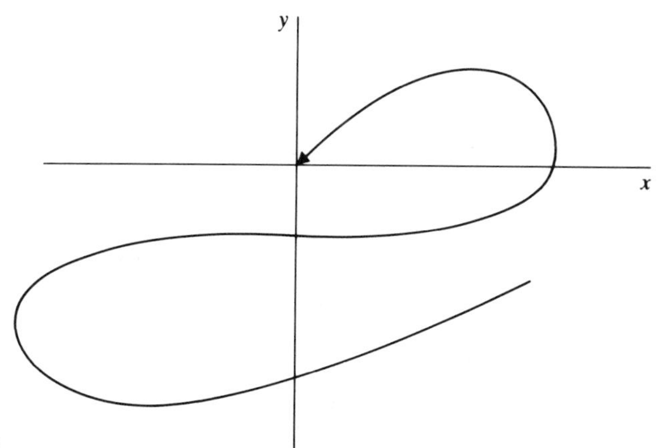

FIGURE S12.2

12.3. We apply Rouché's theorem with $2iz^2$ as f and $\sin z$ as g. For $z = x \pm i$, $|\sin z| = (\sin^2 x \cosh^2 1 + \cos^2 x \sinh^2 1)^{1/2} < \cosh 1 < 1.5431$. For $z = \frac{1}{2}\pi + iy$, $|y| \leq 1$, $|\sin z| = \cosh y \leq \cosh 1$. On the other hand, $2|z^2| \geq 2$. Consequently, $|\sin z| < |2iz^2|$ on the boundary of the rectangle, and there are two zeros inside (one of which is obviously at 0).
12.4. Repeat the first proof of Rouché's theorem, with f and g meromorphic, and observe that everything is the same except for what the changes in $\arg f$ and $\arg(f+g)$ represent. Example: $f = 1$, $g(z) = z/(4z^2 - 1)$, $C =$ unit circle.
12.5. If $P(z) = z^n + a_1 z^{n-1} + \cdots + a_0$, we have

$$P(z) = z^n + g(z),$$

where $|g(z)| < |z|^n$ on a sufficiently large circle C with center at 0 (Exercise 1.23). Rouché's theorem says that $P(z)$ has just as many zeros inside C as z^n does, namely, n.
12.6. Proof by contradiction. Suppose that $f(z_0) = 0$ for some $z_0 \in D$. Then by Sec. **7G** there is a punctured disk Δ, centered at z_0, in which $f(z) \neq 0$. Let C be the boundary of Δ and let $|f(z)| \geq m$ on C. Since $f_n(z) \to f(z)$ uniformly on C, we have $|f_n(z) - f(z)| < m/2$ on C if n is sufficiently large. For such an n take $g(z) = f_n(z) - f(z)$. Then $|f(z) + g(z)| = |f_n(z)| < |f(z)|$ on C, and Rouché's theorem says that f_n and f have the same number of zeros in Δ, namely, 1, contradicting the hypothesis that $f_n(z) \neq 0$.
12.7. For example, $f_n(z) = 1 - (z^n/n) \to 1$ uniformly for $|z| \leq 1$, but $f_n(z) = 0$ at the n nth roots of n, which have all points of the unit circle as limit points because $n^{1/n}$ (principal value) $\to 1$.
12.8. Let $\delta_n \to 0$. Choose $\varepsilon_n \to 0$ so that $|f(z)| < \delta_n$ in the disk D_n: $|z - z_n| < \varepsilon_n$. If we look at a particular n, we have, in D_n, $T_m(z) \to f(z)$ uniformly as $m \to \infty$, so by Hurwitz's theorem infinitely many T_m have zeros in D_n. Let T_{m_k} be one such partial sum, and let $T_{m_k}(w_n) = 0$, $w_n \in D_n$. Since $D_n \to 0$, we have $w_n \to 0$.

Section 13

13.1. Examples: e^z in the whole plane (or in any disk of radius greater than π); or $f(z)=z^2$ in the cut disk $0<|z|<1$, $0<\arg z<2\pi$.
13.2. Let z_1 and z_2 be points in D; then, with integration along the line segment connecting z_1 and z_2, we have

$$f(z_2)-f(z_1)=\int_{z_1}^{z_2} f'(w)\,dw=\int_{z_1}^{z_2}\{[f'(w)-f'(z_0)]+f'(z_0)\}\,dw$$

$$=f'(z_0)(z_2-z_1)+\int_{z_1}^{z_2}[f'(w)-f'(z_0)]\,dw,$$

$$|f(z_2)-f(z_1)|\geq|f'(z_0)|\,|z_2-z_1|-\int_{z_1}^{z_2}|f'(w)-f'(z_0)|\,|dw|$$

$$\geq|f'(z_0)|\,|z_2-z_1|-|z_2-z_1|\max|f'(w)-f'(z_0)|,$$

where the maximum is taken for w on the line segment (z_1, z_2). But $|f'(w)-f'(z_0)|<|f'(z_0)|$ by hypothesis, so

$$|f(z_2)-f(z_1)|>0,$$

that is, $f(z_2)\neq f(z_1)$.
13.3. The only univalent polynomials are of the form $az+b$, since polynomials of higher degree take values more than once. A transcendental entire function has an essential singularity at ∞, and univalence contradicts the Casorati–Weierstrass theorem (Sec. **8E**).

Several other proofs are given by R. K. Williams, "On the Linearity of One-to-One Entire Functions," *Delta* 6 (1976): 72–77; most of these proofs depend on material beyond the scope of this book.
13.4. Let $f(z)=\lim f_n(z)$, where f_n are analytic and univalent and converge uniformly in D. Let z_0 be any point of D and consider the sequence $\{g_n(z)\}=\{f_n(z)-f(z_0)\}$, which converges to $f(z)-f(z_0)$. By Hurwitz's theorem (Sec. **12C**; Exercise 12.6) either $f(z)-f(z_0)\equiv 0$ or $f(z)-f(z_0)$ is zero-free. Since this is true for every z_0 in D, either f takes no value twice or f is constant.
13.5. Let $f(z)=z^2$ in the half disk $|z|<1$, $0\leq\arg z\leq\pi$. We can have $z_1^2=z_2^2$ only if $z_1=\pm z_2$. If z_1 is in the interior of the half disk, $-z_1$ is outside, so f is univalent inside the half disk. On the other hand, if z is on the bounding diameter, then $f(z)=f(-z)$.

Section 14

14.1. On C_2,

$$\frac{1}{w-z}=\frac{w^{-1}}{1-(z/w)}=w^{-1}\sum_{n=0}^{\infty}\left(\frac{z}{w}\right)^n,$$

with uniform convergence with respect to z; also, $f(w)$ is bounded for w on C_2. Hence we can integrate term by term to get

$$f_2(z)=\frac{1}{2\pi i}\sum_{n=0}^{\infty} z^n\int_{C_2}\frac{f(w)\,dw}{w^{n+1}}=\sum_{n=0}^{\infty} c_n z^n.$$

Similarly, on C_1,

$$\frac{1}{w-z}=\frac{z^{-1}}{(w/z)-1}=-z^{-1}\sum_{n=0}^{\infty}\left(\frac{w}{z}\right)^n=-\sum_{n=1}^{\infty}w^{n-1}z^{-n},$$

so

$$f_1(z)=\frac{1}{2\pi i}\sum_{n=1}^{\infty}z^{-n}\int_{C_1}w^{n-1}f(w)\,dw,$$

with integration in the positive direction.

14.2. If f has a pole of order n at z_0, then

$$f(z)=(z-z_0)^{-n}\phi(z),$$

where ϕ is analytic and not 0 at z_0. Hence

$$\begin{aligned}c_{-n}&=\frac{1}{2\pi i}\int_C f(w)(w-z_0)^{n-1}\,dw\\&=\frac{1}{2\pi i}\int_C (w-z_0)^{-n}(w-z_0)^{n-1}\phi(w)\,dw\\&=\frac{1}{2\pi i}\int_C\frac{\phi(w)\,dw}{w-z_0}=\phi(z_0)\neq 0;\end{aligned}$$

whereas when $m>n$,

$$\begin{aligned}c_{-m}&=\frac{1}{2\pi i}\int_C (w-z_0)^{-n}(w-z_0)^{m-1}\phi(w)\,dw\\&=\frac{1}{2\pi i}\int_C (w-z_0)^{n-m-1}\phi(w)\,dw=0\end{aligned}$$

because the integrand is analytic at z_0.

14.3. The residue of f at ∞ is $-(2\pi i)^{-1}\int_C f(z)\,dz$, where C surrounds all the finite singular points of f and C is described counterclockwise. Since $\int_C z^n\,dz=0$ for $n\neq -1$, and $\int_C z^{-1}\,dz=2\pi i$, the result follows.

14.4. If f had two Laurent series with a common annulus of convergence, the difference of these series would have sum 0. The integral formula for the coefficients would then show that the difference has all its coefficients zero.

14.5. $\frac{1}{2z}-\frac{1}{z}\left(1+\frac{1}{z}+\frac{1}{z^2}+\frac{1}{z^3}+\dots\right)+\frac{1}{2z}\left(1+\frac{2}{z}+\frac{4}{z^2}+\frac{8}{z^3}+\dots\right)=\frac{1}{z^3}+\frac{3}{z^4}+\dots.$

14.6. The residue of

$$\frac{1}{(w-1)(w-2)w^{n+2}}$$

at $w=2$ is 2^{-n-2}. Hence the third Laurent series is the second Laurent series plus

$$\sum_{n=-\infty}^{\infty}2^{-n-2}z^n,$$

that is,

$$-\frac{1}{2z}-\sum_{n=2}^{\infty}z^{-n}-\sum_{n=0}^{\infty}2^{-n-2}z^n+\sum_{-\infty}^{\infty}2^{-n-2}z^n=-\sum_{n=2}^{\infty}z^{-n}+\sum_{n=2}^{\infty}2^{n-2}z^{-n}.$$

14.7. (*a*)

$$\frac{z+1}{z^3(z-2)} = \frac{1}{z^3}\left(\frac{z-2+3}{z-2}\right) = \frac{1}{z^3}\left(1+\frac{3}{z-2}\right)$$
$$= \frac{1}{z^3}\left\{1 - \frac{3}{2}\sum_{n=0}^{\infty}\left(\frac{z}{2}\right)^n\right\}$$
$$= \frac{-1}{2z^3} - \frac{3}{2}\sum_{n=1}^{\infty} 2^{-n}z^{n-3},$$

when $0<|z|<2$.

When $2<|z|$, we add to the coefficient of z^n the residue of

$$\frac{w+1}{(w-2)w^{n+4}}$$

at $w=2$, namely,

$$\frac{3}{2^{n+4}}.$$

(*b*)

$$\frac{z^3}{(2z+1)(3z-2)} = \frac{-z^3}{2(1+2z)(1-\frac{3}{2}z)}$$
$$= -\frac{z^3}{2}\left(\frac{\frac{2}{3}}{1+2z} + \frac{\frac{1}{2}}{1-\frac{3}{2}z}\right)\cdot\frac{6}{7}$$
$$= -\frac{3z^3}{7}\left[\frac{2}{3}\sum_{n=0}^{\infty}(-1)^n 2^n z^n + \frac{1}{2}\sum_{n=0}^{\infty}\left(\frac{3}{2}\right)^n z^n\right]$$
$$= \frac{1}{7}\sum_{n=3}^{\infty}\left[(-2)^{n-2} - \left(\frac{3}{2}\right)^{n-2}\right]z^n$$

for $0<|z|<\frac{1}{2}$.

For $\frac{1}{2}<|z|<\frac{2}{3}$ we must add to the coefficient of z^n the residue of

$$\frac{w^{-n+2}}{6(w+\frac{1}{2})(w-\frac{2}{3})}$$

at $w=-\frac{1}{2}$, namely,

$$-\frac{1}{7}(-2)^{n-2}.$$

For $|z|>\frac{2}{3}$ we must further add to the coefficient of z^n the residue at $w=\frac{2}{3}$, namely,

$$\frac{1}{7}(3/2)^{n-2}.$$

14.8. $(\log i)/(2i) = \pi/4$ and $\log(-i)/(-2i) = -3\pi/4$. These do not have to be complex conjugates, because $\log z$ is not real on the negative real axis.

14.9. The series converges only for $|z|>1$. It is therefore not the Laurent series for the function in a neighborhood of 0. The residue is 1.

14.10. To find the residue, write

$$f(z) = z\left(1-\frac{a}{z}\right)^{1/2}\left(1-\frac{b}{z}\right)^{1/2}$$

and expand the square roots by the binomial theorem to get the Laurent series of

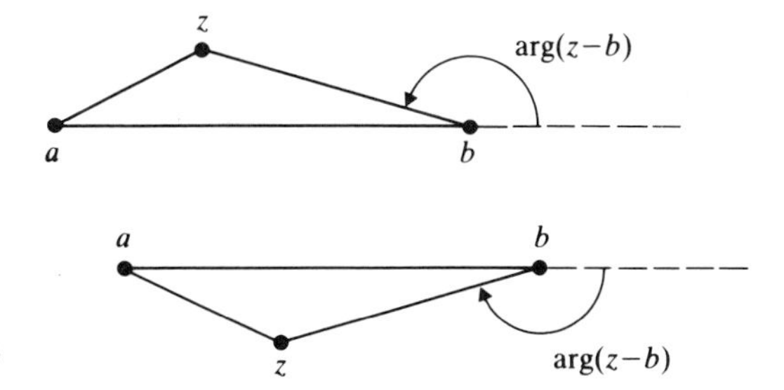

FIGURE S14.9

f about ∞:

$$f(z) = z\left\{1 - \frac{a}{2z} - \frac{a^2}{8z^2} + O(z^{-3})\right\}\left\{1 - \frac{b}{2z} - \frac{b^2}{8z^2} + O(z^{-3})\right\}.$$

The coefficient of $1/z$ is $\frac{1}{4}ab - \frac{1}{8}(a^2 + b^2) = -\frac{1}{8}(a-b)^2$; the residue is the negative of this.

Now let C be a contour that surrounds the interval $[a, b]$, described *clockwise*; we have

$$\int_C f(z)\,dz = 2\pi i \text{ times the residue at } \infty$$

$$= \tfrac{1}{4}\pi i(a-b)^2.$$

If we shrink C down to the interval itself, we have the upper side of the interval described from left to right, plus the lower side from right to left. As z approaches a point x of the interval from above, it is clear from Fig. S14.9 that we should take $\arg(x-a) = 0$, $\arg(x-b) = \pi$, so that $(x-b)^{1/2} = i\sqrt{(b-x)}$. For x on the lower side, $\arg(x-b) = -\pi$ and $(x-b)^{1/2} = -i\sqrt{(b-x)}$. Hence we get

$$\lim \int_C f(z)\,dz = 2i\int_a^b \sqrt{(x-a)(b-x)}\,dx.$$

Since $\int_C f(z)\,dz = (4\pi i)^{-1}(a-b)^2$, we finally have

$$\int_a^b \sqrt{(x-a)(b-x)}\,dx = \frac{1}{8}\pi(a-b)^2.$$

Section 15

15.1 For the final result, see Sec. **15C**.

15.2.

$$\frac{e^{z^2}}{z^{-3}\sin(z^3)} = \frac{1 + z^2 + \frac{1}{2}z^4 + \frac{1}{6}z^6 + \cdots}{1 - \frac{1}{6}z^6 + \cdots}.$$

$$\begin{array}{r|l} & 1 + z^2 + \tfrac{1}{2}z^4 + \cdots \\ \hline 1 - \dfrac{z^6}{6} + \cdots & 1 + z^2 + \tfrac{1}{2}z^4 + \tfrac{1}{6}z^6 + \cdots \\ & 1 \qquad\qquad\qquad - \tfrac{1}{6}z^6 \\ \hline & z^2 + \tfrac{1}{2}z^4 + \tfrac{1}{3}z^6 + \cdots \end{array}$$

15.3. (*a*) Let $u = \sin z$. Then

$$
\begin{aligned}
e^u &= 1 + u + \frac{u^2}{2} + \frac{u^3}{6} + \frac{u^4}{4!} + \frac{u^5}{5!} + \cdots \\
u &= z \qquad - \frac{z^3}{6} \qquad + \frac{z^5}{5!} \qquad + O(z^7) \\
u^2 &= \qquad z^2 \qquad - \frac{2z^4}{6} \qquad + O(z^6) \\
u^3 &= \qquad z^3 \qquad - \frac{3z^5}{6} \qquad + O(z^7) \\
u^4 &= \qquad z^4 \qquad + O(z^6) \\
u^5 &= \qquad z^5 \qquad + O(z^7)
\end{aligned}
$$

Hence

$$
\begin{aligned}
e^u = 1 + z \qquad &- \frac{z^3}{6} \qquad + \frac{z^5}{5!} + \cdots \\
+ \frac{z^2}{2} \qquad &- \frac{z^4}{6} + \cdots \\
&+ \frac{z^3}{6} \qquad - \frac{z^5}{12} + \cdots \\
&+ \frac{z^4}{4!} + \cdots \\
&+ \frac{z^5}{5!} + \cdots \\
&+ \cdots \\
= 1 + z + \frac{z^2}{2} &- \frac{z^4}{8} - \frac{z^5}{15} + O(z^6).
\end{aligned}
$$

(*b*) Let $u = \cos z - 1$. Then

$$
\begin{aligned}
u &= -\frac{z^2}{2} + \frac{z^4}{4!} - \frac{z^6}{6!} + O(z^8) \\
&= z^2\left(-\frac{1}{2} + \frac{z^2}{4!} - \frac{z^4}{6!} + \cdots\right) + O(z^8) \\
u^2 &= z^4\left[\frac{1}{4} - \frac{z^2}{4!} + O(z^4)\right] \\
u^3 &= -\frac{z^6}{8} + O(z^8) \\
e^u &= 1 - \frac{z^2}{2} + \frac{z^4}{4!} - \frac{z^6}{6!} + O(z^8) \\
&\quad + \frac{z^4}{8} - \frac{z^6}{48} + O(z^8)
\end{aligned}
$$

$$-\frac{z^6}{48}+O(z^8)$$

$$=1-\frac{z^2}{2}+\frac{z^4}{6}-\frac{31}{720}z^6+O(z^8)$$

$$e^{\cos z}=e\cdot e^{\cos z-1}=e\left[1-\frac{z^2}{2}+\frac{z^4}{6}-\frac{31}{720}z^6+O(z^8)\right].$$

(*c*) Let $u=ze^z$.

$$\log(1+u)=u-\frac{1}{2}u^2+\frac{1}{3}u^3-\frac{1}{4}u^4+\frac{1}{5}u^5-\cdots$$

$$u=ze^z=z+z^2+\frac{1}{2}z^3+\frac{1}{6}z^4+\frac{1}{24}z^5+O(z^6)$$

$$-\frac{1}{2}u^2=-\frac{1}{2}z^2e^{2z}=-\frac{1}{2}\left(z^2+2z^3+2z^4+\frac{4}{3}z^5\right)+O(z^6)$$

$$\frac{1}{3}u^3=\frac{1}{3}z^3e^{3z}=\frac{1}{3}\left(z^3+3z^4+\frac{9}{2}z^5\right)+O(z^6)$$

$$-\frac{1}{4}u^4=-\frac{1}{4}z^4e^{4z}=-\frac{1}{4}(z^4+4z^5)+O(z^6)$$

$$\frac{1}{5}u^5=z^5e^{5z}=\frac{1}{5}z^5+O(z^6)$$

$$\log(1+u)=z+z^2+\frac{1}{2}z^3+\frac{1}{6}z^4+\frac{1}{24}z^5$$

$$-\frac{1}{2}z^2-z^3-z^4-\frac{2}{3}z^5$$

$$+\frac{1}{3}z^3+z^4+\frac{3}{2}z^5$$

$$-\frac{1}{4}z^4-z^5$$

$$+\frac{1}{5}z^5+O(z^6)$$

$$=z+\frac{1}{2}z^2-\frac{1}{6}z^3-\frac{1}{12}z^4+\frac{3}{40}z^5+O(z^6).$$

Section 16

16.1.

$$[f(z)]^m=\frac{1}{2\pi i}\int_C\frac{[f(w)]^m\,dw}{w-z}.$$

Let L be the length of C and $\delta = \min |w - z|$ for w on C. Then

$$|f(z)|^m \le \frac{1}{2\pi}\frac{L}{\delta}\max_{w\in C}|f(w)|^m,$$

$$|f(z)| \le \left(\frac{L}{2\pi\delta}\right)^{1/m}\max_{w\in C}|f(w)| \to \max_{w\in C}|f(w)|$$

as $m \to \infty$.

16.2. $e^{f(z)}$ is analytic wherever f is; we have $|e^{f(z)}| = e^{\operatorname{Re} f(z)}$, and a maximum for $\operatorname{Re} f(z)$ would correspond to a maximum for $|e^{f(z)}|$. Similarly for $\operatorname{Im} f(z)$, using $|e^{-if(z)}| = e^{\operatorname{Im} f(z)}$.

16.3. Apply the maximum principle to $1/f(z)$.

16.4. Apply Exercise 16.3 to $e^{f(z)}$ or $e^{-if(z)}$.

16.5. Consider $g(z) = f(z)/(z - z_0)$ in Δ. Then $|g(z)| \le M$ on $\partial\Delta$ and g has a removable singularity at z_0, which we may suppose has been removed. The maximum principle for g says that $|g(z)| \le M$ in Δ, whence $|f(z)| \le M\,|z - z_0|$.

16.6. $|f'(z_0)| \le M$ because $f'(z) = \lim_{z\to z_0} f(z)/(z - z_0)$.

16.7. If $f'(z_0) = 0$, then in a neighborhood U: $|z - z_0| < \varepsilon$ of z_0, with ε so small that f is analytic in U, we can represent f by its Taylor series with center z_0. If $f \not\equiv 0$, we have

$$f(z) - f(z_0) = (z - z_0)^k\left[a_k + \sum_{n=k+1}^{\infty} a_n(z - z_0)^{n-k}\right], \qquad a_k \ne 0, \qquad k \ge 2.$$

When ε is small, the part of U inside $|z| = 1$ is nearly half a disk, and $\arg z$ varies by nearly π on the part S of ∂U that is inside $|z| = 1$. Consequently, $\arg[f(z) - f(z_0)]$ varies by nearly $k\pi$ on S ($k \ge 2$), and this makes $w = f(z)$ go outside the disk $|w| \le |f(z_0)|$. Consequently, $|f|$ takes values greater than $|f(z_0)|$ at some points z with $|z| < 1$, contradicting the hypothesis that $|f|$ has its maximum at z_0 for $|z| \le 1$.

Second proof (suggested by R. B. Burckel). We may assume that $z_0 = 1 = |f(z_0)|$. Since f is not a constant, the maximum principle shows that $|f(z)| < 1$ in $|z| < 1$. First suppose that $f(0) = 0$. Then, by Schwarz's lemma, $|f(z)| \le |z|$ ($|z| < 1$), so for $0 < x < 1$ we have

$$\frac{|f(1) - f(x)|}{1 - x} \ge \frac{|f(1)| - |f(x)|}{1 - x} \ge \frac{1 - x}{1 - x} = 1.$$

Letting $x \to 1$, we have $|f'(1)| \ge 1$.

If, however, $f(0) = a \ne 0$, apply the preceding argument to

$$g(w) = \frac{f(w) - a}{1 - \bar{a}f(w)},$$

which has $g(0) = 0$, $|g(1)| = 1$ because $|f(1)| = 1$, and

$$g'(1) = \frac{1 - |a|^2}{[a - \bar{a}f(1)]^2} f'(1) \ne 0$$

by the previous argument applied to g. Hence $f'(1) \ne 0$ in any case.

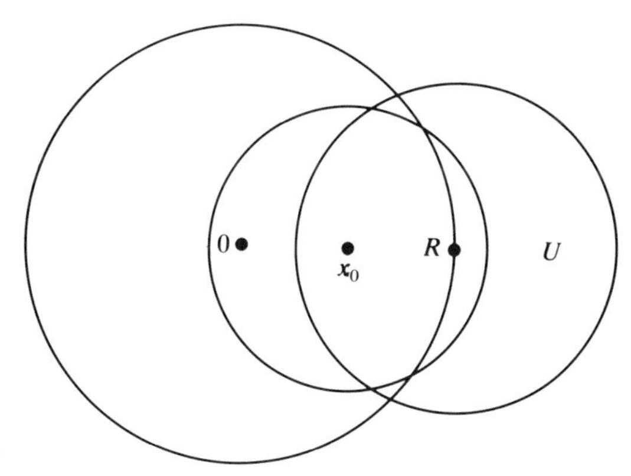

FIGURE S17.3

16.8. Let $\operatorname{Re} f \le M$. Then $|e^{f(z)}| = e^{\operatorname{Re} f(z)} \le M$. By Liouville's theorem, e^f is a constant and therefore f is a constant.
16.9. Let $R = 2r$. Then Carathéodory's inequality gives us $|f(re^{i\theta})| = O[(2r)^\lambda]$. The conclusion follows from Exercise 7.8.

Section 17

17.1. The highest power of z in $P_k(z)$ has exponent $pn_k + n_k$; the lowest power in $P_{k+1}(z)$ has exponent pn_{k+1}. The difference of these exponents is $pn_{k+1} - (p+1)n_k > 0$, because $n_{k+1}/n_k > \lambda > (1/p) - 1 = (p-1)/p$.
17.2. If $p = 1$ and $\lambda > 2$, we have $p > 1/(\lambda - 1)$, and we want to show that $\sum a_k P_k(z) = \sum a_k [z(1-z)]^{n_k}$ converges when $|1 - z| < 1$. We know that the series converges for $|z| < 1$; since the series is unchanged if we interchange z and $1 - z$, it also converges for $|1 - z| < 1$.
17.3. If f is analytic at $z = R$, then f is analytic in some neighborhood U of the point R. If we expand f about any point x_0 with $0 < x_0 < R$, then a disk $|z - x_0| < R - x_0 + \delta$, with a sufficiently small positive δ, will be in the region of analyticity of f. (See Fig. S17.3.)
17.4. If f is analytic in $|z| \le R$, it is analytic in an open disk about each point of the circumference C: $|z| = R$. By the Heine–Borel theorem, a finite number of the disks cover the circumference; their union is an open set containing the trace of C and therefore contains a circle, centered at 0, of radius greater than R.
17.5. Since $a_n \ge 0$ and $\sum a_n$ diverges, if we are given a (large) number K, we can make $\sum_{n=0}^M a_n > K$, if M is large enough. Now, if $0 < x < 1$,

$$f(x) = \sum_{n=0}^{\infty} a_n x^n \ge \sum_{n=0}^{M} a_n x^n \ge x^M \sum_{n=0}^{M} a_n > Kx^M.$$

By taking x sufficiently close to 1, we have $f(x) > K/2$. But K is arbitrary, so $f(x) \to \infty$.
17.6. The series $\sum (\operatorname{Re} a_n) z^n$ cannot be continued past $z = R$. If $f(z) = \sum a_n z^n$ could, then the Taylor expansion of f about $z = R/2$ would converge at some real points $x = R + \varepsilon$, with $\varepsilon > 0$. Then its real and imaginary parts would have to

converge separately, and the real part is just the Taylor expansion about $x = R/2$ of $\sum (\text{Re } a_n)x^n$, because for real x we have $\text{Re} f(x) = \sum x^n \text{ Re } a_n$.
17.7. When n is sufficiently large, $n!p/q$ is an even integer and $e^{in!\pi p/q} = 1$. Hence the power series of $f(re^{in\pi p/q})$ has positive coefficients when n is sufficiently large, and consequently $f(ze^{in\pi p/q})$ cannot be continued past 1 on the radius $z = x$. This is the same as saying that f cannot be continued along the ray $\arg z = \pi p/q$. Since these rays intersect the unit circumference in a dense set, f cannot be analytic at any point of the circumference.
17.8. Every partial sum of the power series in Hadamard's theorem is equal to a partial sum whose index is some n_k, so the convergence of a subsequence of partial sums is equivalent to the convergence of the entire sequence of partial sums.

Section 18

18.1. Imitate the argument at the beginning of the proof of Abel's theorem to get, with $x = 1$, $B_0 = 0$,

$$B_n = b_1 + \cdots + b_n,$$

$$\sum_{n=M}^{N} a_n b_n = \sum_{n=M}^{N} a_n(B_n - B_{n-1}) = a_M(B_M - B_{M-1}) + \cdots + a_N(B_N - B_{N-1})$$

$$= -a_M B_{M-1} + B_M(a_M - a_{M+1}) + \cdots + B_{N-1}(a_{N-1} - a_N) + a_N B_N.$$

Since we assumed B_k bounded, say $|B_k| \le P$, and $a_k - a_{k-1} \ge 0$, we have

$$\left| \sum_{n=M}^{N} a_n b_n \right| \le Pa_M + P[(a_M - a_{M+1}) + \cdots + (a_{N-1} - a_N)] + Pa_N$$

$$\le Pa_M + P(a_M - a_N) + Pa_N.$$

Since $a_M \to 0$, it follows that $\sum_{n=M}^{N} a_n b_n \to 0$ as M and $N \to \infty$; that is, Cauchy's criterion for convergence is satisfied.
18.2. Consider the functions f, g, h defined by the series

$$f(x) = \sum_{n=0}^{\infty} a_n x^n,$$

$$g(x) = \sum_{n=0}^{\infty} b_n x^n,$$

$$h(x) = \sum_{n=0}^{\infty} c_n x^n,$$

which converge for $|x| < 1$ because the series of coefficients were assumed to converge. Since the series A and B converge, Abel's theorem says that $\lim_{x\to1-} f(x) = \sum_{n=0}^{\infty} a_n$ and $\lim_{x\to1-} g(x) = \sum b_n$. But $h(x) = f(x)\, g(x)$ for $|x| < 1$, and since $\sum c_n$ converges, $\lim_{x\to1-} h(x) = \sum_{n=0}^{\infty} c_n$. The theorem follows.
18.3. By Exercise 4.17, $\sum \cos nx$ and $\sum \sin nx$ have bounded partial sums for each

x in $0 < x \leq \pi$ and $0 \leq x \leq \pi$, respectively (since the terms of the sine series are 0 at $x = 0$ or π). Now apply Exercise 18.1.

18.4. Imitate the proof of Abel's theorem, taking ε to be an upper bound for $|s_k|$ (hence not necessarily small). We have

$$f(x) = (1 - x) \sum_{k=1}^{M} s_k x^k + (1 - x) \sum_{k=M+1}^{\infty} s_k x^k = S_1 + S_2.$$

Then the argument used for Abel's theorem shows that $|S_2| \leq \varepsilon$ for $0 < x < 1$. For a fixed M, the sum S_1 still approaches 0 as $x \to 1-$, so $|S_1| < \varepsilon$ if $x \geq 1 - \delta$, with some (small) positive δ. It remains to show that S_1 is bounded on $0 < x < 1 - \delta$. But

$$|S_1| \leq \sum_{k=1}^{M} \varepsilon(1-\delta)^k \leq \varepsilon \sum_{k=0}^{\infty} (1-\delta)^k = \frac{\varepsilon}{\delta}.$$

Hence S_1 is indeed bounded for $0 < x < 1$.

18.5. We may suppose that $R = 1$ and $s_n \to +\infty$. Take an arbitrary (large) number Q and write

$$f(x) = (1 - x) \sum_{k=1}^{M} s_k x^k + (1 - x) \sum_{k=M+1}^{\infty} s_k x^k = S_1 + S_2,$$

where M is chosen so that $s_k > Q$ for $k > M$. Now fix M; then S_1 is bounded, and

$$S_2 \geq (1 - x)Q \sum_{k=M+1}^{\infty} x^k = Qx^{M+1},$$

so that $S_2 \geq Q/2$ if x is close enough to 1. Since Q is arbitrary, $\lim_{x\to 1-} f(x) = \infty$.

18.6.

$$\frac{d}{dx} \sum_{n=0}^{\infty} (-1)^n x^n = \sum_{n=1}^{\infty} n x^{n-1}(-1)^n.$$

Hence we have

$$\sum_{n=1}^{\infty} n x^n (-1)^n = x \frac{d}{dx} \frac{1}{1+x} = -\frac{x}{(1+x)^2} \to -\frac{1}{4}, \qquad x \to 1-.$$

18.7. Let $\sum a_n x^n$ have radius of convergence 1. If the series were Abel summable at z with $|z| > 1$, we would have to have $\sum a_n z^n x^n$ convergent for this z and $0 < x < 1$. But then $\sum a_n w^n$ would converge for some w with $|w| > 1$, which would imply that the radius of convergence of $\sum a_n z^n$ is greater than 1.

18.8. Here the partial sums are $s_0 = 1$, $s_1 = 0$, $s_2 = 1$, $s_3 = 0$, and so on, that is, $s_{2k} = 1$, $s_{2k+1} = 0$. Hence all the terms of the sequence $\{\frac{1}{2}(s_n + s_{n+1})\}$ are equal to $\frac{1}{2}$ and the Y sum of the series is $\frac{1}{2}$.

18.9. It is no harder to prove the *general form of Kronecker's lemma*, namely, that if $s_n \to L$ and w_k are positive numbers such that $W_n = \sum_{k=1}^{n} w_k \to \infty$, then the weighted averages

$$\frac{1}{W_n} \sum_{k=1}^{n} w_k s_k \to L.$$

If $s_n \to L$, then $L+\varepsilon > s_n > L-\varepsilon$ for $n>N$, and $|s_n| <$ some number A for all n. Write

$$\frac{w_1s_1+w_2s_2+\cdots+w_ns_n}{w_1+\cdots+w_n}=\frac{w_1s_1+\cdots+w_Ns_N}{W_n}+\frac{w_{N+1}s_{N+1}+\cdots+w_ns_n}{W_n}.$$

With N fixed, the second fraction exceeds $(L-\varepsilon)(W_n-W_N)/W_n$, which is arbitrarily close to $L-\varepsilon$ if n is large enough; and the absolute value of the first fraction does not exceed NA/W_n, which is arbitrarily small if n is large enough. On the other hand, the second fraction does not exceed $(L+\varepsilon)(W_n-W_N)/W_n$ for large n.

18.10. The partial sums of the first series are $s_{2k+1}=1$, $s_{2k}=0$. Then

$$\frac{s_1+s_2+\cdots+s_{2n}}{2n}=\frac{n}{2n}\to\frac{1}{2}$$

and

$$\frac{s_1+s_2+\cdots+s_{2n+1}}{2^{n+1}}=\frac{n+1}{2n+1}\to\frac{1}{2}.$$

If $(s_1+\cdots+s_n)/n\to L$, then $(s_1+\cdots+s_{n+1})/(n+1)\to L$ and therefore $(s_1+\cdots+s_{n+1})/n\to L$ also. Consequently,

$$\frac{s_1+\cdots+s_{n+1}}{n}-\frac{s_1+\cdots+s_n}{n}\to 0,$$

which is to say that $s_{n+1}\to 0$ if $\sum a_n$ is summable $(C, 1)$. For the series $\sum(-1)^n n$, s_{n+1}/n does not approach 0.

18.11. The partial sums are $s_{3k+1}=1$, $s_{3k-1}=s_{3k}=0$, and so $(s_1+\cdots+s_n)/n\to\frac{1}{3}$. The series is $(C, 1)$ summable, but to $\frac{1}{3}$ instead of $\frac{1}{2}$.

The same series is not summable by the Y method at all, since the running averages equal $\frac{1}{2}$ two-thirds of the time and 0 the rest.

18.12. Repeat the proof of Tauber's theorem with a large number ε instead of a small number.

18.13. $\sum_{n=0}^{\infty} a_nx^n > Q$ if x is close enough to 1. Let $|na_n| < M$, Q large.

$$\sum_{n=0}^{N} a_nx^n + \sum_{N+1}^{\infty} a_nx^n > Q,$$

where $1-x$ is nearly $1/N$.

$$\left|\sum_{N+1}^{\infty} a_nx^n\right| = \left|\sum_{N+1}^{\infty} na_n\frac{x^n}{n}\right| < \frac{M}{N+1}\sum_{N+1}^{\infty} x^n < M.$$

But

$$\sum_{n=0}^{N} a_nx^n > Q - \sum_{N+1}^{\infty} a_nx^n > Q - M$$

independently of x. Letting $x\to 1$, we have $\sum_{n=0}^{N} a_n > Q - M$, so $\sum_{n=0}^{N} a_n \to \infty$.

Section 19

19.1.

$$\left.\frac{\partial v}{\partial x}\right|_{(x,y)} = \lim_{h\to 0}\frac{1}{h}\int_{(x,y)}^{(x+h,y)} [-u_2(s,y)\,ds] = -u_2(x,y),$$

because we can adjust the path of integration to have a horizontal segment from (x, y) to $(x+h, y)$. Similarly for $\partial v/\partial y$.

19.2. (*a*) $3x^2y - y^3$

(*b*) $e^{-y}\sin x$

(*c*) $2\tan^{-1}\left(\frac{y}{x}\right)$

(*d*) $\dfrac{x-1}{(1-x)^2+y^2}$

19.3. If $f(z)=u(x,y)+iv(x,y)$ is analytic at $z_0=x_0+iy_0$, $g(\zeta)=f(\zeta+z_0)$ is analytic at $\zeta=0$.

$$\begin{aligned} f(z) = f(\zeta+z_0) &= f(\xi+x_0+i(\eta+y_0)) \\ &= u(\xi+x_0,\ \eta+y_0)+iv(\xi+x_0,\ \eta+y_0) \\ &= u(\zeta+x_0,\ y_0)+iv(\zeta+x_0,\ y_0) \\ &= u(z-iy_0,\ y_0)+iv(z-iy_0,\ y_0). \end{aligned}$$

Illustration:

$$u=\frac{x}{x^2+y^2},\qquad v=\frac{-y}{x^2+y^2}.$$

$$\frac{z-iy_0}{(z-iy_0)^2+y_0^2}-\frac{iy_0}{(z-iy_0)^2+y_0^2}=\frac{z-2iy_0}{z^2-2izy_0}=\frac{1}{z}.$$

19.4. (*a*) By Rule *B*, $2\left[\left(\frac{z}{2}\right)^3-3\frac{z}{2}\left(\frac{z}{2i}\right)^2\right]=2\left(\frac{z^3}{8}+\frac{3z^3}{8}\right)=z^3.$

(*b*)

$$\begin{aligned} f(z) &= 2\cos\frac{z}{2}e^{iz/2}-1 \\ &= 2\cos\frac{z}{2}\left(\cos\frac{z}{2}+i\sin\frac{z}{2}\right)-1 \\ &= 2\left(\cos^2\frac{z}{2}+\frac{i}{2}\sin z\right)-1=\cos z+i\sin z=e^{+iz}. \end{aligned}$$

(*c*) $2\log z$.

(*d*) $\dfrac{-i}{1-z}$.

(*e*) Best to use polar coordinates: $r^{1/2}\cos\theta/2=\operatorname{Re} z^{1/2}$.

19.5. Let f be analytic at $z_0 = x_0 + iy_0$. Set $z = \zeta + z_0 = \xi + i\eta + z_0$. Then

$$\begin{aligned} f(z) &= f(\zeta + z_0) = f[\xi + x_0 + i(\eta + y_0)] \qquad \text{(analytic at } \zeta = 0) \\ &= u(\xi + x_0, \eta + y_0) + iv(\xi + x_0, \eta + y_0) \\ &= 2u\left(\frac{\zeta}{2} + x_0, \frac{\zeta}{2i} + y_0\right) - u(x_0, y_0) \\ &= 2u\left(\frac{z - z_0}{2} + x_0, \frac{z - z_0}{2i} + y_0\right) - u(x_0, y_0) \\ &= 2u\left(\frac{z + \overline{z_0}}{2}, \frac{z - \overline{z_0}}{2i}\right) - u(x_0, y_0). \end{aligned}$$

Illustration:

$$u = \frac{x}{x^2 + y^2}.$$

$$2\frac{(z + \overline{z_0})/2}{[(z + \overline{z_0})/2]^2 + [(z - \overline{z_0})/(2i)]^2} = 4\frac{z + \overline{z_0}}{4z_0\overline{z_0}} = \frac{z + \overline{z_0}}{z\overline{z_0}} = \frac{1}{\overline{z_0}} + \frac{1}{z},$$

$$\begin{aligned} f(z) &= \frac{1}{z} + \frac{1}{\overline{z_0}} - \frac{x_0}{z_0\overline{z_0}} \\ &= \frac{1}{z} + \frac{1}{|z_0|^2}(z_0 - x_0) \\ &= \frac{1}{z} + \frac{1}{|z_0|^2}(iy_0) \\ &= \frac{1}{z} + \text{pure imaginary}. \end{aligned}$$

19.6. Rule B gives

$$f(z) = 8\left[\left(\frac{z}{2}\right)^2\left(\frac{z}{2i}\right)\right] = -iz^3,$$

but $\operatorname{Re}(-iz^3) = 3x^2y - y^3 \neq 4x^2y$.

Section 20

20.1

$$e^{1/z} = \exp\left[\frac{x}{(x^2 + y^2)}\right]\exp\left[\frac{-iy}{(x^2 + y^2)}\right].$$

Then

$$\operatorname{Im} e^{1/z} = -\exp\left[\frac{x}{(x^2 + y^2)}\right]\sin\left[\frac{y}{(x^2 + y^2)}\right]$$

is clearly 0 when $y = 0$ and harmonic for $y > 0$.

When $x = 0$ and $y > 0$, $\operatorname{Im} e^{1/z} = -\sin(1/y)$, which is equal to ± 1 when $y = 2/[(2k + 1)\pi]$.

20.2. If $U(z)$ is harmonic for $|z|<R$, then $U(Rz)=u(z)$ is harmonic for $|z|<1$;

$$u(z)=\frac{1}{2\pi}\int_{-\pi}^{\pi} u(e^{i\phi})\frac{(1-|z|^2)}{1+|z|^2-2\,|z|\cos(\theta-\phi)}\,d\phi.$$

$$U(re^{i\theta})=\frac{1}{2\pi}\int_{-\pi}^{\pi} U(Re^{i\phi})\frac{1-|z/R|^2}{1-|z^2/R^2|-2\,|z/R|\cos(\theta-\phi)}\,d\phi$$

$$=\frac{1}{2\pi}\int_{-\pi}^{\pi} U(Re^{i\phi})\frac{R^2-|z|^2}{R^2-|z|^2-2R\,|z|\cos(\theta-\phi)}\,d\phi.$$

Section 21

21.1.

$$c_n=\frac{1}{2\pi}\int_{-\pi}^{0} e^{-in\theta}\,d\theta,\qquad c_0=\frac{1}{2}.$$

For $n\neq 0$,

$$c_n=-\frac{1}{2\pi n i}[1-(-1)^n].$$

$$\psi(\theta)=\frac{1}{2}-\frac{1}{\pi i}\sum_{\text{odd }n}\frac{1}{n}e^{in\theta}$$

$$=\frac{1}{2}-\frac{1}{\pi i}\sum_{\text{odd }n\geq 1}\frac{1}{n}(e^{in\theta}-e^{-in\theta})$$

$$=\frac{1}{2}-\frac{2}{\pi}\sum_{\text{odd }n}\frac{\sin n\theta}{n}.$$

$$u(r,\theta)=\frac{1}{2}-\frac{2}{\pi}\sum_{k=0}^{\infty}\frac{\sin(2k+1)\theta}{(2k+1)}r^{2k+1}.$$

$$u\left(\frac{1}{2},\frac{\pi}{2}\right)=\frac{1}{2}-\frac{2}{\pi}\sum_{k=0}^{\infty}\frac{(-1)^k}{2^{2k+1}(2k+1)}$$

$$=\frac{1}{2}-\frac{2}{\pi}\tan^{-1}\left(\frac{1}{2}\right)$$

$$=0.2048\cdots.$$

In fact, the series for $u(r,\theta)$ can be summed in closed form. We have

$$\frac{1}{2}-\frac{2}{\pi}\sum_{\text{odd }k}\frac{\sin k\theta}{k}r^k=\frac{1}{2}-\frac{2}{\pi}\operatorname{Im}\sum_{\text{odd }k}\frac{z^k}{k},\qquad z=re^{i\theta}.$$

But

$$2\sum_{\text{odd }k}\frac{z^k}{k}=\log\frac{1+z}{1-z}\qquad(|z|<1)$$

and

$$\operatorname{Im}\log\frac{1+z}{1-z}=\tan^{-1}\frac{2r\sin\theta}{1-r^2}.$$

This can also be obtained more directly; compare Secs. **24B** and **26C**.

For the beer bottle problem, we want (assuming the radius of the bottle to be 1 unit) a solution of Laplace's equation for the unit disk with boundary values 100 on the upper semicircle and 32 on the lower semicircle. We can get this from the harmonic function $u(r, \theta)$ that has the values 0 (up) and 1 (down) by taking a linear combination of 1 and $u(r, \theta)$, because constants are solutions of Laplace's equation and a linear combination of solutions is a solution. The result is given in the text (Sec. **20A**).

21.2.

$$\sum_{k=-n}^{n} c_k e^{ikx} = \frac{1}{2\pi} \sum_{k=-n}^{n} e^{ikx} \int_{-\pi}^{\pi} e^{-ikt} f(t)\, dt$$
$$= \frac{1}{2\pi} \int_{-\pi}^{\pi} f(t) \sum_{k=-n}^{n} e^{ik(x-t)}\, dt.$$

By the formula for the sum of a geometric series,

$$\sum_{k=0}^{n} e^{iku} = \frac{e^{i(n+1)u} - 1}{e^{iu} - 1},$$
$$\sum_{k=-n}^{n} e^{iku} = \frac{e^{i(n+1)u} - 1}{e^{iu} - 1} + \frac{e^{-i(n+1)u} - 1}{e^{-iu} - 1} - 1$$
$$= \frac{\cos nu - \cos(n+1)u}{1 - \cos u} = \frac{\sin(n+\frac{1}{2})u}{\sin \frac{1}{2}u}.$$
$$\sum_{k=-n}^{n} c_k e^{ikx} = \frac{1}{2\pi} \int_{-\pi}^{\pi} \frac{\sin(n+\frac{1}{2})(x-t)}{\sin \frac{1}{2}(x-t)} f(t)\, dt.$$

21.3.

$$\frac{\partial v}{\partial \theta} = r \frac{\partial u}{\partial r}.$$
$$u = \sum_{-\infty}^{\infty} c_n r^{|n|} e^{in\theta}.$$
$$\frac{\partial v}{\partial \theta} = \sum_{-\infty}^{\infty} c_n\, |n|\, r^{|n|} e^{in\theta}.$$
$$v = \sum_{\substack{-\infty \\ n \neq 0}}^{\infty} c_n \frac{|n|}{in} r^{|n|} e^{in\theta}$$
$$= -i \sum_{-\infty}^{\infty} c_n (\operatorname{sgn} n) r^{|n|} e^{in\theta}.$$

21.4. Try for a solution

$$c \ln r + \sum_{k=-\infty}^{\infty} (\lambda_k r^{|k|} + \mu_k r^{-|k|}) e^{ik\theta}$$

that reduces to $\omega(\theta)$ for $r = \sigma$, and to $\phi(\theta)$ for $r = \rho$. Then

$$c \ln \sigma + \lambda_0 + \mu_0 = \frac{1}{2\pi} \int_0^{2\pi} \omega(\theta)\, d\theta = C_0,$$
$$c \ln \rho + \lambda_0 + \mu_0 = \frac{1}{2\pi} \int_0^{2\pi} \phi(\theta)\, d\theta = c_0;$$

and for $k = \pm 1, \pm 2, \ldots,$

$$\lambda_k \rho^{|k|} + \mu_k \rho^{-|k|} = \frac{1}{2\pi} \int_0^{2\pi} e^{-ik\theta} \phi(\theta)\, d\theta = c_k,$$

$$\lambda_k \sigma^{|k|} + \mu_k \sigma^{-|k|} = \frac{1}{2\pi} \int_0^{2\pi} e^{-ik\theta} \omega(\theta)\, d\theta = C_k.$$

Consequently, $c \ln(\sigma/\rho) = C_0 - c_0$, determining c; and λ_k and μ_k are determined by the equations above.

Section 22

22.1. For $\phi[u(x, y), v(x, y)]$, we compute

$$\frac{\partial \phi}{\partial x} = \phi_1 u_1 + \phi_2 v_1$$

$$\frac{\partial \phi}{\partial y} = \phi_1 u_2 + \phi_2 v_2$$

$$\frac{\partial^2 \phi}{\partial x^2} = \phi_{11} u_1^2 + \phi_1 u_{11} + \phi_{22} v_1^2 + \phi_2 v_{11} + 2\phi_{12} u_1 v_1$$

$$\frac{\partial^2 \phi}{\partial y^2} = \phi_{11} u_2^2 + \phi_2 u_{22} + \phi_{22} v_2^2 + \phi_2 v_{22} + 2\phi_{12} u_2 v_2$$

$$\frac{\partial^2 \phi}{\partial x^2} + \frac{\partial^2 \phi}{\partial y^2} = \phi_{11}(u_1^2 + u_2^2) + \phi_{22}(v_1^2 + v_2^2)$$

(by Laplace's equation for u and v, and the Cauchy–Riemann equations)

$$= 0$$

(by the Cauchy–Riemann equations, and Laplace's equation for ϕ).

Section 23

23.1. We have

$$\mathbf{v} = \nabla \phi = \mathbf{i} \frac{\partial \phi}{\partial x} + \mathbf{j} \frac{\partial \phi}{\partial y}.$$

But

$$\frac{\partial \phi}{\partial y} = -\frac{\partial \psi}{\partial x},$$

so

$$\mathbf{v} = \mathbf{i} \frac{\partial \phi}{\partial x} - \mathbf{j} \frac{\partial \psi}{\partial x} = \mathbf{i} \operatorname{Re} f' - \mathbf{j} \operatorname{Im} f'.$$

23.2. If C is a single streamline, ψ is constant on C. But a harmonic function that is constant on a closed curve is constant inside, by the maximum and minimum principles, and so is the analytic function of which it is the imaginary part (Exercise 4.9).
23.3. Let $w = f(z)$ map a region D, bounded by a curve C, to a region Δ, bounded by a curve Γ, in the w plane, so that $w = u + iv$, and let $\phi(u, v)$ be harmonic in Δ. We want to show that if $\partial\phi/\partial\mathbf{n} = 0$ on an arc Γ_1 of Γ, then $(\partial/\partial\mathbf{n})\phi[u(x, y), v(x, y)] = 0$ on the corresponding arc C_1 of C. The condition $\partial\phi/\partial\mathbf{n} = 0$ means that no heat flows *across* Γ_1, so that Γ_1 is an isothermal. Identify the problem with a fluid flow problem with Γ_1 as a streamline and ϕ as velocity potential. The conformal map carries ϕ to a harmonic function (Sec. **22C**), and its harmonic conjugate ψ to a streamline, hence carries the orthogonal trajectory ϕ = constant to an equipotential, on which $\partial\phi/\partial\mathbf{n} = 0$.

Notice that we did *not* say that the value of $\partial\phi/\partial\mathbf{n}$ is preserved under a conformal map, only that the condition $\partial\phi/\partial\mathbf{n} = 0$ is preserved. This is an important point, because asking for a harmonic function with $\partial\phi/\partial\mathbf{n}$ given on the boundary of a region is a **Neumann problem**, which we cannot solve just like a Dirichlet problem.

Section 24

24.1. The complex potential is $az + b\log z$; to keep the formulas simple, take $a = b = 1$. We need only consider the upper half z plane, since the flow in the lower half plane is the reflection of the flow in the upper half plane with respect to the real axis, by the Schwarz reflection principle. Since

$$w = u + iv = x + \log|z| + i(y + \arg z),$$

the images of lines $v = c$ are curves $y + \tan^{-1}(y/x) = c$.

First consider the image of the half plane $v > \pi$. When $c > \pi$, $v = y + \tan^{-1}(y/x) = c$, $u = x + \frac{1}{2}\log(x^2 + y^2)$, we have $du/dx > 0$, $dy/dx > 0$, and so the flow is from left to right and upward. When $u \to -\infty$, $x \to -\infty$, and when $u \to +\infty$, $x \to +\infty$. When $c = \pi$, however, we can have $y = 0$ and $\arg z = \pi$, or $y > 0$ and $y + \tan^{-1}(y/x) = \pi$. Consider first the case $y = 0$. Then $u = x + \log|x|$, which starts at $x = -\infty$ with $u = -\infty$; as x increases, $du/dx = 1 + (1/x)$, which is positive up to $x = -1$, where the derivative of the complex potential has a simple zero. Consequently, the streamline follows the x axis to $x = -1$, but then turns 90° to the left and follows the heavy line in Fig. S24.1 since du/dx and dy/dx are both still positive. Thus the half plane $v \geq \pi$ in the w plane maps to the region above the heavy line. To discuss the strip $0 \leq v \leq \pi$, let us start at $v = 0$; then we have to have $\arg z = 0$, $y = 0$, and $u = x + \log x$ (with $x > 0$). Since $du/dx > 0$ and $u \to +\infty$ as $x \to +\infty$, the line $v = 0$ corresponds to the positive x axis. When $0 < x < \pi/2$, we have $\arg z = \pi/2 - y$ and so $\arg z < \pi/2$; the streamlines remain in the right-hand half plane. For $\pi/2 < c < \pi$, the streamlines begin by going into the left-hand half of the z plane but cannot cross the streamline corresponding to $v = \pi$. Finally, when $c = \pi$, we again have either $y = 0$, $\arg z = \pi$, or $y > 0$ and $y + \tan^{-1}(y/x) = \pi$. The first case now gives us the segment $-1 < x < 0$ traced from right to left, and the second case gives us the upper part of the heavy curve again.

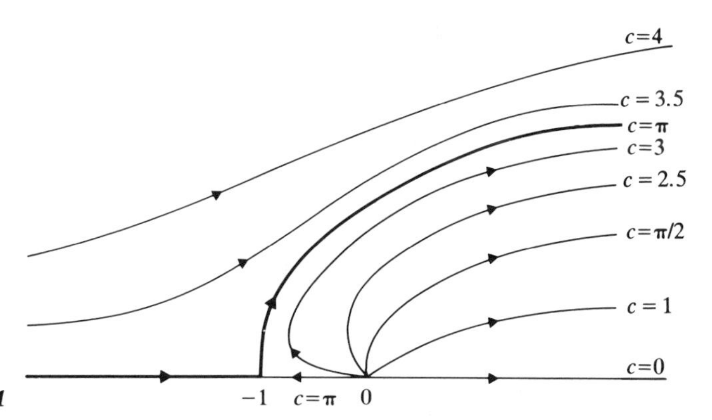

FIGURE S24.1

We can think of the picture, in a vertical plane, for $v > \pi$, as representing the flow of an (incompressible) wind blowing from left to right against a hill, whose profile is the heavy line. Notice that the point $z = -1$ is a stagnation point, where the wind velocity is zero; that would be a good place to stay if the wind were very strong. If we reflect the part of the figure with $v > \pi$ across the real axis and interpret it in a horizontal plane, it represents the flow in a wide river encountering the end of a long island.

24.2. Let ψ_1 be the function considered in Exercise 21.1. Then

$$\begin{aligned}\psi &= 1 - 2\psi_1\\ &= \frac{4}{\pi}\sum_{k=0}^{\infty}\frac{\sin(2k+1)\theta}{2k+1}r^{2k+1} = \operatorname{Im}\frac{4}{\pi}\sum_{k=0}^{\infty}\frac{e^{(2k+1)i\theta}}{2k+1}r^{2k+1}\\ &= \operatorname{Im}\frac{4}{\pi}\sum_{k=0}^{\infty}\frac{z^{2k+1}}{2k+1} = \operatorname{Im}\frac{4}{\pi}\cdot\frac{1}{2}\log\frac{1+z}{1-z}\\ &= \frac{2}{\pi}\operatorname{Im}\log\frac{1+z}{1-z}.\end{aligned}$$

Hence ψ is $(2/\pi)$ times the stream function for the complex potential

$$\log\frac{1+z}{1-z} = \log(1+z) - \log(1-z),$$

which is the potential of a source at $z = -1$ together with a sink at $z = +1$.

24.3. To keep the formulas simple, take $z_0 = 0$. The complex potential is $1/z = \bar{z}/|z|^2$, so the streamlines are the curves $-y/(x^2+y^2) = c$. When $c = 0$, this equation becomes $y = 0$; otherwise, it represents a family of circles with centers at $(0, -1/(2c))$ and passing through the origin. When $c > 0$, the circles are in the lower half plane; when $c < 0$, in the upper half plane. Thus the uniform flow from left to right in the upper half of the w plane maps to a flow in the lower half z plane. The complex velocity is

$$\frac{-1}{z^2} = \frac{y^2 - x^2 + 2ixy}{(x^2+y^2)^2},$$

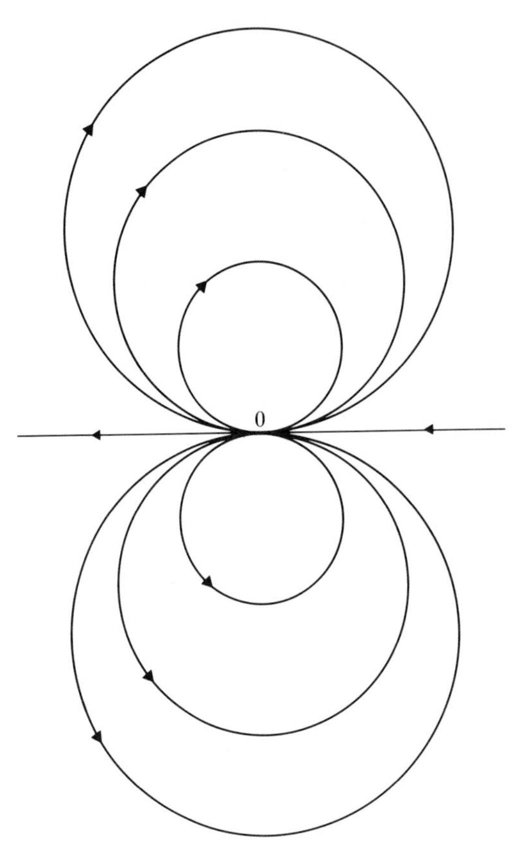

FIGURE S24.3

and its imaginary part is

$$\frac{2xy}{(x^2+y^2)^2};$$

the velocity vector has components

$$\left(\frac{y^2-x^2}{(x^2+y^2)^2}, \frac{-2xy}{(x^2+y^2)^2}\right).$$

For $y<0$ the flow is upward when $x>0$ and to the left or right according as $|y|<x$ or $|y|>x$, and the opposite in the upper half plane. When $c=0$, $y=0$, and the velocity vector is along the real axis in the negative direction, becoming infinite at 0. See Fig. S24.3.

24.4. For the complex potential e^z, we have

$$e^z = e^x(\cos y + i \sin y).$$

Streamlines $v=c$ correspond to streamlines $e^x \sin y = c$. If $0<y<\pi/2$ and $-\infty<x<\infty$, then both u and v take all positive values, so that the strip

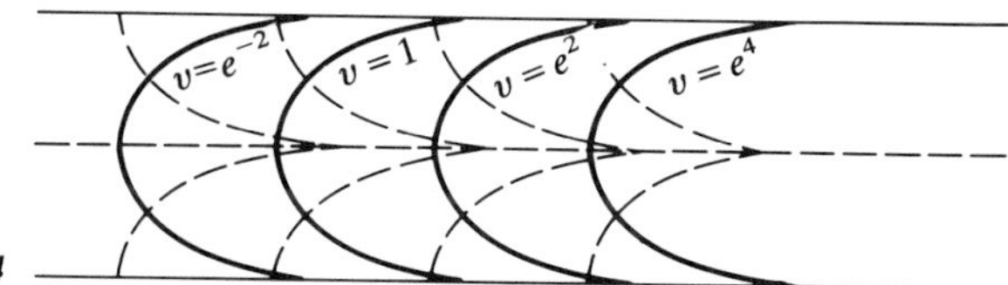

FIGURE S24.4

$0<y<\pi/2$ in the z plane maps to the first quadrant in the w plane. Similarly, the strip $\pi/2<y<\pi$ maps to the second quadrant. Hence the strip $0<y<\pi$ maps to the upper half of the z plane. Similarly, $\pi<y<2\pi$ maps to the lower half plane. Some curves $v=c$ are sketched in Fig. S24.4. The curves are all congruent because their equations can be written as $x+\log\sin y=\log c$, so that a shift in $\log c$ corresponds to a shift in x.

Since

$$\frac{dx}{dy}+\frac{\cos y}{\sin y}=0$$

on a streamline, the leftmost point is at $y=\pi/2$, that is, in the middle of the strip; $y\to 0$ or π as $x\to+\infty$, and $dy/dx\to 0$ as $y\to 0$ or π, so that the streamlines have the lines $y=0$ and $y=\pi$ as asymptotes. Because of the periodicity of e^z, congruent strips in the z plane each correspond to upper or lower halves of the w plane. The equipotentials $e^x\cos y=c$ are simply the streamlines translated by $\pi/2$ in the y direction, and of course are orthogonal to the streamlines (some equipotentials are shown as dotted lines in the figure).

24.5. The stream function for $w=z+z^{-1}$ is $v=\operatorname{Im} w=y-y/(x^2+y^2)$. The streamline $v=0$ appears at first sight to consist of both the real axis ($y=0$) and the circle $x^2+y^2=1$. This cannot be a single streamline, so we must study the function $w=z+z^{-1}$ more carefully. When $y=0$, $w=x+x^{-1}$. Let z start at $-\infty$ on the x axis and move to the right. Then $dw/dz=1-x^{-2}$ is positive as long as $x<-1$. At $x=-1$, $dw/dz=0$; and angles are doubled going from the z plane to the w plane, or halved going the other way. Hence the image of $\operatorname{Im} w=0$ turns 90° to the left at $z=-1$ and follows the upper semicircle. At $z=+1$ it turns left again and continues along the real axis. Thus the upper half w plane corresponds to the part of the z plane above the real axis and above the circle. (See Fig. S24.5a, p. 322.)

Now consider what happens inside the circle. Let z start at -1 and move to the right along the real axis; then w starts at -2, $dw/dx<0$; so $w\to-\infty$ as $x\to 0$ from the left. Starting over at $x=0$ and going to the right, w starts from $+\infty$ and decreases to 2 at $z=1$. Then w goes to -2 but z follows the circle back to -1. The streamlines then look like Fig. S24.5b. We see that the lower half w plane is mapped to the inside of the semicircular area.

If we reflect the whole pattern into the lower half plane, we get a picture like Fig. S24.5c in the z plane. It is clear that the circle $|z|=1$ can be inserted as a boundary, and the flow outside it is then a uniform flow disturbed by a circular barrier.

The complex velocity is $1-1/z^2$, which has absolute value less than 2 when $|z|>1$ and 2 when $z=i$, so that the speed is greatest at the top of the circle.

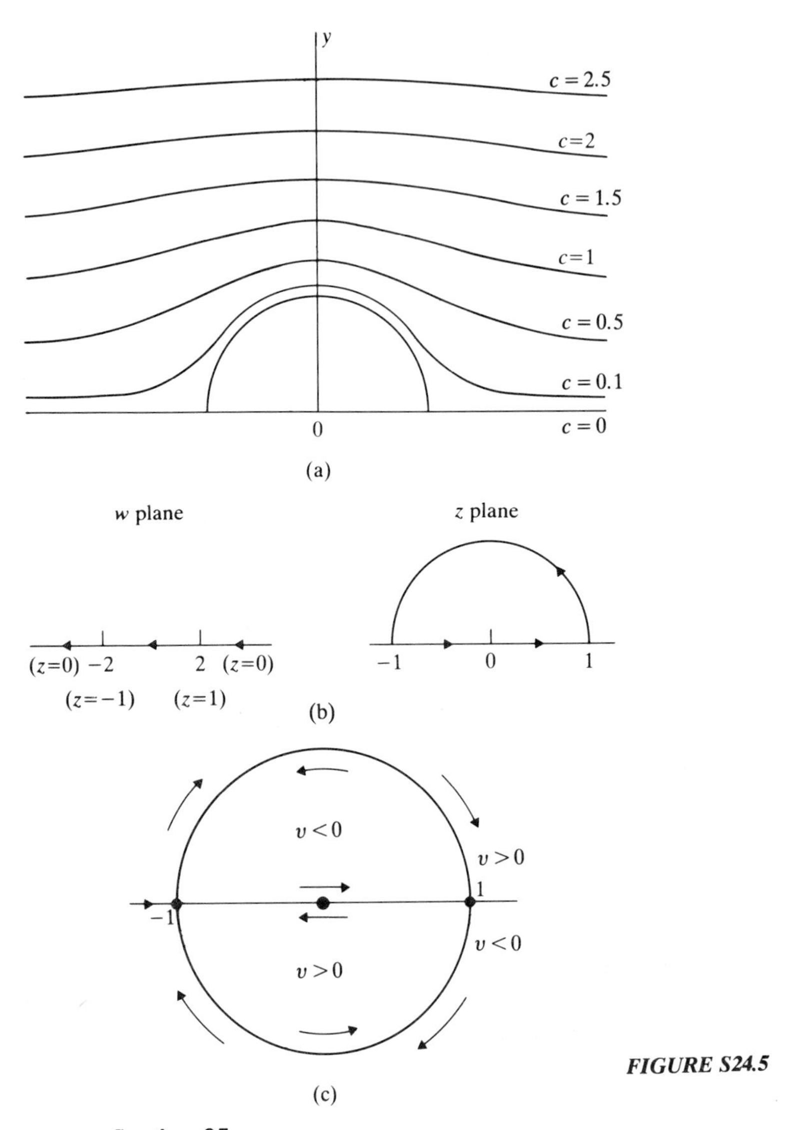

FIGURE S24.5

Section 25

25.1. Let the transformation be $w = (az + b)/(cz + d)$. The inverse image of 0 is real, so b/a is real; the image of ∞ is real, so d/c is real. (I assume that none of the coefficients is zero; if any of them are zero, the problem becomes simpler.) Now we can write the transformation as

$$w = \frac{(a/b)z + 1}{(c/d)z + 1} = \frac{\alpha z + 1}{\beta z + 1},$$

where α and β are real.

25.2. If $b = d = 0$, w is constant and the transformation degenerates. If $d \neq 0$, then since 0 corresponds to 0, $b/d = 0$ and therefore $b = 0$, and $w = az/(cz + d)$. Here $a \neq 0$, since otherwise the transformation degenerates. If ∞ corresponds to ∞, then $c = 0$ and $w = az/d$. If 1 corresponds to 1, the transformation becomes $w = z$.

25.3. The transformation must carry $z = 0$ to $w = e^{i\lambda}$ (λ real), and (because it carries symmetric points to symmetric points) it carries an arbitrary point μ in the upper half plane, and its complex conjugate $\bar{\mu}$, to 0 and ∞. If $w = (az + b) \div (cz + d)$, from $z = 0$ we get $b/d = e^{i\lambda}$. From μ and $\bar{\mu}$ we get $a\mu + b = 0$ and $c\bar{\mu} + d = 0$. Since $d = be^{-i\lambda}$, we have $a\mu + b = 0$, $c\bar{\mu} + be^{-i\lambda} = 0$. Hence $a = -b/\mu$, $c = -be^{-i\lambda}/\bar{\mu}$, $d = be^{-i\lambda}$. The transformation becomes

$$\frac{(-b/\mu)z + b}{(-be^{-i\lambda}/\bar{\mu})z + be^{-i\lambda}} = \frac{\bar{\mu}}{\mu} e^{i\lambda} \frac{-z + \mu}{-z + \bar{\mu}} = e^{i\theta} \frac{\mu - z}{\bar{\mu} - z},$$

with $\theta = \lambda - 2 \arg \mu$.

25.4. Let $w = (az + b)/(cz + d)$. Since $w = 0$ and ∞ correspond to symmetric points $z = \alpha$ ($|\alpha| < 1$) and $1/\bar{\alpha}$, we get $b = -a\alpha$, $c = -\bar{\alpha}d$,

$$w = -\frac{a}{d} \frac{z - \alpha}{\bar{\alpha}z - 1}.$$

But when $|z| = 1$, $|(z - \alpha)/(\bar{\alpha}z - 1)| = 1$ (Exercise 1.22), and so $-a/d = e^{i\lambda}$.

25.5. We know from Exercise 25.1 that a Möbius transformation that carries the real axis to the real axis can be written with real coefficients. Conversely, a transformation with real coefficients carries the real axis to the real axis. Hence our transformation must be $w = (az + b)/(cz + d)$ with real coefficients. To make the upper half plane correspond to the upper half plane, make $z = i$ correspond to a point w with $\operatorname{Im} w > 0$; then

$$\frac{ai + b}{ci + d} = \frac{ac + bd - bci + adi}{c^2 + d^2}$$

must have positive imaginary part, so $ad - bc > 0$.

To transform the right-hand half plane to the right-hand half plane, first carry the right-hand half plane to the upper-half ζ plane by $\zeta = iz$; then

$$w = \frac{a\zeta + b}{c\zeta + d}$$

carries the upper half plane to the upper half plane, and multiplication by $-i$ carries this back to the right-hand half plane. Consequently,

$$w = -i\frac{aiz + b}{ciz + d} = \frac{az - bi}{ciz + d}, \qquad ad - bc > 0,$$

must be the general Möbius transformation from the right-hand half plane to the right-hand half plane. We can write it more compactly as

$$w = i\frac{az - bi}{cz - di}.$$

25.6. By Exercise 25.4, we see that

$$g(z)=\frac{f(z)-f(0)}{\overline{f(0)}f(z)-1}$$

has $g(0)=0$, $|g(z)|\le 1$, so $|g'(0)|\le 1$ by Schwarz's lemma. Now, by the maximum principle, $|f(0)|<1$ unless f is constant, so we may compute

$$g'(0)=\frac{[\overline{f(0)}f(0)-1]f'(0)}{[\overline{f(0)}f(0)-1]^2}=\frac{f'(0)}{|f(0)|^2-1}.$$

Then

$$1\ge |g'(0)|=\frac{|f'(0)|}{1-|f(0)|^2},\qquad |f'(0)|\le 1-|f(0)|^2\le 1.$$

25.7. Let $F(0)=\alpha$. Then $w=F(z)$ maps the z disk $|z|<1$ into the left-hand w plane, and 0 to α. We can map the half plane to the unit disk in a t plane by

$$t=g(w)=\frac{\alpha-w}{\bar{\alpha}+w}.$$

Then $g[F(z)]$ maps the unit disk into the unit disk and 0 to 0. By Schwarz's lemma,

$$\left|\frac{d}{dz}g[F(z)]\right|\le 1,$$

that is,

$$\left|g'[F(0)]F'(0)|_{z=0}\right|\le 1.$$

But

$$g'(w)=\frac{-(\bar{\alpha}+w)-(\alpha-w)}{(\bar{\alpha}+w)^2}=\frac{-(\bar{\alpha}+\alpha)}{(\bar{\alpha}+w)^2},$$

$$g'(\alpha)=-\frac{1}{\bar{\alpha}+\alpha},\qquad \alpha=F(0),$$

so

$$\left|\frac{F'(0)}{\bar{\alpha}+\alpha}\right|\le 1,$$

$$|F'(0)|\le|\bar{\alpha}+\alpha|=|2\,\mathrm{Re}\,F(0)|=-2\,\mathrm{Re}\,F(0).$$

25.8. Since $f(z)\ne 0$ in $|z|\le 1$, there is an analytic branch of $F(z)=\log f(z)$ in this disk. Then $\mathrm{Re}\,F(z)<0$ because $|f(z)|<1$, and the preceding exercise shows that

$$|F'(0)|\le -2\log|f(0)|,$$

$$\left|\frac{f'(0)}{f(0)}\right|\le -2\log|f(0)|,$$

$$|f'(0)|\le -2\,|f(0)|\log|f(0)|.$$

The maximum of $-u\log u$ for $0<u<1$ is $1/e$ at $u=1/e$, so $|f'(0)|\le 2/e$.

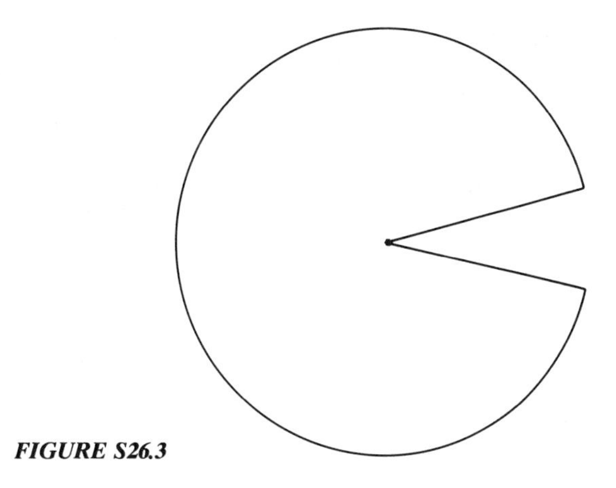

FIGURE S26.3

Section 26

26.1. For $w = \log z = u + iv = re^{i\theta}$, we have $v = 0$ on the positive real z axis, $v = \pi$ on the negative real axis, and $0 < \operatorname{Im} w < \pi$ for z in the upper half plane. The correspondence is clearly one-to-one on the boundary.

26.2. After what we have just done in the text, the shortest method is to invert the mapping from the half plane to the half disk, so that we have a mapping from the half disk to the half plane, and then map the half plane to the disk. In fact, we had the map

$$w = \left(\frac{\zeta + 1}{1 - \zeta}\right)^2$$

from the ζ half disk to the w half plane; and in Exercise 25.3 we had (with a change of notation) the map

$$z = \frac{w - i}{w + i}$$

from the half plane to the unit disk. The required transformation is the composition of these.

26.3. We may think of the region as the limit of a disk with a small sector removed. The transformation $\zeta = \sqrt{z}$ will carry this to a half disk, which was mapped to a disk in the preceding exercise. See Fig. S26.3.

26.4. The Möbius transformation $\zeta = (z + 1)/(z - 1)$ sends -1 to 0 and 1 to ∞; when $-1 < z < 1$, we have ζ real and negative. Hence the complement of the interval $[-1, 1]$ is mapped to the plane cut along the negative real axis. This can be thought of as an angle with opening 2π. Then $w = \sqrt{\zeta}$ with $\operatorname{Re} w > 0$ takes this to an angle of opening π, the right-hand half plane; and a Möbius transformation will take this to the disk $|w| < 1$ (see Exercises 25.3 and 25.5).

Section 27

27.1. If a region D has no boundary points, it is the extended plane (or the finite plane if we do not consider ∞ as a boundary point). A function mapping D to the unit disk would be bounded and analytic in the whole finite plane, hence a constant by Liouville's theorem.

A region D with only one boundary point in the extended plane is either the finite plane, or the extended plane with a finite point p deleted. In the latter case, suppose that $w = \phi(z)$ maps D to the unit disk Δ; the map $\zeta = 1/(z-p) = \psi(z)$ sends D to the finite ζ plane; the inverse map $z = \psi^{-1}(\zeta)$ sends the finite ζ plane to D; $w = \phi[\psi^{-1}(\zeta)]$ sends the finite ζ plane to D and thence to Δ. We have already seen that this last map is impossible.

27.2. The transformation $w = (z-z_0)/(z-\bar{z}_0)$, $\operatorname{Im} z_0 > 0$, maps the upper half plane to the unit disk and z_0 to 0. The Poisson kernel is then

$$\frac{1}{2\pi}\frac{\partial}{\partial y}\operatorname{Re}\left(\log\frac{z-z_0}{z-\bar{z}_0}\right)_{y=0} = \frac{1}{2\pi}\operatorname{Re}\frac{\partial}{\partial y}[\log(z-z_0)-\log(z-\overline{z_0})]_{y=0}$$

$$= \frac{1}{2\pi}\operatorname{Re}\left(\frac{i}{x-z_0}-\frac{i}{x-\overline{z_0}}\right)$$

$$= \frac{1}{\pi}\frac{y_0}{(x-x_0)^2+y_0^2}.$$

See also Note 4 at the end of Sec. **27**.

Section 28

28.1. Provided that D does not contain ∞, we can form an increasing sequence of regions D_n' that exhausts D. We can, for example, cover the plane with a network of squares of side 1 and take D_1' to consist of all the squares in this network that are completely in D and are connected (in D) to some given square; then refine the network so that it consists of squares of side $\frac{1}{2}$ and let D_2' consist of D_1' together with the new squares that are in D and connected (in D) to the given square; and so on. For each D_n' there is a sequence $\{f_m^{(n)}\}$ that converges uniformly on $D_1' \cup D_2' \cup \cdots \cup D_n'$. The diagonal sequence $\{f_n^{(n)}\}$ converges uniformly on each D_n', and hence uniformly on each compact subset of D.

Now suppose that D contains ∞. If D is the extended plane, all the f_n are constants and there is nothing to prove. If D contains ∞ but is not the extended plane, D has a boundary point b, and we can consider the transformation $w = 1/(z-b)$. The image of D in the w plane is a region that does not contain ∞. A function f becomes $g(w) = f[(1+bw)/w]$, and these functions form a normal family. Hence $\{f\}$ is also a normal family.

28.2. We have to show that if w_n approaches a boundary point w_0 of D, then every convergent subsequence of $\{f(w_n)\}$ must have its limit on the boundary of Δ. The limit of such a subsequence, if not on the boundary of Δ, must be inside; suppose that it is inside. If $f(w_{n_k}) \to z_1$ then, since the inverse of f is continuous, $w_{n_k} \to f^{-1}(z_1)$, which would be an interior point of D (because the inverse image of an open neighborhood of z_1 is an open set in D). But $w_{n_k} \to w_0$, a boundary point of D. Notice that Δ could have been any region, and that conformality played no role: f only needed to be a homeomorphism.

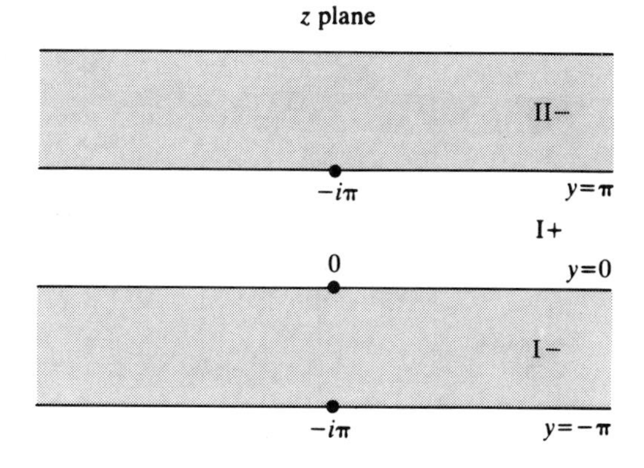

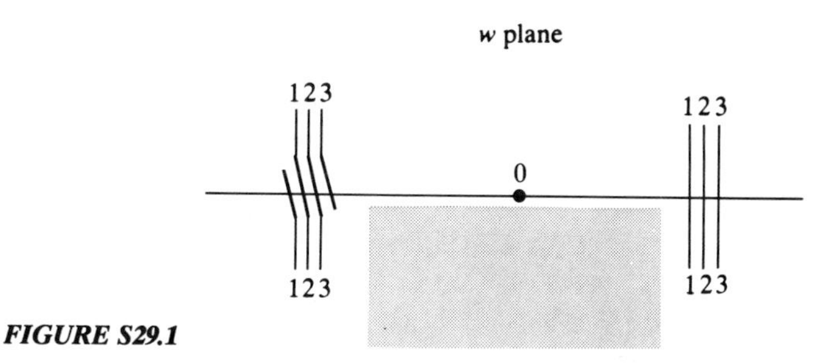

FIGURE S29.1

Section 29

29.1. $w = e^z$. We have $w = e^x(\cos y + i \sin y)$. When $v = 0$, $e^x \sin y = 0$, so $y = k\pi$, $k = 0, \pm 1, \pm 2, \ldots$, and $u = e^x \cos y$. Take $y = 0$ and let u start at $+\infty$ and move to the left; then $u = e^x$, and z moves monotonically to the left until u reaches 0, when $x = -\infty$. If u continues to the left, $e^x \cos y$ is negative and y can be π. Then z starts at $-\infty$ and x increases as u decreases, so z describes the line $y = \pi$. The region labeled I + in Fig. S29.1 is on the right of the path of z, so I + corresponds to an upper half plane. Similarly, I − corresponds to a lower half plane. Then I + and I − join along the positive u axis, so along the negative real axis I + must join to a different lower half plane (II −); and so on. The sheets of the Riemann surface are connected cyclically around 0. Since e^z is never 0, the point 0 is not a point of the surface. Notice that if w makes a circuit around 0, it cannot return to the sheet it started from unless it makes the same number of turns in the opposite direction.

29.2. $w = \sin z = \sin x \cosh y + i \cos x \sinh y$. We saw in Sec. **26B** that the half strip $-\pi/2 \le x \le \pi/2$, $0 \le y < \infty$, corresponds to an upper half w plane. Similarly, the strip $\pi/2 \le x \le 3\pi/2$, $0 \le y < \infty$, corresponds to a lower half plane, and so does $-\pi/2 \le x \le \pi/2$, $-\infty < y \le 0$.

To verify the connections between the different z regions, notice in Fig. 29.2 (p. 328) that when z crosses the interval $-\pi/2 < x < \pi/2$, $y = 0$, then $\operatorname{Im} w = 0$

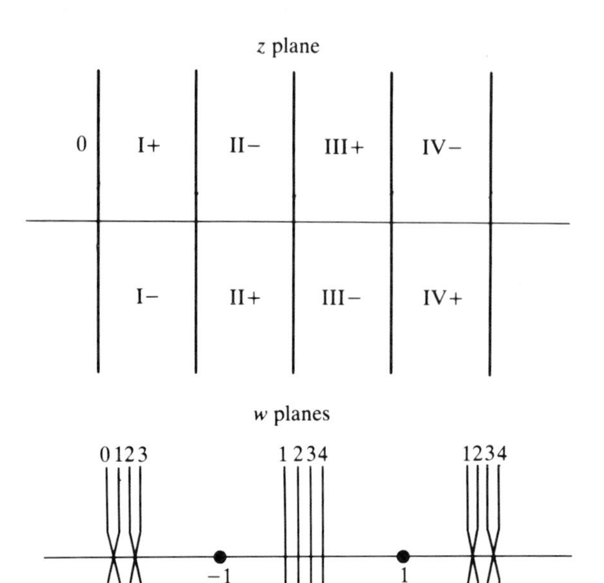

FIGURE S29.2

and $w = \sin x$ is between -1 and 1, so we can join the half strips I− and I+ to correspond to a whole w plane. When z crosses from I+ to II−, $x = \pi/2$, $0 < y < \infty$, $\operatorname{Re} w = \cosh y$ with $\operatorname{Im} w = 0$. Then w crosses the real axis between 1 and $+\infty$, and II− must correspond to the lower half of a second sheet of the Riemann surface, II+ to the upper half. Hence we connect the upper sheet of plane 1 to the lower sheet of plane 2 along the ray $(1, \infty)$. Continuing in this way, we get the indicated correspondence. Notice that the sheets connect differently around $+1$ and -1.

Section 30

30.1. The center of the orthogonal circle is the intersection of the tangents to C at P and Q, as shown in Fig. S30.1.

30.2. As Fig. S30.2 shows, the sides of a non-Euclidean triangle are concave outward since their centers are outside the bounding circle. Hence a non-Euclidean triangle is inside the Euclidean triangle with the same vertices, and therefore has smaller angles. Consequently the sum of the angles of a non-Euclidean triangle is less than 180°. The sum can be arbitrarily close to 0, since a triangle with all its vertices on the "line at infinity" would have all its angles zero.

Section 31

31.1. If $\sum \operatorname{Log}(1 + a_n)$ converges with some choice of each logarithm, then

$$\sum [\operatorname{Log}(1 + a_n) + 2\pi i h_n]$$

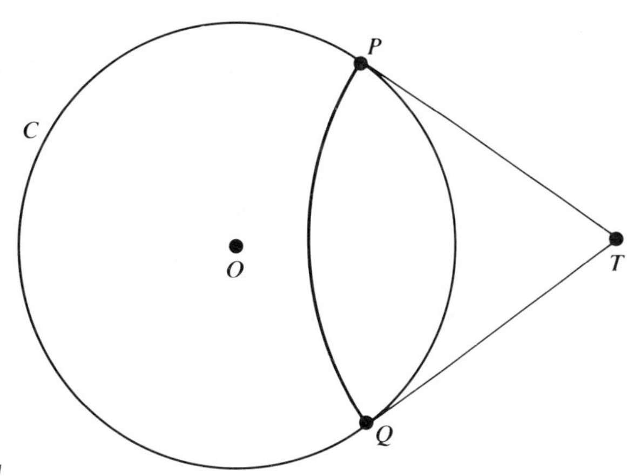

FIGURE S30.1

converges, where each h_n is an integer. Then the general term of this series tends to zero. But $\text{Log}(1 + a_n) \to 0$ since $1 + a_n \to 1$. Consequently, $h_n \to 0$, but since each h_n is an integer, $h_n = 0$ for all sufficiently large n.

31.2. If $|z| < \frac{1}{2}$,

$$\text{Log}(1+z) = z - \frac{z^2}{2} + \frac{z^3}{3} - \frac{z^4}{4} + \cdots.$$

Hence

$$|\text{Log}(1+z) - z| \le |z|^2 \left|\frac{1}{2} - \frac{z}{3} + \frac{z^2}{4} - \cdots\right|$$

$$\le |z|^2\left(\frac{1}{2} + \frac{1}{4} + \frac{1}{8} + \cdots\right) = |z|^2,$$

$$\left|\text{Log}(1+z) - z + \frac{z^2}{2}\right| \le |z|^3 \left|\frac{1}{2} + \frac{1}{4} + \frac{1}{8} + \cdots\right| = |z|^3,$$

and so on.

FIGURE S30.2

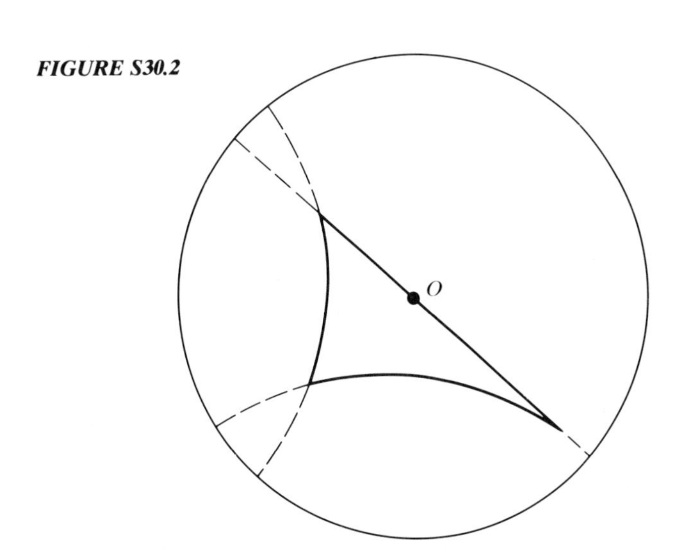

31.3. $|\text{Log}(1+a_k)-a_k|\le|a_k|^2$ by Exercise 31.2. Hence if $\sum|a_k|^2$ converges, so does $\sum[\text{Log}(1+a_k)-a_k]$; and then if $\sum a_k$ converges, so does $\sum \text{Log}(1+a_k)$.

31.4. (*a*)

$$\left|\text{Log}(1+a_k)-a_k+\frac{a_k^2}{2}-\frac{a_k^3}{3}\right|<|a_k|^4=\frac{1}{k^{4/3}},$$

so

$$\sum\left[\text{Log}(1+a_k)-a_k+\frac{a_k^2}{2}-\frac{a_k^3}{3}\right]$$

converges (absolutely). Now $\sum a_k$ and $\sum a_k^3$ converge (conditionally), but $\sum a_k^2$ diverges and hence so must $\sum \text{Log}(1+a_k)$.

(*b*)

$$\sum\left[\text{Log}(1+a_k)-a_k+\frac{a_k^2}{2}\right]$$

converges since $\sum|a_k^3|$ converges. According as k is odd or even,

$$a_k+\frac{a_k^2}{2}=\frac{1}{k^{1/2}}-\frac{1}{2k}\quad\text{or}\quad-\frac{1}{(k-2)^{1/2}}-\frac{1}{2(k-2)},$$

so $\sum(a_k+\frac{1}{2}a_k^2)$ diverges; hence $\sum \text{Log}(1+a_k)$ diverges.

(*c*)

$$|\text{Log}(1+a_k)-a_k|\le\frac{|z|^2}{k^2},$$

so $\sum[\text{Log}(1+a_k)-a_k]$ converges. But since $\sum a_k$ diverges, the product diverges.

(*d*) $\sum|a_k|$ converges and $\sum[\text{Log}(1+a_k)-a_k]$ converges, so the product converges.

(*e*) A partial product is either one of the partial products of the product in (*d*), or one of these times a factor whose limit is 1. Hence the product converges.

(*f*) Let PQ be a side of the n-gon and let r_n be the radius of C_n. In Fig. S31.4 (schematic), PQ is tangent to C_{n-2} at T; then $OP=r_{n-1}=$

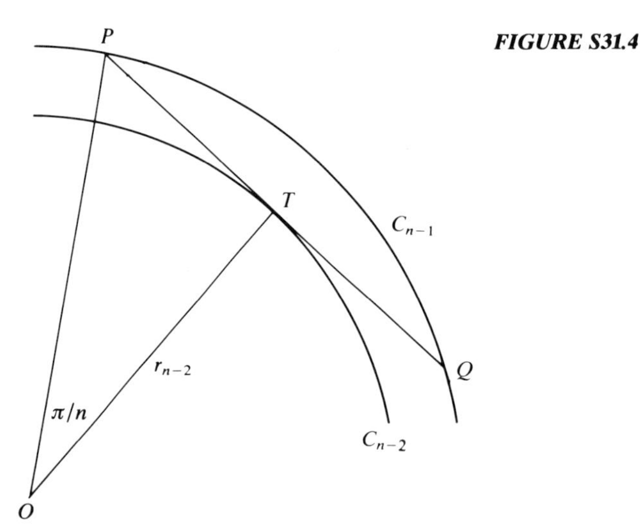

FIGURE S31.4

$r_{n-2}\sec(\pi/n)$. Since $r_1 = 1$, $r_2 = \sec(\pi/3)$, and

$$r_n = \prod_{k=3}^{n+1} \sec\frac{\pi}{k}.$$

The figure is inside a circle of radius

$$R = \prod_{k=3}^{\infty} \sec\frac{\pi}{k},$$

which is finite if the product for $1/R$ converges. Now

$$\cos\frac{\pi}{k} = 1 - \frac{\pi^2}{2k^2} + O\left(\frac{1}{k^4}\right),$$

and the product converges by Exercise 31.3.

31.5. Let z belong to a compact set K that contains no point $\pm n$. Since K is bounded, when n is sufficiently large we have $|z^2/n^2| < \frac{1}{2}$ and

$$\left|\operatorname{Log}\left(1 - \frac{z^2}{n^2}\right) + \frac{z^2}{n^2}\right| \le \frac{1}{2}\left|\frac{z^2}{n^2}\right|^2 \le \frac{c}{n^4}$$

for some number c. Since $\sum z^2/n^2$ converges uniformly on K (by the same reasoning), $\sum \operatorname{Log}(1 - z^2/n^2)$ converges uniformly on K.

31.6. We must show that $\cot\pi(n + \frac{1}{2} + iy)$, $-n \le y \le n$, and $\cot(\pi x + n\pi i)$, $-(n - \frac{1}{2}) \le x \le (n + \frac{1}{2})$, are bounded. Since

$$\left|\cot\pi\left(n + \frac{1}{2} + iy\right)\right| = \left|\frac{\cos\pi(n + \frac{1}{2} + iy)}{\sin\pi(n + \frac{1}{2} + iy)}\right| = \left|\frac{\sin iy}{\cos iy}\right| = \left|\frac{\sinh y}{\cosh y}\right|,$$

it is enough to consider positive y. Then an upper bound is

$$\frac{1 - e^{-2\pi y}}{1 + e^{-2\pi y}} \le 1.$$

This disposes of the vertical sides of C_n. On the horizontal sides,

$$\cot(\pi x + n\pi i) = \frac{\cos(\pi x + n\pi i)}{\sin(\pi x + n\pi i)} = \frac{\cos\pi x\cosh n\pi - i\sin\pi x\sinh n\pi}{\sin\pi x\cosh n\pi + i\cos\pi x\sinh n\pi},$$

$$|\cot(\pi x + n\pi i)|^2 = \frac{\cos^2\pi x\cosh^2 n\pi + \sin^2\pi x\sinh^2 n\pi}{\sin^2\pi x\cosh^2 n\pi + \cos^2\pi x\sinh^2 n\pi}$$

$$\le \frac{\cosh^2 n\pi}{\sinh^2 n\pi}$$

We may suppose $n > 0$, since the last fraction is an even function. Then

$$\frac{\cosh^2 n\pi}{\sinh^2 n\pi} = \left(\frac{e^{n\pi} + e^{-n\pi}}{e^{n\pi} - e^{-n\pi}}\right)^2 \le \left(\frac{2e^{n\pi}}{e^{n\pi} - 1}\right)^2,$$

which is clearly bounded. Now we have

$$|I_n| \le \int_{C_n} \frac{O(1)\,|dw|}{|w|\,|w - z|} \le O(1)\cdot(\text{length of } C_n)\cdot\max_{C_n}\frac{1}{|w|\,|w - z|},$$

and this approaches 0 for each fixed z.

31.7.

$$f(w) = \frac{d}{dw}(\log \sin w - \log w),$$

$$f(w) = \frac{\cos w}{\sin w} - \frac{1}{w} = \frac{\cos w}{\sin w} - \frac{\sin w}{w} \cdot \frac{1}{\sin w}$$

$$= \frac{1}{\sin w}\left(\cos w - \frac{\sin w}{w}\right)$$

$$= \frac{1}{\sin w}\left[\left(1 - \frac{w^2}{2} + \cdots\right) - \left(1 - \frac{w^2}{3!} + \cdots\right)\right]$$

$$= \frac{O(w^2)}{\sin w} \to 0.$$

Section 32

32.1. Suppose that $f(z) \not\equiv 0$. From Jensen's formula we have, when $f(0) \neq 0$,

$$\int_0^R \frac{n(t)}{t}\, dt = \frac{1}{2\pi}\int_0^{2\pi} \log |f(Re^{i\theta}| \, d\theta - \log |f(0)|,$$

where $n(t)$ is the number of zeros z_k with $|z_k| \le t$. If $f(0) = 0$, we can consider $f(z)/z^p$, where p is the order of the zero at 0. Then Jensen's formula gives us

$$\int_0^R \frac{n(t)}{t}\, dt = \frac{1}{2\pi}\int_0^{2\pi} \log |f(Re^{i\theta})| \, d\theta - p \log R - \log |f(0)|.$$

Since f has at least one zero of modulus k, $k = 1, 2, \ldots, [R]$, we may argue loosely that $n(t)$ is at least t, so

$$\int_1^R t^{-1} n(t)\, dt \ge R$$

and

$$\frac{1}{2\pi}\int_0^{2\pi} \log |f(Re^{i\theta})| \, d\theta \le \log A + BR,$$

so that we have a contradiction for large R if $B < 1$. A careful use of the same ideas would go as follows.

We have $n(t) = 0$ for $0 \le t < 1$, $n(t) \ge 1$ for $1 \le t < 2$, $n(t) \ge 2$ for $2 \le t < 3$, and so on. Hence $n(t) \ge [t]$ (the integral part of t), and

$$[R] - 1 - \log[R] = \int_1^{[R]} \frac{t-1}{t}\, dt \le \int_0^R \frac{n(t)}{t}\, dt$$

$$\le \frac{1}{2\pi}\int_0^{2\pi} \log |f(Re^{i\theta})| \, d\theta - p \log R - \log |f(0)|.$$

But $\log |f(Re^{i\theta})| \le \log A + BR$, so we have

$$[R] - 1 - \log[R] \le \log A + BR - p \log R - \log f(0),$$

which is impossible for large R if $B < 1$.

32.2. As in Exercise 32.1, if we assume that $f(0) \neq 0$, we have

$$n(t) \geq 1 + 2 + \cdots + [t] = \frac{[t]([t]+1)}{2}$$

or about $t^2/2$. Hence $\int_0^R t^{-1} n(t)\, dt$ is approximately $\frac{1}{2}\int_0^R t\, dt = R^2/4$. However,

$$\frac{1}{2\pi}\int_0^{2\pi} \log|f(Re^{i\theta})|\, d\theta \leq \log A + BR^2,$$

so we will get a contradiction if $B < \frac{1}{4}$.

More precisely,

$$n(t) \geq \frac{(t-1)t}{2}$$

for $R \geq t \geq 1$,

$$\int_0^R \frac{n(t)}{t}\, dt \geq \frac{1}{2}\int_1^{[R]} (t-1)\, dt \geq \frac{1}{4}([R]-1)^2;$$

whereas

$$\frac{1}{4}([R]-1)^2 \leq \frac{1}{2\pi}\int_0^{2\pi} \log|f(Re^{i\theta})|\, d\theta \leq \log A + BR^2.$$

Section 33

33.1. If D is unbounded and has an exterior point z_1, then $|z - z_1|^{-1}$ is bounded in D and $\delta\,|z - z_1|^{-1} < 1$ for sufficiently small δ. We may take $g_\varepsilon(z) = [\delta/(z - z_1)]^\varepsilon$.

Section 34

34.1. Suppose that $f(z) \sim \sum a_n z^{-n}$ and also $f(z) \sim \sum b_n z^{-n}$. Then

$$\sum_{k=0}^{n} a_k z^{-k} - \sum_{k=0}^{n} b_k z^{-k} = o(z^{-n}),$$

uniformly in an angle in which both series are asymptotic representations of f. Then

$$\sum_{k=0}^{n} (a_k - b_k) z^{-k} = o(z^{-n}).$$

Let $n = 0$; then $a_0 - b_0 = o(1)$, which is to say that $a_0 = b_0$. Now, with $a_0 = b_0$, let $n = 1$; we then have

$$(a_1 - b_1) z^{-1} = o(z^{-1}).$$

And so on.

34.2. We have

$$z^n \sum_{k=1}^{n} a_k z^{-k} \to 0 \quad \text{and} \quad z^n \sum_{j=1}^{n} b_j z^{-j} \to 0,$$

uniformly as $|z| \to \infty$ in an angle. If we form the product

$$\sum_{k=1}^{n} a_k z^{-k} \sum_{j=1}^{n} b_j z^{-j} = \sum_{k=1}^{n} \sum_{j=1}^{n} a_k b_j z^{-(k+j)},$$

then the terms with $k+j=m$, $m \le n$, yield

$$\sum_{m=0}^{n} c_m z^{-m},$$

where $c_m = \sum_{j=0}^{m} a_j b_{m-j}$. The other terms all involve powers z^{-m} with $m > n$. Hence

$$z^n\left[f(z)g(z) - \sum_{m=0}^{n} c_m z^{-m}\right] = z^n\left[f(z)g(z) - \sum_{k=0}^{n} a_k z^{-k} \sum_{j=0}^{n} b_j z^{-j} + \phi_n(z)\right],$$

where $\phi_n(z)$ is a finite sum of terms involving z^{-p} with $p \ge n+1$. Consequently, $z^p \phi_n(z) \to 0$ uniformly.

34.3. The function has an asymptotic series for real positive x because

$$|e^{-x} \sin(e^{2x})| \le e^{-x} = o(x^{-n})$$

for each n as $x \to +\infty$. However, the derivative of $e^{-z} \sin(e^{2z})$ is

$$-e^{-z} \sin(e^{2z}) + 2e^{z} \cos(e^{2z}),$$

which does not approach zero in any angle containing the positive real axis. Indeed,

$$\begin{aligned} 2\cos(e^{2z}) &= \exp(ie^{2z}) + \exp(-ie^{2z}) \\ &= \exp(ie^{2x} \cos 2y - e^{2x} \sin 2y) + \exp(-ie^{2x} \cos 2y + e^{2x} \sin 2y); \\ |2\cos(e^{2z})| &\ge \exp(e^{2x} \sin 2y) - o(1) \qquad (x>0,\ \sin 2y > 0) \\ &= \exp[e^{2r\cos\theta} \sin(2r \sin\theta)] - o(1); \end{aligned}$$

and

$$2\,|e^{z} \cos(e^{2z})| \to \infty$$

on any sequence $|z| = r_n$ such that $2r_n \sin\theta \equiv \pi/2 \pmod{2\pi}$. There evidently is such a sequence $\{r_n\}$ for each θ, $0 < \theta < \pi/2$.

34.4. We have

$$(1+t^2)^{-1/2} = \sum_{n=0}^{\infty} (-1)^n t^{2n} \frac{(2n)!}{2^{2n}(n!)^2}.$$

Then

$$\int_0^{\infty} e^{-zt}(t^2+1)^{-1/2}\, dt \sim \sum_{n=0}^{\infty} (-1)^n \frac{[(2n)!]^2}{2^{2n}(n!)^2} z^{-2n-1}.$$

34.5. The graph of $P(t)$ looks like Fig. S34.5, with $P(n+0) = \frac{1}{2}$, $P(n-0) = -\frac{1}{2}$. Hence

$$\begin{aligned} \int_{k+0}^{k+1-0} f'(t)P(t)\, dt &= P(t)f(t)\Big|_{k+0}^{k+1-0} + \int_k^{k+1} f(t)\, dt \\ &= -\frac{1}{2}f(k+1) - \frac{1}{2}f(k) + \int_k^{k+1} f(t)\, dt. \end{aligned}$$

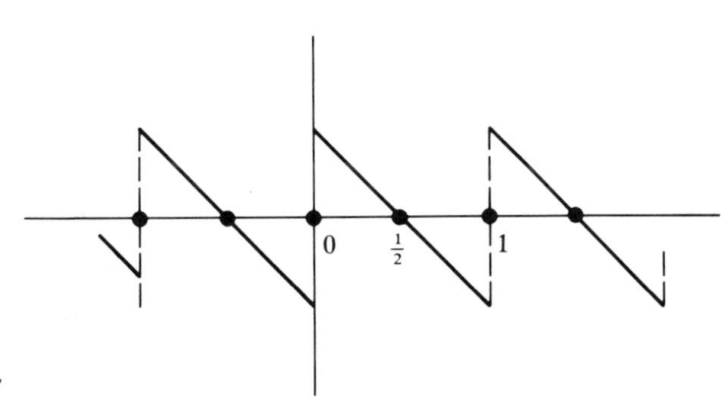

FIGURE S34.5

Then

$$\begin{aligned}\int_1^n f'(t)P(t)\,dt &= -\frac{1}{2}\sum_{k=1}^{n-1} f(k+1) - \frac{1}{2}\sum_{k=1}^{n-1} f(k) + \int_1^n f(t)\,dt \\ &= \frac{1}{2}f(n) + \frac{1}{2}f(1) - \sum_{k=1}^{n} f(k) + \int_1^n f(t)\,dt \\ &= -\sum_{k=1}^{n} f(k) + \frac{1}{2}[f(n)+f(1)] + \int_1^n f(t)\,dt,\end{aligned}$$

so

$$\sum_{k=1}^{n} f(k) = \int_1^n f(t)\,dt + \frac{1}{2}[f(n)+f(1)] - \int_1^n f'(t)P(t)\,dt.$$

34.6. It is enough to show that

$$\lim_{n\to\infty}\int_1^n t^{-1}P(t)\,dt$$

exists. If we write

$$\int_1^n t^{-1}P(t)\,dt = \int_1^{3/2} + \int_{3/2}^{2} + \cdots + \int_{n-1/2}^{n},$$

it is evident from the graph of P (and easily verified analytically) that the successive integrals alternate in sign, decrease in absolute value, and approach zero. Hence the series of integrals converges.

34.7. (*a*) $1 - e^{-t} = \int_0^t e^{-u}\,du < \int_0^t du = t.$

(*b*) $1 - e^{-t} = t - \dfrac{t^2}{2!} + \dfrac{t^3}{3!} - \cdots.$

If $0 < t < 1$, the absolute values of the terms decrease and the signs alternate, so that $1 - e^{-t} > t - \frac{1}{2}t^2$.

(*c*) For $0 < t < 1$, we have $0 < 1 - \frac{1}{2}t < 1$, so $1 - \frac{1}{2}t > (1 - \frac{1}{2}t)^2$.

34.8. We saw in Sec. **31C** that Wallis's product can be written as

$$\left(\frac{2}{\pi}\right)^{1/2} = \lim_{N\to\infty}\frac{(2N)!(2N+1)^{1/2}}{2^{2N}(N!)^2}.$$

Applying Stirling's formula (with C unknown), we get

$$\left(\frac{2}{\pi}\right)^{1/2} = \lim_{N\to\infty} \frac{2^{2N}N^{2N}(2N)^{1/2}e^{-2N}C(2N+1)^{1/2}}{2^{2N}N^{2N}e^{-2N}C^2N} = \frac{2}{C},$$

$$C = 2\left(\frac{\pi}{2}\right)^{1/2} = (2\pi)^{1/2}.$$

Section 35

35.1. First, $g(z_1) = g(z_2)$ implies $f(z_1^2) = f(z_2^2)$, hence $z_1^2 = z_2^2$, $z_1 = \pm z_2$. However, g is an odd function, so $z_1 = -z_2$ would imply that $g(z_1) = -g(z_2)$. But then $g(z_1) = g(z_2) = -g(z_1)$, whence $g(z_1) = 0$, because the only zero of g is at 0. Thus $g(0) = 0$.

Next,

$$g(z) = z(1 + a_2z^2 + \cdots)^{1/2} = z[1 + (a_2z^2 + \cdots)]^{1/2}$$
$$= z\left(1 + \frac{1}{2}(a_2z^2 + \cdots) + \left(\frac{1}{2}\right)^2\left(-\frac{1}{2}\right)(a_2z^2 + \cdots)^2 + \cdots\right)$$

by the binomial theorem; compare Sec. **15E**. Hence

$$g(z) = z + \frac{1}{2}a_2z^3 + O(z^5),$$

and the power series of g converges in the unit disk because $f(z^2)$ has no zeros except for the double zero at $z = 0$, and hence has an analytic square root for $|z| < 1$. Therefore $g \in S$.

35.2. None of a, b, c is necessarily in S.

(*a*) g could be $-f$.

(*b*) If $f(z) = z/(1-z)^2 = z + 2z^2 + \cdots$, $zf'(z) = z + 4z^2 + \cdots$. Here the coefficient of z^2 is too large.

(*c*) As for (*b*), $f(z)^2/z = z + 4z^2 + \cdots$.

*References**

L. V. Ahlfors. *Complex Analysis.* 3d ed. New York: McGraw-Hill, 1979.

M. L. Boas. *Mathematical Methods in the Physical Sciences.* 2d ed. New York: Wiley, 1983; *Solutions of Selected Problems,* New York: Wiley, 1984.

R. P. Boas. *A Primer of Real Functions.* 3d ed. Washington, D.C.: Mathematical Association of America, 1981.

R. C. Buck and E. F. Buck. *Advanced Calculus.* 3d ed. New York: McGraw-Hill, 1978.

R. B. Burckel. *An Introduction to Classical Complex Analysis,* vol. 1. New York: Academic Press, 1979.

G. F. Carrier, M. Krook, and C. E. Pearson. *Functions of a Complex Variable: Theory and Technique.* New York: McGraw-Hill, 1966.

H. Cartan. *Elementary Theory of Analytic Functions of One or Several Complex Variables.* Reading, Mass.: Addison-Wesley, 1963.

E. F. Collingwood and A. J. Lohwater. *The Theory of Cluster Sets.* Cambridge: Cambridge University Press, 1966.

L. Comtet. *Advanced Combinatorics.* Dordrecht, The Netherlands: Reidel, 1974.

E. T. Copson. *An Introduction to the Theory of Functions of a Complex Variable.* Oxford: Clarendon Press, 1935.

P. Dienes. *The Taylor Series.* Oxford: Clarendon Press, 1931. Reprint. New York: Dover Publications, 1957.

H. Goldstein. *Classical Mechanics.* Reading, Mass.: Addison-Wesley, 2d ed. 1980.

G. H. Hardy. *Divergent Series.* Oxford: Clarendon Press, 1949.

P. Henrici. *Applied and Computational Complex Analysis,* vol. 3. New York: Wiley, 1986.

E. Hille. *Analytic Function Theory.* Boston: Ginn & Co., 1959; (vol. 1), 1962; (vol. 2). Reprint. Chelsea, New York, 1974.

J. D. Jackson. *Classical Electrodynamics.* 2d ed. New York: Wiley, 1975.

H. Jeffreys and B. S. Jeffreys. *Methods of Mathematical Physics.* Cambridge: Cambridge University Press, 1950.

H. Kober. *Dictionary of Conformal Representations.* New York: Dover Publications, 1952.

N. Levinson. *Gap and Density Theorems.* New York: American Mathematical Society, 1940.

J. E. Littlewood. *Lectures on the Theory of Functions.* Oxford: Oxford University Press, 1944.

A. I. Markushevich. *Theory of Functions of a Complex Variable.* vol. 1. Translated by R. A. Silverman. Englewood Cliffs, N.J.: Prentice-Hall, 1965.

A. I. Markushevich. *Selected Chapters in the Theory of Analytic Functions.* In Russian. (*Izbrannye glavy teorii analiticheskih funktsiĭ,* Moscow: Nauka, 1976.

L. M. Milne-Thomson. *Theoretical Hydrodynamics.* 2d ed. New York: Macmillan, 1950.

D. S. Mitrinović and J. D. Kečkić. *The Cauchy Method of Residues, Theory and Applications.* Dordrecht, The Netherlands: Reidel, 1984.

* This list includes only books that are cited in the text.

W. F. Osgood. *Lehrbuch der Funktionentheorie,* vol. 1. 5th ed. Leipzig: Teubner, 1982. Reprint. New York: Chelsea, 1966.

H. R. Pitt. *Tauberian Theorems.* Oxford: Clarendon Press, 1958.

G. Pólya. *Collected Mathematical Papers,* vol. 1. Cambridge, Mass.: M.I.T. Press, 1974.

G. Pólya and G. Szegő. *Problems and Theorems in Analysis,* vol. 1. New York: Springer-Verlag, 1972.

A. G. Postnikov. Tauberian Theory and Its Applications. *Proc. Steklov Inst. Math.,* no. 2 (1980). [Translation of *Trudy Mat. Inst. Steklov.,* no. 144 (1979).]

R. Remmert. *Funktionentheorie I.* New York: Springer-Verlag, 1984.

H. Sagan. *Advanced Calculus.* Boston: Houghton-Mifflin, 1974.

W. Sibagaki. *Theory and Applications of the Gamma Function* (in Japanese). Tokyo: Iwanami Syoten, 1952.

O. Szász. *Introduction to the Theory of Divergent Series.* New York: Hafner, 1948.

E. C. Titchmarsh. *The Theory of Functions.* Oxford: Clarendon Press, 1932.

D. V. Widder. *Advanced Calculus.* 2d ed. Englewood Cliffs, N.J.: Prentice-Hall, 1961.

Index

Index of Symbols